FUNDAMENTALS OF MOLECULAR BIOLOGY

JAYANTA K. PAL

Professor, Department of Biotechnology
University of Pune, Pune

SAROJ S. GHASKADBI

Head of Department, Department of Zoology
University of Pune, Pune

OXFORD
UNIVERSITY PRESS

OXFORD
UNIVERSITY PRESS

YMCA Library Building, Jai Singh Road, New Delhi 110001

Oxford University Press is a department of the University of Oxford.
It furthers the University's objective of excellence in research, scholarship,
and education by publishing worldwide in

Oxford   New York
Auckland   Cape Town   Dar es Salaam   Hong Kong   Karachi
Kuala Lumpur   Madrid   Melbourne   Mexico City   Nairobi
New Delhi   Shanghai   Taipei   Toronto

With offices in
Argentina   Austria   Brazil   Chile   Czech Republic   France   Greece
Guatemala   Hungary   Italy   Japan   Poland   Portugal   Singapore
South Korea   Switzerland   Thailand   Turkey   Ukraine   Vietnam

Oxford is a registered trade mark of Oxford University Press
in the UK and in certain other countries.

Published in India
by Oxford University Press

© Oxford University Press 2009

ISBN-13: 978-0-19-569781-0
ISBN-10: 0-19-569781-2

Typeset in Times
by Tej Composers, New Delhi
Printed in India by Radha Press, Delhi 110031
and published by Oxford University Press
YMCA Library Building, Jai Singh Road, New Delhi 110001

Dedicated to
the teachers who taught us
and the students whom we taught and learned from

Preface

Molecular biology, which has its root in biochemistry, has emerged as a main discipline in life sciences, particularly with the pioneering discovery of the DNA structure by Watson and Crick in 1953. Our understanding on the flow of genetic information from DNA to RNA to protein, called the central dogma of molecular biology, began around this time. Not only did this discipline enhance our understanding of biology, it also led to the development of several advanced tools including recombinant DNA technology that forms the basis of the contemporary biotechnology.

In the recent times, research and development in this area of molecular biology and recombinant DNA technology has made an extensive impact on medicine, agriculture, and environment. Thus, in the post-genomic era, any research seeking a deeper understanding of biology and its applications demands an understanding of the basic concepts in molecular biology. Every student in biology today therefore has to master this subject.

About the Book

This book, *Fundamentals of Molecular Biology*, is an outcome of several years of our experience in teaching molecular biology and related subjects in the University of Pune. Many excellent books are available in this and related fields. However, we have felt the need of a concisely written textbook that provides a clear idea on the concepts, rationale, methodologies, and applications of molecular biology. We hope that this book will serve our basic purpose of ensuring that students learn the basics of molecular biology, a necessary prerequisite to understand and appreciate life processes which are intricately linked to human health and disease.

The contents of this book have been structured keeping in mind the requirements of both undergraduate and postgraduate students of biology and biotechnology. In addition, this book will also meet the requirements of students pursuing advance level post M.Sc. diploma courses in genetic engineering and biotechnology, and also of the research students interested in molecular biology.

We have provided an exhaustive account of various methodologies including biochemical and biophysical ones that have aided in unravelling the molecular mystery of cells in terms of their structure-function relationship. The style and language of description has been kept simple, and to an extent, it mimics the style of our routine classroom teaching. Each chapter, barring Chapter 1, is organized uniformly with an overview of the chapter, the main text, followed by a summary, study questions, and further reading. A number of figures have been produced in colour for better clarity in reading and understanding.

Content and Structure

While emphasizing the basics of molecular biology, we have provided up-to-date information on all the topics. These have been dealt in the form of 11 chapters.

Chapter 1 introduces the concept of molecular biology. It also emphasizes the paradigm change in our concept of central dogma. Further, a number of model organisms used to understand various molecular phenomena have been highlighted.

Chapter 2 depicts the molecular composition of a cell, emphasizing the macromolecules in particular.

Chapter 3 describes various DNA modifying enzymes that are instrumental in regulating various biological processes. Usage of these enzymes as important tools in molecular biology research has been elaborated.

Chapter 4 deals with a diverse array of biophysical and biochemical techniques including those used for recombinant DNA research. These techniques which are a combination of classical and advanced type are essential for molecular biology research. This chapter is therefore the largest of the book. The readers, particularly the young researchers, may acquaint themselves with a variety of techniques that may be useful for their research.

Chapter 5 throws light on the concept of genome, its structure and complexity across species with various levels of evolution. It also emphasizes the major differences of the genetic material between a prokaryote and a eukaryote. In eukaryotes, the genome organization gets more elaborate due to the presence of chromatin.

Chapters 6 to 8 describe the functions of genome in relation to its perpetuation (DNA replication), maintenance and variation (damage repair and recombination), and gene expression (transcription).

Chapters 9 and 10 explain the second process of gene expression (translation) along with various post-synthetic modifications including folding of the protein phenotypes for making them functional. Chapter 10 also includes the mechanisms of protein breakdown.

Chapter 11 introduces two newly emerging branches of biology in the post-genomic era, genomics and proteomics. These are the thrust areas of present day research in molecular biology.

A detailed glossary of important terminology is also given at the end of the book. We will be very happy and receptive to suggestions and criticisms from readers.

Acknowledgements

We express our gratitude and thanks to those who have helped in completion of the book. First and foremost, we deeply appreciate the continuous help obtained from many of our current M.Sc. and Ph.D. students. Among them, Dr Pal is deeply indebted to Smriti Mittal for her untiring help in preparing a number of figures included in the book, and to Neah Likhite, S. Manjutha, Abhijeet Kulkarni, Sunil Berwal, Parijat Senapati, and Sakshi Arora for proofreading the manuscript. Dr Ghaskadbi would like to thank Swati Khole, Acharya Zanhar, Rakhi Wishwakarma, and Sheetal Kumar for their help in drawing several figures.

The authors have applied for permissions to use the various images featured in the text. Due acknowledgement will be made in future editions of the book when the permissions are granted.

Dr Pal is grateful to Anita (wife) and Joeeta (daughter) for their continuous support and for tolerating and bearing with his continuous preoccupation with the writing work. Dr Ghaskadbi is indebted to Surendra (husband) and Pallavi (daughter) for their constant support and encouragement.

We would like to thank our colleagues for their suggestions and criticisms, and many others for allowing us to adapt as well as to cite some of their work in this book. Lastly, we would like to thank the editorial team of Oxford University Press, India, for their cooperation and help.

Jayanta K. Pal
Saroj S. Ghaskadbi

Contents

Preface   v
List of Colour Plates   xiii

1. The Concept   1

1.1 Definition of Molecular Biology and its Importance   2
1.2 Central Dogma of Molecular Biology   2
1.3 Model Organisms for Studying Molecular Biology   3

2. Molecules of a Cell   5

2.1 Small Inorganic Molecules   6
2.2 Small Organic Molecules and Macromolecules   6
     2.2.1 Carbohydrates   6
     2.2.2 Lipids   8
     2.2.3 Nucleic acids   12
     2.2.4 Proteins   17

3. Enzymes in Molecular Biology Research   27

3.1 DNA Modifying Enzymes   28
     3.1.1 Polymerases   28
     3.1.2 Ligases   30
     3.1.3 Phosphatases   32
     3.1.4 Polynucleotide kinases   32
3.2 DNA Degrading Enzymes   33
     3.2.1 Nucleases: endo- and exonucleases   33
     3.2.2 Restriction endonucleases   37

4. Techniques in Molecular Biology   43

4.1 Requirements of a Molecular Biology Laboratory   44
     4.1.1 Construction   44

4.1.2 Containment facilities   44

4.1.3 Equipments   44

4.1.4 Special practices   45

4.2 Biophysical Techniques   45

4.2.1 Microscopy   45

4.2.2 Centrifugation   58

4.2.3 Spectroscopy   61

4.2.4 X-ray crystallography   67

4.3 Biochemical Techniques   69

4.3.1 Electrophoresis   69

4.3.2 Chromatography   81

4.3.3 Radiolabelling and detection   86

4.3.4 Sequencing techniques   91

4.3.5 Polymerase chain reaction   96

4.3.6 Methods for determining DNA-protein interactions   105

4.4 Recombinant DNA Techniques   109

4.4.1 Molecular cloning   109

4.4.2 Cloning vectors   111

4.4.3 Transfer of recombinant DNA to the host cells   115

4.4.4 Hybridization techniques   115

4.4.5 Molecular probes   117

5. Concept of Genome and its Organization   **131**

5.1 Nucleic Acids as the Genetic Material   *132*

5.2 DNA Content of a Cell and C-value Paradox   *134*

5.3 Genome and its Organization   *140*

5.3.1 Non-coding DNA sequences within genes   142

5.3.2 Extra-genic sequences   143

5.3.3 Gene families   145

5.4 Packaging of DNA   *153*

5.4.1 Viral DNA   154

5.4.2 Bacterial DNA   154

5.4.3 Eukaryotic DNA   154

5.4.4 Chromatin organization   156

5.4.5 Chromatin higher order structure   158

5.5 Organelle Genome   *160*

5.6.1 Mitochondrial genome   162

5.6.2 Chloroplast genome   165

6. DNA Replication   **171**

6.1 General Features of DNA Replication   *172*
6.2 Replication in Prokaryotes   *173*
    6.2.1 Initiation   174
    6.2.2 Elongation   175
    6.2.3 Maturation of Okazaki fragments   178
6.3 Replication in Eukaryotes   *180*
6.4 Termination of Replication   *180*
6.5 Regulation of Replication   *184*

7. DNA Damage, Repair, and Recombination   **187**

7.1 Types of DNA Damages   *188*
7.2 Multiple Repair Pathways   *189*
    7.2.1 Nucleotide excision repair   191
    7.2.2 Base excision repair   192
    7.2.3 Mismatch repair system   192
    7.2.4 Recombination repair   194
    7.2.5 Double-strand break repair   195
7.3 DNA Recombination   *196*
    7.3.1 Homologous genetic recombination   196
    7.3.2 Meiotic recombination   198
    7.3.3 Other recombination events   200

8. Transcription   **205**

8.1 Transcription Machinery in Prokaryotes   *207*
    8.1.1 Initiation and elongation of transcription   208
    8.1.2 Termination of transcription   210
    8.1.3 Regulation of transcription   211
8.2 Transcription in Eukaryotes   *224*
    8.2.1 Promoters of eukaryotic polymerases   225
    8.2.2 Transcription factors in eukaryotes   227
    8.2.3 Transcription activators   231
    8.2.4 Chromatin and transcription   236
8.3 Post-transcriptional Events   *236*
    8.3.1 Splicing   237
    8.3.2 RNA editing   243
    8.3.3 Processing of mRNA at 3' end   245
    8.3.4 Processing of mRNA at 5' end   247
    8.3.5 Production of mature rRNA   248
    8.3.6 Production of mature tRNA   249

9. Protein Synthesis **257**

9.1 Components of Protein Synthesis Machinery *258*
    9.1.1 Messenger RNA 258
    9.1.2 Transfer RNA 263
    9.1.3 Ribosome 266
9.2 Genetic Code *268*
9.3 Mechanism of Protein Synthesis *272*
    9.3.1 Initiation 273
    9.3.2 Elongation 278
    9.3.3 Termination 281
9.4 In vitro Cell-free Translation Systems *283*
9.5 Regulation of Protein Synthesis *286*
    9.5.1 Global regulation 286
    9.5.2 mRNA-specific regulation 288

10. Protein Folding and Modifications **299**

10.1 Protein Folding *300*
10.2 Protein Processing *303*
10.3 Protein Modifications *305*
    10.3.1 Glycosylation 305
    10.3.2 Attachment of lipids 307
    10.3.3 Attachment of glycolipids 310
    10.3.4 Protein phosphorylation 311
10.4 Protein Degradation *313*
    10.4.1 Lysosomal pathway 313
    10.4.2 Ubiquitin-proteasome pathway 314

11. Genomics and Proteomics **319**

11.1 Genomics *320*
    11.1.1 The human genome project 320
    11.1.2 An overview of the human genome 323
    11.1.3 Applications of genomics 326
11.2 Proteomics *332*
    11.2.1 Objectives of proteomics 332
    11.2.2 Methodologies for proteomics 333
    11.2.3 Protein-protein interactions and interactome 340

Glossary   *348*
Index   *371*

List of Colour Plates

Colour Plate 1
- A, B, and Z forms of DNA. (Chapter 2, p. 15)
- Tertiary structure of H119A variant of ribonuclease. (Chapter 2, p. 22)
- Quaternary structure of haemoglobin. (Chapter 2, p. 22)

Colour Plate 2
- Cells stained with a coloured dye Giemsa for morphological observation through light microscopy. (Chapter 4, p. 47)
- Image of melanotic melanoma cells by immunofluorescence microscopy. (Chapter 4, p. 50)
- (a) Double labeled immunofluorescence. (b) Schematic representation of the excitation and emission wavelength of some fluorochromes. (Chapter 4, p. 51)

Colour Plate 3
- Fluorescence microscopy. (Chapter 4, p. 51)
- Fura-2, a Ca^{2+} sensitive fluorochrome, can be used to monitor the relative concentrations of cytosolic Ca^{2+} in different regions of live cells. (Chapter 4, p. 52)
- Fluorescence microscope image of mitotic cells. (Chapter 4, p. 53)
- A deconvolved reconstructed image of live bovine pulmonary artery endothelial cells stained with LysoTracker Red DND-99, dihydrorhodamine 123 and Hoechst 33258. (Chapter 4, p. 54)

Colour Plate 4
- (a) The procedure for casting agarose gel. (b) The difference in the DNA samples run in different concentrations of agarose gel. (Chapter 4, p. 70)
- Schematic representation of the gel showing 100 bp DNA ladder and the position of xylene cyanol and bromopheol blue after agarose gel electrophoresis. (Chapter 4, p. 70)
- The procedure followed while performing SDS-PAGE. (Chapter 4, p. 74)

Colour Plate 5
- The SDS-PAG stained with silver stain and coomassie blue stain. (Chapter 4, p. 76)
- Principle of isoelectric focusing. (Chapter 4, p. 78)
- Ponceau red S stained membrane. (Chapter 4, p. 80)

Colour Plate 6
- Affinity chromatography. (Chapter 4, p. 85)
- Ion-exchange chromatography. (Chapter 4, p. 84)

Colour Plate 7
- Polymerase Chain Reaction (PCR). (Chapter 4, p. 97)
- TA cloning. (Chapter 4, p. 102)

Colour Plate 8
- Real time PCR. (Chapter 4, p. 104)

- Chromatin immunoprecipitation (ChIP). (Chapter 4, p. 108)

Colour Plate 9

- Structure of a eukaryotic gene. (Chapter 5, p. 141)
- Structure of haemoglobin. (Chapter 5, p. 148)
- Dual staining of a *Euglena gracilis* cell. (Chapter 5, p. 163)

Colour Plate 10

- Endosymbiont theory of evolution of mitochondria and chloroplast. (Chapter 5, p. 160).
- Organization of the liverwort chloroplast genome. (Chapter 5, p. 165)

Colour Plate 11

- Synthesis of the leading and lagging strand at the replication fork. (Chapter 6, p. 177)

Colour Plate 12

- Maturation of Okazaki fragments. (Chapter 6, p. 179)
- Initiation of DNA replication. (Chapter 6, p. 175)
- Different types of DNA damages. (Chapter 7, p. 189)

Colour Plate 13

- 3-dimensional high resolution X-ray crystal structures of prokaryotic ribosomal subunits. (Chapter 9, p. 268)
- Termination of translation. (Chapter 9, p. 284)
- Elongation during translation. (Chapter 9, p. 279)

Colour Plate 14

- Overview of translation. (Chapter 9, p. 272)
- Initiation of translation in prokaryotes. (Chapter 9, p. 274)

Colour Plate 15

- Recycling of EF-Tu.GTP during elongation of translation. (Chapter 9, p. 280)
- Summarized scheme of translation regulation in eukaryotic cells. (Chapter 9, p. 288)

Colour Plate 16

- Chaperones during protein transport into mitochondria. (Chapter 10, p. 300)
- Structure of chaperonin. (Chapter 10, p. 301)
- The Ubiquitin-Proteasome pathway of protein degradation. (Chapter 10, p. 314)

Colour Plate 17

- Sequential involvement of Hsp70 and Hsp60 chaperones in protein folding. (Chapter 10, p. 302)
- Cyclin degradation by Ubiquitin-Proteasome pathway during cell cycle. (Chapter 10, p. 315)

Colour Plate 18

- Synthesis of N-linked glycoproteins. (Chapter 10, p. 306)
- Microarray and DNA chips. (Chapter 11, p. 328)

Colour Plate 19

- Schematic representation of the procedure followed during 2-D polyacrylamide gel electrophoresis. (Chapter 11, p. 334)
- Analysis and identification of proteins in a sample containing a mixture of proteins by LC-MS/MS. (Chapter 11, p. 338)

Colour Plate 20

- Analysing gene expression by microarray. (Chapter 11, p. 329)
- The principle of a basic yeast two-hybrid system. (Chapter 11, p. 341)
- Principle of GST pulldown assay. (Chapter 11, p. 342)

The Concept

OVERVIEW

- A clear concept in molecular biology is essential for unraveling the mystery of a cell.

- The central dogma of molecular biology guides us to learn molecular details of flow of genetic information for functioning of a cell.

- Experiments in molecular biology have been aided by a number of model organisms as representatives of various grades of evolution.

1.1 DEFINITION OF MOLECULAR BIOLOGY AND ITS IMPORTANCE

Molecular Biology deals with our detailed understanding at the molecular level, of various biological processes that govern the existence of living organisms and their perpetuation. In order to have a detailed understanding at the molecular level, a panoply of highly sophisticated tools and techniques is used, and these are collectively called as *molecular biology techniques*. With the advent of various advanced and highly efficient techniques, now it is possible to readdress the mechanisms of various biological processes and to obtain a global understanding of them. Such a synthesis allows us to rebuild the existing models and to seek answers to further queries, resulting in a comprehensive understanding of various biological processes. Thus, Molecular Biology has a pivotal role in unravelling the mystery of living cells/organisms.

Molecular Biology, although initially learned as a part of Biochemistry, became an independent subject after the discovery of the DNA structure by Watson and Crick in 1953. Since then, till late 60s, a great deal of astounding discoveries led to a clear concept of genes, their structure, organization and function, which constitute the core of Molecular Biology. The broader concept of gene function is popularly called as the *central dogma of molecular biology*.

Considering the above, this subject therefore, is of tremendous importance for learning the fundamental concepts of biology at a greater depth. As a follow-up of such understanding, a variety of applications in various areas have emerged. In association with recombinant DNA techniques in particular, Molecular Biology has been strengthening the output of Biotechnology. Thus, this subject has a great promise to mankind, and it needs to be mastered appropriately to reap all the endless benefits that are hidden behind the widely diverse organisms on earth.

1.2 CENTRAL DOGMA OF MOLECULAR BIOLOGY

The central dogma of molecular biology deals with the detailed flow of genetic information from DNA to polypeptide. It is an outcome of extensive research and discoveries by several scientists in the 1950s and 1960s. Therefore, the central dogma changed over this period based on more and more understanding of various mechanisms of gene expression. It states that the genes present in the genome (DNA) are transcribed into mRNAs, which are then translated into polypeptides or proteins, which are the phenotypes. It also incorporates the reverse flow of information from mRNA to DNA by reverse transcription in case of virus, in particular (Fig. 1.1A).

Central dogma served as the central concept in Molecular Biology. Our understanding of the process of genetic information flow which holds the secret of life, reproduction and death of a cell has been principally guided by the central dogma. Further, with the advent of more sophisticated techniques, the study of flow

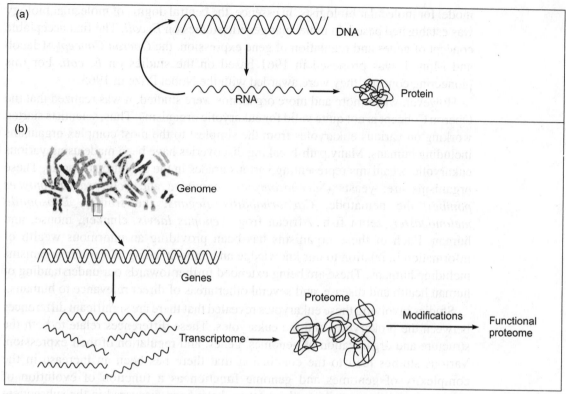

Figure 1.1 **(a)** Central dogma of molecular biology. **(b)** Central dogma, a global view (in the context of post genomic era).

of genetic information has been more global (Fig. 1.1B). Simultaneously, there has been continuous addition of vocabulary in Modern Molecular Biology, such as genomics, transcriptomics, proteomics, metabolomics, etc. The most important lesson that we have learnt is that in order to understand the main function of genome (i.e., flow of genetic information), we need to understand the organization of genome first. Therefore, in the recent past, genome sequencing and analysis were emphasized in such a fabulous scale. This has therefore made the subject of Molecular Biology more magnanimous and has broaden its scope in terms of various applications. Some of its applications in various areas, such as in human health, disease, and agriculture, etc. have given birth to Modern Biotechnology, a subject of immense potential to mankind.

1.3 MODEL ORGANISMS FOR STUDYING MOLECULAR BIOLOGY

Most of the molecular biology principles including the flow of genetic information have been learnt using *Escherichia coli* (*E. coli*) as the model organism. Due to its simple genome, very short duplication time, etc., it has been the most favourite model for molecular biologists. In essence, the central dogma of molecular biology was established based on molecular genetic studies on *E. coli*. The first acceptable concept of genes and regulation of gene expression, the *Operon Concept* of Jacob and Monod, was proposed in 1961 based on the studies on *E. coli*. For this pioneering concept, they were awarded with the Nobel Prize in 1965.

However, when more and more organisms were studied, it was realized that the Operon Concept is not quite valid for eukaryotic organisms. Thus, scientists started working on various eukaryotes from the simplest to the most complex organisms including humans. Many path-breaking discoveries have been made using various eukaryotic organisms representing various grades in the evolutionary scale. These organisms are: yeasts (*Saccharomyces cerevisiae* and *Schizosaccharomyces pombe*), the nematode, *Caenorhabditis elegans*, the fruitfly, *Drosophila melanogaster*, zebra fish, African frog, *Xenopus laevis*, chicken, mouse, and human. Each of these organisms has been providing an enormous wealth of information in relation to our knowledge and understanding of various organisms including humans. These are being extended further towards our understanding of human health and disease, and several other areas of direct relevance to humans.

Studies involving these eukaryotes revealed that there are significant differences between the prokaryotes and the eukaryotes. These differences relate to both the structure and organization of genomes, genes, and regulation of their expression. Various studies lead to the conclusion that there has been an increase in the complexity of genomes and genome function as a function of evolution of organisms. Further details on these topics have been discussed in the subsequent chapters.

Molecules of a Cell

2

OVERVIEW

- The main chemical constituents of a cell are: small molecules, macromolecules, and water.

- Water has been selected as the suitable medium and it contributes to almost 70% of the mass of a cell.

- The various polymers, namely lipids, polysaccharides, nucleic acids, and proteins are essential for building the structure and imparting functions of a cell.

The basic chemical constituents of a cell are: various types of small molecules, polymers of certain small molecules, and water as the solvent. Among these, water contributes to the bulk, about 70% of the total mass. Water is a polar molecule containing hydrogen atoms with a slight positive charge and oxygen with a negative charge. This property of water is central to most of the biological reactions and processes in a cell. Because of its polar nature, it can interact with other polar molecules and also with charged ions, making them hydrophilic and soluble in water. At the same time, it keeps the hydrophobic non-polar molecules away, and this is the basis of the formation of membranes in a cell.

2.1 SMALL INORGANIC MOLECULES

Among the small molecules, the inorganic ions contribute to approximately 1% of the cell mass. These include sodium, potassium, magnesium, calcium, phosphate, chloride, and bicarbonate ions. As evident from the chemistry of a cell, these ions constitute a very important part of various cell metabolism including enzyme activities and other functions in a cell.

The other small molecules that carry out essential functions of the cell belong to energy storing molecules like ATP and small signaling molecules, such as hormones, neurotransmitters, and small gases (e.g., nitric oxide).

2.2 SMALL ORGANIC MOLECULES AND MACROMOLECULES

The other major group of small molecules is organic in nature and these also constitute the larger molecules, namely carbohydrates, lipids, nucleic acids, and proteins. Most of the carbohydrates (polysaccharides), nucleic acids, and proteins are referred to as the macromolecules. These polymers are made up of their small organic precursors, namely sugars, nucleotides, and amino acids. Macromolecules contribute to the bulk of the cell's dry weight (approximately 80 – 90%).

2.2.1 Carbohydrates

Carbohydrate is the general name for simple sugars as well as complex polysaccharides. The name *carbohydrate* derives from: C = carbo and H_2O = hydrate. Simple sugars or *monosaccharides* are the units or building blocks of a polysaccharide. When a few sugars are joined together in a molecule, it is called an *oligosaccharide*. A polysaccharide which could be a linear or a branched type of molecule is made up of repeating units of the same identical sugar or of different sugars.

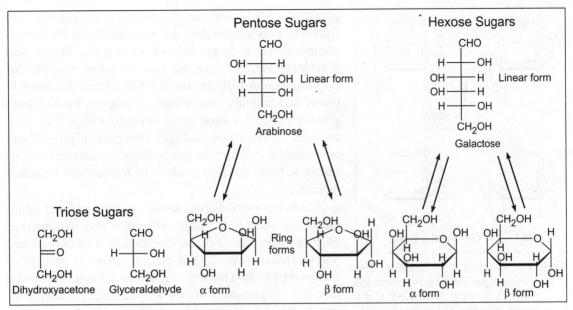

Figure 2.1   Basic structure of sugars. Trioses contain 3 carbons, pentoses contain 5 carbons, and hexoses contain 6 carbons. The sugars are shown in its linear and ring form.

The basic structure of a representative simple sugar is presented in Figure 2.1. The structural formula that is used to represent such a molecule is $(CH_2O)_n$. It is a covalent combination of carbon and water in a one-to-one ratio, and n could be 3, 4, 5, 6, or 7. Among sugars, hexoses and pentoses are the most common. Many hexoses including glucose, mannose, and galactose are biologically important sugars. These sugars are identical to each other except that the orientation of the groups bonded to carbon 2 or 4 differs (Fig. 2.2). For example, in mannose and glucose, the group bonded to carbon 2 is reversed. And in galactose and glucose, the orientation of the groups attached to carbon 4 differs. They can be interconverted to one another by an enzyme called *epimerase*. Sugars with 5 or more carbons can cyclize to form ring structures and they are prevalent in the cells. These forms of sugars occur in two alternative forms, and they are called α and β based on the configuration of the carbon 1 (Fig. 2.1).

In oligosaccharides or polysaccharides, monosaccharides are joined together by dehydration reactions: removal of H_2O and joining of sugars by a glycosidic bond between two of their carbons (Fig. 2.3). Disaccharides are composed of two sugars. For example, lactose is composed of glucose and

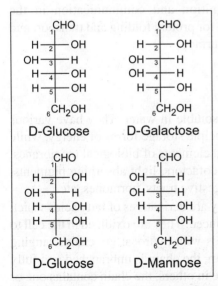

Figure 2.2   Epimers. Glucose, mannose and glucose, galactose are epimers of each other.

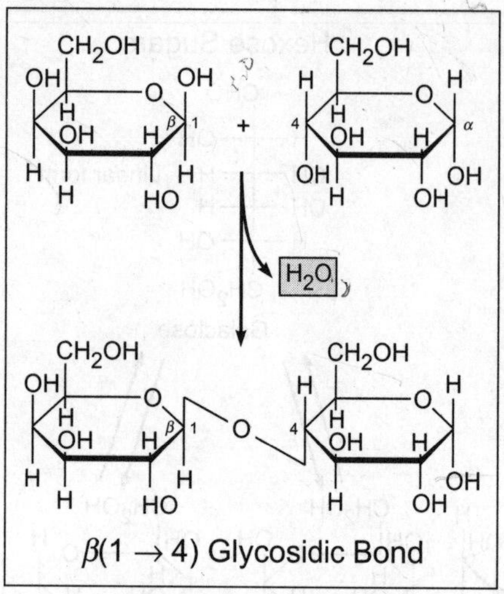

Figure 2.3 Glycosidic bond formation. Two simple sugars are joined by dehydration reaction as water is removed. In the figure, galactose and glucose residues are joined to form a disaccharide, lactose through $\beta\,(1 \rightarrow 4)$ glycosidic linkage.

galactose, while sucrose is made up of glucose and fructose. Polysaccharides are macromolecules/polymers composed of a large number of sugars. Starch and glycogen (Fig. 2.4) are the two common examples of polysaccharides that are stored forms of carbohydrates in plants and animals, respectively. These are formed from glucose molecules alone in the α configuration (Fig. 2.4), through (1-4) carbon linkage. However, glycogen and amylopectin, one form of starch contain occasional $\alpha\,(1\text{-}6)$ linkages. These linkages result in the formation of branched molecules.

Another polysaccharide, cellulose, the principal structural component of the plant cell wall is also composed of glucose molecules of β conformation. It is an unbranched polysaccharide (Fig. 2.4) of great mechanical strength as seen in the plants. This structure is so rigid that it cannot be digested as opposed to starch. Polysaccharides often contain modified sugars linked to various small groups such as amino, sulfate, and acetyl groups. Glycosaminoglycans which are the major polysaccharides of extracellular matrix in animal cells are examples of such modified polysaccharides.

Carbohydrates of the mono- and polysaccharide types are not only important as the energy sources in cells, but are also important as parts of glycoproteins. These molecules are involved in cell-cell recognition and communication in the multicellular context. They are also important for protein folding and transport and targeting of proteins into organelles and cell surface.

2.2.2 Lipids

Lipids are a diverse group of compounds, insoluble in water. They have various biological functions. Fats and oils are the principle storage forms of energy, while phospholipids and sterols are, major structural elements of biological membranes. Other lipids have functions, such as enzyme cofactors, light absorbing pigments, electron carriers, emulsifying agents in the digestive tracts, hormones, etc.

The fats and oils stored in the form of energy are derivatives of fatty acids which are hydrocarbon derivatives and are highly reduced. They are oxidized in the cell to CO_2 and H_2O. Fatty acids are carboxylic acids with hydrocarbon chains ranging from 4–36 carbon atoms long. In some of them, the chain is unbranched and fully saturated, i.e. without any double bonds; while in others, the chain contains one or more double bonds and are referred to as being unsaturated (Fig. 2.5).

Some fatty acids contain three carbon rings, hydroxyl groups, or methyl group branches. The nomenclature of these compounds specifies the chain length and the

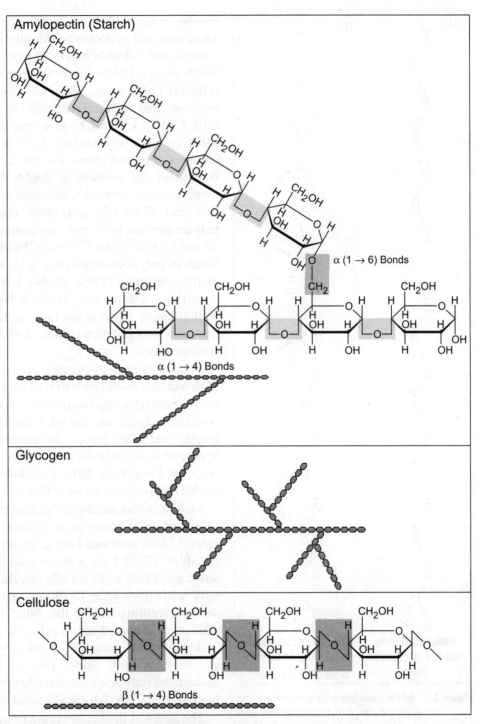

Figure 2.4   Structure of polysaccharides. Polyaccharides are made up of many simple sugars. In the figure, three polysaccharide structures are shown: amylopectin, glycogen, and starch.

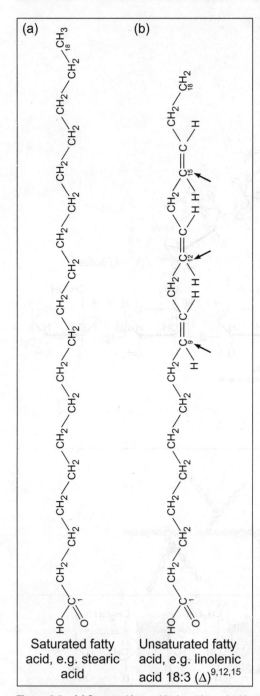

(a) Saturated fatty acid, e.g. stearic acid

(b) Unsaturated fatty acid, e.g. linolenic acid 18:3 $(\Delta)^{9,12,15}$

Figure 2.5 (a) Saturated fatty acids do not have double bonds along the length (b) Unsaturated fatty acids contain at least one double bond along the length

number of double bonds, e.g. palmitic acid 16:0 (with 16 carbons and no double bond) and oleic acid 18:1 (18 carbons and 1 double bond). The positions of double bonds are specified by superscript numbers following Δ (delta). For example, 18 carbon long alpha fatty acid linolenic acid with three double bonds between C9-C10, C12-C13, C15-C16 is written as $18:3(\Delta)^{9,12,15}$. Most commonly occurring fatty acids have even numbers of carbon atoms. Further, a common pattern is seen in the location of double bonds. In most monosaturated fatty acids, the double bond is between C-9 and C-10$(\Delta)^9$, and other double bonds of polyunsaturated fatty acids are generally between C-12 and C13 (Δ^{12}) and C15 and C16 (Δ^{15}). The double bonds of polyunsaturated fatty acids are separated by methylene group. These double bonds of naturally occurring unsaturated fatty acids are in *cis* configuration, whereas the fatty acids obtained from dairy products and meat contain double bonds in *trans* configuration.

Trans fatty acids correlate with increased levels of LDL and decreased levels of HDL in blood. Therefore it is generally recommended that one avoids consuming large amount of these fatty acids. The length and the degree of unsaturation of the hydrocarbon chain in the fatty acid decide its physical property. Longer the fatty acid chain and fewer the double bonds, low is the solubility in water.

Melting points are also influenced by the length and degree of unsaturation of the hydrocarbon chain. At around 25°C, saturated fatty acids between the chain length of 12–24 have a waxy consistency, whereas unsaturated fatty acids are oily liquids. Fully saturated fatty acids have great flexibility and have the most stable conformation in the fully extended form, whereas unsaturated fatty acids, due to the presence of double bonds, cannot be packed together as tightly as saturated fatty acids. Due to loose packing, unsaturated fatty acids also have lower melting points. In vertebrates, free fatty acids circulate in the blood.

The simplest lipid constructed from fatty acids is a *triacylglycerol*. Triacylglycerols are composed of three fatty acids each in ester linkage with a single glycerol. Those with the same fatty acids in all three positions are called *simple triacylglycerols* and are named after the fatty acid content, e.g. tripalmitin, triolein, etc. Most naturally occurring triacylglycerols are derived from mixed fatty acids. These are non-polar hydrophobic molecules, insoluble in water. In most eukaryotic cells, triacylglycerols can be seen as oily droplets in the aqueous cytosol. Specialized cells called *adipocytes* store large amounts of triacylglycerols as fat droplets that nearly fill the cell. Triacylglycerols are also stored in seeds of many plants providing energy and biosynthetic precursors during seed germination. Adipocytes and germinating seeds contain lipases which hydrolyze stored triacylglycerols releasing fatty acids.

Using triacylglycerols as stored fuels is advantageous rather than using polysaccharides such as glycogen and starch. This is because the carbon atoms of fatty acids are more reduced and therefore their oxidation yields more than twice as much energy as compared to carbohydrates. Secondly, triacylglycerols are hydrophobic and therefore unhydrated. Therefore, the organism carrying fat as a fuel does not have to carry the extra weight of water. In some animals, e.g. seals, penguins, walruses, etc., triacylglycerols stored under the skin serve as insulation against low temperature. During hibernation, the huge fat reserves serve the dual purpose of insulation and energy storage. Most vegetable oils, dairy products, and animal meat contain complex mixtures of simple and mixed triacylglycerols.

Waxes are esters of long chain saturated and unsaturated fatty acids with long chain alcohols. They have higher melting points than triacylglycerols. They have diverse functions. They are the chief storage forms of metabolic fuels. Leaves of some plants are coated with a thick layer of waxes that prevent excess evaporation of water. Biological waxes are commonly used in manufacturing of lotions, ointments, and polishes.

Lipids are an extremely important component of biological membrane which acts as a barrier to the passage of polar molecules and ions. One end of the membrane lipids is hydrophobic, while the other is hydrophilic. The most abundant of these membrane lipids are the *glycerophospholipids* in which hydrophobic regions are composed of two fatty acids joined to glycerol. Common glycerophospholipids are phosphatidylethanolamine and phosphatidylcholine. Other lipids found in the membrane are galactolipids, sulpholipids which lack the phosphate group of phospholipids, tetraether lipids (found in archaebacterial membrane), sphingolipids containing sphingosine, a long chain aliphatic amino alcohol but no glycerol, and sterols which have four fused rings and hydroxyl group. *Cholesterol* which is a major sterol in animals is a structural component of membrane and serves as a precursor to a wide variety of steroids.

Another group of lipids present in small amounts acts as *messengers*. Some serve as potent signals, e.g. hormones carried in the blood from one tissue to another or as intracellular messengers generated in response to extracellular signaling. Others function as *enzyme cofactors* in electron transport chain reaction or in transfer of sugar moieties in a glycosylation reaction. A third group consists of pigment molecules that absorb visible light. Phoshatidylinositide and its phosphorylated derivative act at several levels to regulate cell metabolism.

Eicosanoids are paracrine hormones which act only on cells near the point of hormone synthesis instead of being transported to act on cells somewhere else. All eicosanoids are derived from *arachidonic acid*. Three classes of eicosanoids are prostaglandins, thromboxanes, and leukotrienes. *Prostaglandins* act in many tissues by regulating synthesis of the intracellular messenger cAMP. *Thromboxanes* are produced by platelets and are involved in the formation of blood clots and the reduction of blood flow to the site of a clot, whereas *leukotrienes* found in leucocytes act as powerful biological signals.

Steroids are oxidized derivatives of sterols and move through blood stream. On reaching the target cells, they bind to highly specific receptors and trigger changes in gene expression and metabolism. The major groups of steroid hormones are male and female sex hormones produced by the adrenal cortex, e.g. cortisol and aldosterol.

Vitamins A, D, E, and K are fat soluble vitamins which are *isoprenoid compounds* synthesized by the condensation of multiple isoprene units. Vitamin A and D serve as hormone precursors. In addition, Vitamin A also serves as the visual pigment of the vertebrate eye, whereas vitamin D is metabolized by liver and kidney into an active compound that regulates calcium uptake in the intestine and calcium levels in kidney and bones. Vitamin E, also referred to as tocopherol, behaves as a strong antioxidant, protecting fatty acids from oxidation and preventing oxidative damage to membrane lipids. Vitamin K is a crucial component in blood clotting process. Ubiquinones and plastoquinones are isoprenols that function as electron carriers in the oxidation reduction reactions in mitochondria and chloroplast, respectively. Yet another class of lipids involved in activation and addition of sugars on cellular membranes is called *dolichols*.

2.2.3 Nucleic Acids

When DNA is broken down into its component parts, the constituents formed are nitrogenous bases, phosphoric acid, and sugar deoxyribose (Fig. 2.6). Similarly, the constituents of RNA are nitrogenous bases, phosphoric acid, and a ribose sugar. The four nitrogenous bases found in DNA are: Adenine (A), Cytosine (C), Guanine (G), and Thymine (T). RNA contains same bases except that the thymine is replaced by Uracil (U). Among these Adenine and Guanine are the *purine molecules* whereas Cytosine, Thymine and Uracil are *pyrimidines*.

The sugar in RNA is a ribose sugar which differs only in one place from deoxyribose sugar found in DNA (Fig. 2.7). Ribose sugar contains hydroxyl (OH) group in the 2' position, whereas deoxy ribose sugar simply has hydrogen and lacks oxygen. To indicate the positions of the atoms, ordinary numbers are used in the

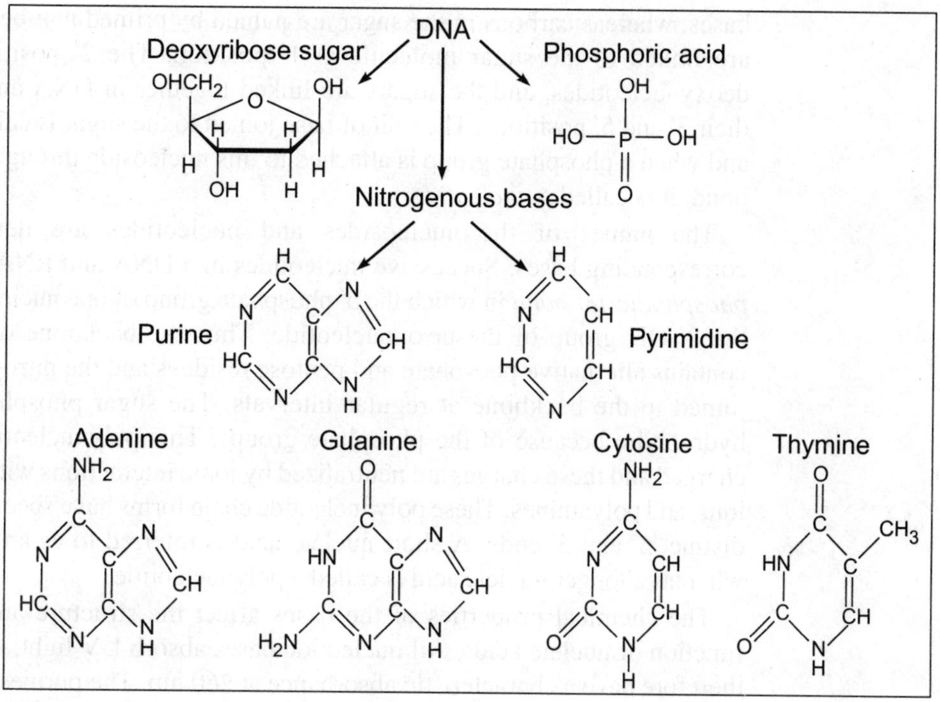

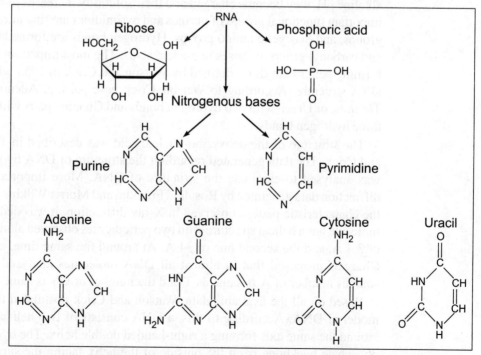

Figure 2.6   Constituents of deoxyribonucleic acid are deoxyribose sugar, nitrogenous bases and phosphoric acid.

Figure 2.7   Constituents of ribonucleic acid are ribose sugar, nitrogenous bases, and phosphoric acid

bases, whereas carbons in the sugar are named by primed numbers. All the bases are linked to the sugar molecule at 1′ position. The 2′ position is deoxy in deoxynucleotides, and the sugars are linked together in DNA and RNA through their 3′ and 5′ positions. The unit of base joined to the sugar is called a *nucleoside* and when a phosphate group is attached to this nucleoside through a phosphoester bond, it is called a *nucleotide*.

The names of the nucleosides and nucleotides are derived from the corresponding bases. Successive nucleotides in a DNA and RNA are linked by a *phosphodiester bond* in which the 5′ phosphate group of one nucleotide is joined to the 3′ OH group of the next nucleotide. Thus the backbone of a nucleic acid contains alternative phosphate and pentose residues and the nitrogenous bases are joined to the backbone at regular intervals. The sugar phosphate backbone is hydrophilic because of the phosphate groups. This polynucleotide is negatively charged and these charges are neutralized by ionic interactions with proteins, metal ions, and polyamines. These polynucleotide chain forms have specific polarity and a distinct 5′ and 3′ ends. A short nucleic acid is referred to as an oligonucleotide, whereas a longer nucleic acid is called a polynucleotide.

The chemical properties of the bases affect the structure and ultimately the function of nucleic acids. All nucleotide bases absorb UV light, and nucleic acids therefore have a characteristic absorbance at 260 nm. The purines and pyrimidines are hydrophobic bases and relatively insoluble in water at neutral pH. At acidic or alkaline pH, they become charged and their solubility increases in water. The most important functional groups of purines and pyrimidines are ring nitrogens, carbonyl groups, and exocyclic amino groups. Hydrogen bonds are formed between amino and carbonyl groups of bases in nucleic acids. The most important hydrogen bond forming pattern are those defined by Watson and Crick in 1953 while elucidating DNA structure. According to Watson–Crick base pairing, Adenosine pairs with Thymine or Uracil with two hydrogen bonds and Guanine pairs with Cytosine with three hydrogen bonds.

The structure of the deoxyribo-nucleic acid was described in 1953 by Watson and Crick. The data generated regarding the structure of DNA by several workers was analysed to elucidate the structure of DNA. More importantly, the X-ray diffraction data generated by Rosalind Franklin and Morris Wilkins was used. From the characteristic pattern observed in X-ray diffraction, it was deduced that DNA molecule has a helical structure with two periodicities observed along their axis, one of 3.4 Å, and the second one of 34 Å. At around the same time, another scientist Chargaff proposed that in almost all DNA molecules analysed from different sources number of A is same as T and the number of G's is same as that of C's.

Based on all the available data, Watson and Crick postulated the double helix model of DNA. According to this, a DNA consists of two helical chains wound around the same axis forming a right-handed double helix. The hydrophilic sugar-phosphate backbone is on the outside of the helix facing the surrounding water, whereas the hydrophobic bases are stacked inside the double helix. The purine and pyrimidine bases of the two strands are base paired with each other through hydrogen bonds and lie perpendicular to the long axis of DNA. The bases are 3.4 Å away from each other and the double helix makes a turn at every 34 Å. The two strands of the DNA are anti-parallel to each other and the base sequence of two strands is complementary to each other. This means that when adenine occurs in one chain, thymine is found in the other. Similarly, when guanine occurs in one chain, cytosine is found in the other. The two strands are held together by hydrogen bonding between complementary base pairs and base stacking interactions.

DNA is a remarkably dynamic molecule. Thermal fluctuations can produce bending, stretching, and melting of the strands. Many variations from Watson–Crick DNA structure are found in cellular DNA. The Watson–Crick model refers to B form of DNA or B-DNA. Under physiological conditions, this is the most stable form of DNA. Many other DNA forms have been discovered among which A and Z forms are well characterized (Fig 2.8). The A form structure is favoured in conditions devoid of water. The DNA is arranged in right-handed helix, but the helix is wider and contains 11 bases per turn. On the other hand, Z-form DNA is a left-handed helical structure. It is more slender and contains 12 base pairs per helical turn. The DNA backbone takes on a zigzag appearance. This structure is more readily formed in sequences containing alternative purines and pyrimidines. Whether or not these alternative forms of DNA occur in cells is uncertain.

A number of structural variations are seen in larger chromosomes which may affect the function of DNA segment. For example, a run of four or more A residues

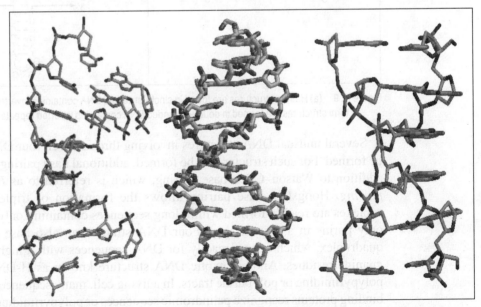

Figure 2.8 A, B, and Z forms of DNA (from left to right). Under physiological conditions, B is the most stable and favourable form of DNA. **(See Colour Plate 1)**

induces a bend in the DNA which may be important for binding of some proteins. The regions of DNA containing inverted repeats of base sequences having two-fold symmetry over two strands of DNA often have the potential to form *hairpin* or *cruciform structures* (Fig 2.9). If the inverted repeats occur within each strand of DNA, the sequence is called a *mirror repeat*. Mirror repeats do not have complementary sequences within the same strand and therefore cannot form hairpin or cruciform structure. On the other hand, palindromic sequences with two-fold symmetry can easily form hairpin or cruciform structures. When only a single strand of DNA is involved, the structure is called hairpin; when both the strands are involved, the structure is called cruciform.

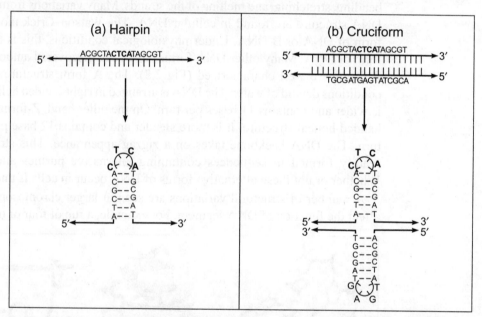

Figure 2.9 **(a)** Hairpin structures are found in single-stranded DNA containing inverted repeats **(b)** Cruciform structures are formed in double-stranded DNA containing inverted repeats.

Several unusual DNA structures involving three or even four DNA strands can be formed. For such structures to be formed, additional base pairing is necessary in addition to Watson–Crick base pairing, which is referred to as *Hoogstein base pairing*. Hoogstein base pairing allows the formation of triplex DNAs. The triplexes are readily formed within long sequences containing only pyrimidine or only purine in a given strand. Four DNA strands can also base pair to form a quadruplex, which occurs readily for DNA sequences with a high proportion of guanine residues. Another exotic DNA structure known as H-DNA is found in polypyrimidine or polypurine tracts. In a living cell, many sequence-specific DNA binding proteins recognize palindromic sequences or polypyrimidine or polypurine sequences that can form triple helices, and therefore, these alternative structures which could be formed in DNA may have a role to play in regulation of gene expression.

RNA has a variety of functions within a cell and for each function, a specific type of RNA is required. RNA molecules differ in chain length and secondary and tertiary structures. mRNA is involved in carrying information from DNA to the site of protein synthesis. rRNA makes the most of the ribosome, whereas small tRNA is involved in transfer of amino acids to the ribosomes. All these RNAs are transcribed as single-stranded molecules and all of these assume different secondary structures. The single strand tends to assume a right-handed helical structure dominated by base stacking interactions. The 3-D structures of these RNAs are complex and unique. The predominant double-stranded structure is an A form right-handed double helix and the B form of RNA has not been observed. Hairpin loops formed between self-complementary sequences are the most common type of secondary structure found in RNA.

2.2.4 Proteins

Proteins (derived from the Greek word 'Proteios' which means 'of the first rank') are the most abundant macromolecules in a cell. As the name indicates, proteins are actually the major players of various biological activity including the cell's survivability, basic metabolic activities, other specific functions (e.g., cell signaling, cellular transport, cell immunity, etc.), and reproduction. Grossly, they are classified as *structural proteins* that are involved in the cell's architecture and organization, and *functional proteins* including the biological catalysts or enzymes. Due to their wide array of activities, they are often referred as the 'doers of the cell'.

The basic constituents of proteins are called amino acids. Twenty different types of amino acid molecules (monomers) polymerize in various combinations and order to form different protein polymers. The various components of an amino acid are: a central α chiral carbon atom ($C\alpha$) bonded to an amino (NH_2) group, a carboxyl (COOH) group, a hydrogen (H) atom, and a variable group called a side chain/R group (Fig. 2.10). It is the side chain that determines the specificity of the amino acid, and thereby its structure and function. Due to the presence of asymmetric carbon atoms in all amino acids but glycine, they can exist in two mirror-image forms, called the *dextro* (D) or *levo* (L) *isomers*. Generally, the L forms of amino acids are part of the proteins in living organisms.

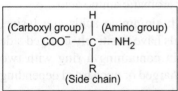

Figure 2.10 Structure of an amino acid. An amino acid consists of a central carbon to which are attached an amino group, a carboxyl group, and a side chain. Side chain is specific for each amino acid

The structure and functions of proteins are largely dependent on the constituent amino acids. Therefore, it is important to know the characteristics of various amino acids, which are imparted largely by their side chains. Amino acids are classified variously based on the size, shape, charge, hydrophobicity, and chemical reactivity of the side chains. However, for our basic understanding, four major categories are to be remembered: *nonpolar*, *polar*, *basic*, and *acidic* amino acids.

There are ten nonpolar amino acids, namely glycine, alanine, valine, leucine, isoleucine, proline, cysteine, methionine, phenylalanine, and tryptophan. Such amino acids with nonpolar side chains are hydrophobic and insoluble in water. The hydrophobicity is proportional to the length of the side chain. Due to their insolubility in water, these hydrophobic amino acids reside in the interior of proteins. Among these, glycine is the smallest and simplest amino acid with only one hydrogen atom as the side chain (Fig. 2.11). It fits into tight spaces. Alanine, valine, leucine, isoleucine, cysteine, and methionine are non-cyclic amino acids, and are with hydrocarbon (up to 4 carbon atoms) side chains, except cysteine and methionine which have one sulfhydryl (SH) group and a sulphur atom, respectively, in addition to hydrocarbon. The SH group can oxidize to form a covalent disulfide bond to a second cysteine. Disulfide bonds stabilize the folded structure of proteins. These amino acids, except cysteine and methionine, are also called *aliphatic amino acids*. Proline has also a hydrocarbon side chain but has a cyclic structure: its side chain is bonded to the nitrogen of the amino group and also to the $C\alpha$. Therefore, proline is very rigid and forms a kink in a protein chain. The remaining two nonpolar amino acids, phenylalanine and tryptophan, have large and bulky side chains containing hydrophobic aromatic rings.

There are five polar amino acids with uncharged and polar side chains (Fig. 2.11). These are: serine, threonine, tyrosine, asparagines, and glutamine. Among these, serine, threonine, and tyrosine have hydroxyl groups on their side chains, whereas asparagine and glutamine have polar amide ($O=C-NH_2$) groups. These amino acids having polar side chains are hydrophilic, and they tend to orient themselves to the exterior of the proteins. It is to be noted that tyrosine having an aromatic ring in its side chain is also grouped under aromatic amino acids.

The three amino acids that are grouped under basic amino acids are: lysine, arginine, and histidine (Fig. 2.11). These amino acids have positively charged side chains. However, histidine which has a side chain containing a ring with two nitrogens called *imidazole* can be either positively charged or uncharged depending on the change of pH in the environment. The activity of some proteins, enzymes in particular, is modulated by the shift in environmental acidity through protonation or deprotonation of histidine side chains.

The remaining two amino acids, aspartic acid and glutamic acid, are grouped under acidic amino acids (Fig. 2.11). They have negatively charged side chains due to the presence of carboxylic acid groups. These amino acids being hydrophilic occur at the exterior of the proteins.

All these 20 amino acids are not uniformly present in all proteins. These vary widely in different proteins. However, in general, leucine, serine, lysine, and glutamic acid are the most abundant (~32%) amino acids in a typical protein; whereas cysteine, typtophan, and methionine are the rare (~5%) amino acids.

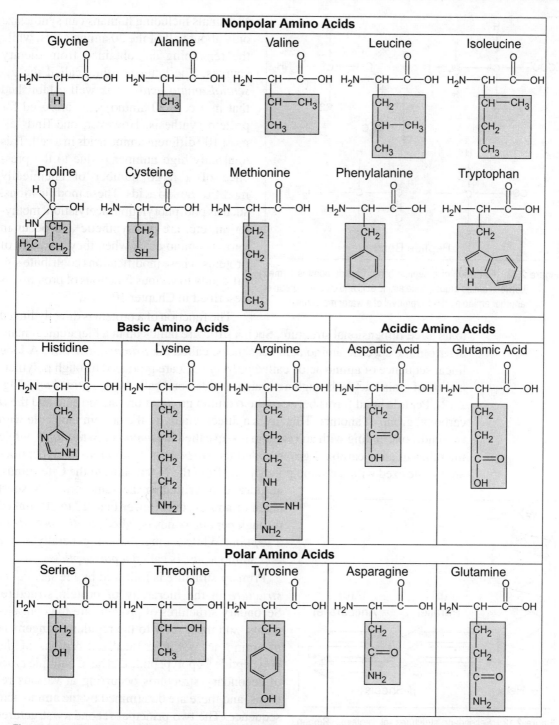

Figure 2.11 Structure of amino acids. The figure divides the amino acids into 4 groups: nonpolar, polar, acidic, and basic amino acids according to the groups which are there in the side chain

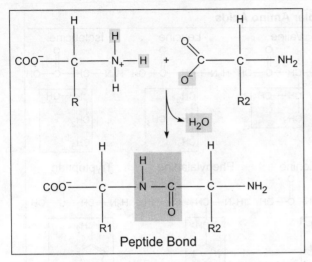

Figure 2.12 Formation of a peptide bond. Peptide bond is formed between amino group of one amino acid and carboxyl group of another amino acid with removal of a water molecule

Mammals including humans can synthesize only about half of the 20 amino acids, while the remaining are obtained from dietary sources. The latter are thus called the *essential amino acids*. It is well established that in a cell, 20 amino acids are used for protein synthesis. However, one finds almost 100 different amino acids in a cell. This unusually high number is due to the presence of a large number of chemically modified amino acids. These modifications, such as phosphorylation, acetylation, methylation, etc. are postsynthetic and occur in various amino acids when they are parts of proteins. These modifications contribute significantly to various functions of proteins, as described in Chapter 10.

The function of a protein is directly linked to its three-dimensional structure. Such a structure builds up in a hierarchical manner from the specific amino acid sequences, called the *primary structure*. A long linear sequence of amino acids called *polypeptides* are produced through polymerization of amino acids through a covalent amide bond called the *peptide bond* (Fig. 2.12). Peptide bond forms between the α amino group of one amino acid and the α carboxyl group of another. This gives a directionality of the protein molecule with two ends, one ending with an α amino group called the *amino* or *N-terminus*, while the other in an α carboxyl group called the *carboxy* or *C-terminus*. Polypeptides are synthesized with the same polarity, i.e. from the N-terminus to the C-terminus, and are also written in the same order. A short chain of amino acids (fewer than 20 to 30) linked through peptide bonds is called an *oligopeptide* or a *peptide*. While a longer chain, generally above 200 amino acids, is called a *polypeptide*.

Primary structure is followed by the *secondary structure* in the hierarchy of protein structure. Secondary structures of proteins are the core elements, and these refer to the regular arrangement of amino acids within localized regions of the polypeptide. A polypeptide can have multiple types of secondary structures occurring at various regions, and these are determined by the amino acid sequence. The two principal secondary structures are α helix and the β sheet (Fig. 2.13). These contribute to almost 60% of a protein, in general.

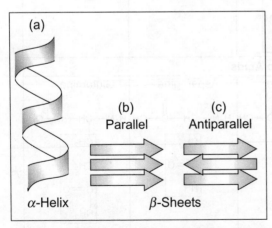

Figure 2.13 Secondary structures of proteins. Ribbon structure of **(a)** α-helix, **(b)** parallel β-sheet, and **(c)** antiparallel β-sheet.

These structures are held together by hydrogen bonds between the CO and NH groups of peptide bonds. In case of α helix, the CO group of one peptide bond forms a hydrogen bond with the NH group of another peptide bond located 4 residues downstream in the polypeptide chain (Fig. 2.14). This results in the coiling of a region of the polypeptide around itself. On the other hand, in case of a β sheet, as the name indicates, two parts of a polypeptide chain lie side by side with hydrogen bonds between them (Fig. 2.14).

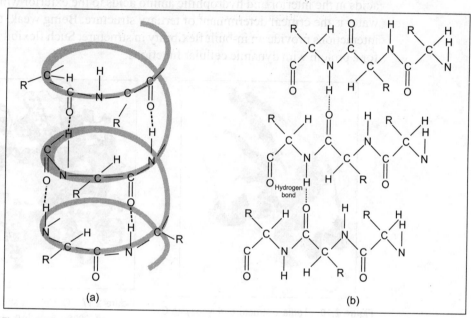

(a)   (b)

Figure 2.14   Secondary structure of **(a)** α-helix showing hydrogen bonds between 1st and 5th amino acids. **(b)** Parallel β-sheet showing interchain hydrogen bonds

Another secondary structure called a short U-shaped β-turn is also present in polypeptides. These are present on the surface of proteins. They form sharp bends which reverse the direction of the backbone of the polypeptide towards the interior of the proteins. These structures are often stabilized by a hydrogen bond between their end residues (Fig. 2.15). The two most frequently occurring amino acids involved in generation of this structure are glycine and proline. β-turn results in folding of large proteins into highly compact structures. The regions of proteins devoid of such secondary structures are referred to have irregular structures.

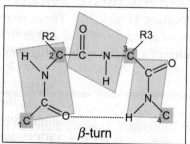

Figure 2.15   Structure of β-turn. In this type of secondary structure, there is hydrogen bonding between 1st and 4th amino acids

The third level of the protein structure hierarchy is the *tertiary structure*. It refers to the overall conformation of a protein resulted due to three-dimensional arrangement of its amino acid residues. It is essentially the folding of the polypeptide chain due to long range interactions between the side chains of amino acids located in different regions of the primary sequence of the polypeptide (Fig. 2.16). The α-helices and β-sheets connected by the loop regions of the polypeptide fold into globular structures called *domains*. These domains are the basic units of the tertiary structure of proteins. In proteins, the tertiary structures are stabilized by hydrophobic interactions between nonpolar side groups along with hydrogen bonds between polar side groups and peptide bonds. Essentially, orientation of the hydrophobic amino acids in the interior and hydrophilic amino acids to the exterior which interact with water is the critical determinant of tertiary structure. Being weak, these stabilizing interactions provide an in-built flexibility in structure. Such flexibility is well-suited for a protein with dynamic cellular functions.

Figure 2.16 Tertiary structure of lysozyme C from *Gallus gallus*. (See Colour Plate 1)

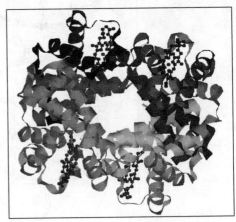

Figure 2.17 Quaternary structure of haemoglobin. It consists of two α-chains and two β-chains. (See Colour Plate 1)

The fourth level of protein structure, called the *quaternary structure*, involves interactions between different polypeptide chains in a protein containing more than one polypeptide. For example, haemoglobin, a tetramer of two types of polypeptide chains, α and β, has a distinctive quaternary structure (Fig. 2.17). These four polypeptides are held together by the similar types of interactions involved in the tertiary structures of proteins, leading to the generation of a functional haemoglobin.

SUMMARY

A cell is chemically composed of small inorganic molecules, organic molecules, polymers of organic molecules, and water. Water contributes to the bulk of the cell, and facilitates macromolecular structure and interactions needed for various metabolic processes. The major part of the organic molecules are polymers which include polysaccharides, lipids, nucleic acids, and proteins. Of these, polysaccharides, nucleic acids, and proteins are referred to as macromolecules.

Carbohydrates

Carbohydrates represent simple sugars and polymers of sugars called the polysaccharides. Polysaccharides often serve as storage forms of sugars, utilized for various structures and functions of a cell, as structural components (cellulose, glycoproteins on membranes), energy source, and cell-cell recognition function.

Lipids

Lipids are water insoluble components of diverse cellular structures. Fats and oils are used universally as stored forms of energy and are derivatives of fatty acids which are either saturated or unsaturated. Triacylglycerols are primary storage fats in specialized cells called adipocytes. Lipids are the major components of membrane, the most abundant amongst them are glycerophospholipids. Some lipids play a major role as signaling molecules, cofactors, and pigments. Phosphatidylinositol bisphosphate is very commonly used as a signaling molecule. Thus, lipids are involved in a variety of functions in a cell. On one hand, they are important structural components, while on the other hand, they play an important role in cell metabolism.

Nucleic acids

The basic constituents of both the nucleic acids (DNA and RNA) are nitrogenous bases, phosphoric acid, and pentose sugar. DNA has a deoxyribose sugar, whereas RNA contains a ribose sugar. Among the nitrogenous bases, DNA has adenine, guanine, cytosine, and thymine, while in RNA, thymine is replaced by uracil. In a polynucleotide chain, the successive nucleotides are linked by a phosphodiester bond. DNA is a double-stranded structure, while RNA is single-stranded. In double-stranded DNA helix, the bases on the opposite strands are hydrogen bonded to each other according to Watson–Crick base pair. The structure of DNA is dynamic and can exist in different forms such as A, B, Z forms. Under physiological conditions, B form is the most stable form of DNA which is a right-handed helical structure. The regions of DNA containing inverted repeats of base sequences often have the potential to form hairpin or cruciform structure. RNA which is transcribed as a single-stranded molecule always assumes different secondary structures. The predominant structure found in RNA is a right-handed double helix.

Proteins

Proteins are the major constituents of a cell. They are also the major functional components of various machinery that control the cell. The monomeric units, amino acids, polymerize in a distinct order and sequence as determined by the genetic code to form polypeptides. The amino acid sequence of a polypeptide is referred to as the primary structure of proteins. The primary structure undergoes folding and compaction to generate a three-dimensional functional structure in a hierarchical manner dictated by the amino acid sequence. The other structures in the hierarchy are called secondary structure (α helix and β sheets), tertiary structure, and quaternary structure.

SHORT ANSWER QUESTIONS

1. Explain why nature has chosen water as the medium for dissolving various constituents in a cell.
2. Why is Vitamin A and Vitamin D supplementation recommended to children?
3. What is the difference between glucose and galactose? Are they interconvertible? If yes, how?
4. Distinguish between amylopectin and cellulose.
5. What are biological polymers? Which ones are referred to as macromolecules and why?
6. Explain why a common procedure for cleaning the grease trapped in a sink is to use a product containing sodium hydroxide.
7. The melting point of oleic acid (18:1) (Δ^9) is 13.4°C while that of lauric acid (12:0) is 44°C. Explain.
8. Draw a general structure of deoxynucleoside triphosphate. Show the structure in detail and indicate the positions of attachment of a base and a phosphate.
9. Why is it said that the structure of DNA is dynamic?
10. What are the characteristic properties of B and Z forms of DNA?
11. What are the roles of hydrophobic amino acids in the formation of 3D-structure of a protein?
12. What are essential amino acids and why are they called so?
13. What do you understand by tertiary structure of a protein?

FURTHER READING

- Berg, J. M., Tymoczko, J.L., and Stryer, L. (2007). *Biochemistry*, 6th ed. Freeman and Co.
- Branden, C. and Tooze, J. (1999). *Introduction to Protein Structure*, 2nd ed. New York: Garland.
- Dowhan, W. (1997). Molecular basis for membrane phospholipid diversity; Why are there so many lipids? *Ann. Rev. Biochem.* **66**, 199–232.
- Gimona, M. (2006). Protein linguistics–A grammer for modular protein assembly. *Nat. Rev. Mol. Cell Biol.* **7**, 68–73.
- Hoekstra, D. (Ed.) (1994). Cell lipids. *Current topics in membranes.* **4**, Academic press Inc. San Diego.

- Kendrew, J.C. (1961). The three-dimensional structure of a protein molecule. *Sci. Amer.* **205**, 96–111.
- Nelson, D.L. and Cox, M.M. (2005). Lehninger Principles of Biochemistry, 4th ed. W.H. Freeman and Co.
- Richardson, J.S. (1981). The anatomy and taxonomy of protein structure. *Adv. Protein chem.* **34**, 167–339.
- Sanger, F. (1988). Sequences, sequences, and sequences. *Ann. Rev. Biochem.* **57**, 1–28.
- Sharon, N. (1980). Carbohydrates. *Sci. Amer.* **243**, 90–116.
- Voet, D. and Voet, J. (2004). *Biochemistry*, 3rd ed. Prentice Hall.
- Watson, J.D. and Crick, F.H.C. (1953). Molecular structures of the nucleic acid: A structure for deoxyribonucleic acid. *Nature* **171,** 737–738.
- Watson, J.D. (1968). The double helix: A personal account of the discovery of structure of DNA. *Atheneun*, New york.
- Wells, R.D. (1988). Unusual DNA structures. *J. Biol. Chem.* **263**, 1095–1098.
- www.rcsb.org/ The protein 3D structure database.
- www.expasy.org/prosite/ PROSITE database of protein families and domains.

3

Enzymes in Molecular Biology Research

OVERVIEW

- The advances in recombinant DNA technology have been possible because of the use of different enzymes to manipulate the DNA molecule in vitro.

- This chapter describes these enzymes as either DNA modifying enzymes or DNA degrading enzymes. The function of most of the enzymes commonly used in molecular biology research is described and their importance explained.

The basis of recombinant DNA technology is the ability to manipulate DNA molecules in the test tube. This, in turn, depends on the availability of purified enzymes whose activities are known and can be controlled. Therefore, these enzymes can be used to make specific changes to the DNA molecules that need to be manipulated. The enzymes used in recombinant DNA technology can be broadly categorized as *nucleic acid degrading* and *nucleic acid modifying enzymes.*

3.1 DNA MODIFYING ENZYMES

These enzymes which are present in the live cells are involved in many crucial functions. Their *in vivo* catalytic characteristics are routinely used to manipulate DNA *in vitro*.

3.1.1 Polymerases

DNA polymerases are a group of enzymes that catalyze the synthesis of polydeoxyribonucleotides from monodeoxyribonucleotide triphosphates (dNTPs), performing the most fundamental function *in vivo* of DNA replication and repair. Different types of polymerases exist in both prokaryotes and eukaryotes. All DNA polymerases for synthesis of DNA require in addition to dNTPs, an initiating oligo-nucleotide called a *primer*, carrying a 3′ end hydroxyl group that is used as a starting point of chain growth. DNA polymerases cannot initiate synthesis *de novo* from mononucleotides. Most of the polymerases are template dependent. The primer, by annealing to the complementary region in the DNA template, provides a double stranded structure to the DNA polymerase. The enzyme then moves along the DNA template extending the primer in 5′-3′ direction (Fig. 3.1) according to the

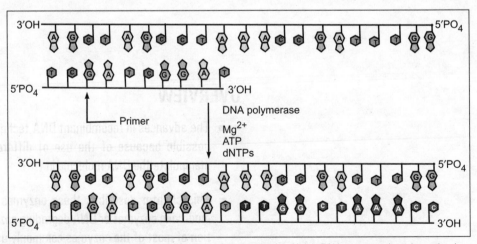

Figure 3.1 DNA polymerase has 5′-3′ polymerization activity. Using DNA as a template, it synthesizes polynucleotide chain by extending the 3′ end of the primer annealed to the DNA template.

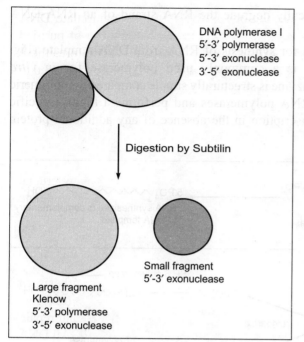

DNA polymerase I
5'-3' polymerase
5'-3' exonuclease
3'-5' exonuclease

Digestion by Subtilin

Small fragment
5'-3' exonuclease

Large fragment
Klenow
5'-3' polymerase
3'-5' exonuclease

Figure 3.2 DNA polymerase has one small and one large sub-unit. These can be separated on cleavage of holoenzyme by protease subtilin giving one large and one small subunit. The large subunit is called as Klenow and retains 5'-3' polymerase and 3'-5' exonuclease activity, while the small subunit has 5'-3' exonuclease activity.

Watson-Crick base pairing rule, which says A pairs with T and C pairs with G. In addition to major 5'-3' polymerase activity, most of the polymerases have 3'-5' proof reading activity responsible for minimizing the error in newly synthesized DNA.

Most commonly used polymerase is *DNA polymerase I* from *E. coli*. This enzyme in addition to the 5'-3' polymerase and 3'-5' exonuclease activities also has 5'-3' exonuclease activity. Its 5'-3' exonuclease activity is of great advantage in the process of nick translation. This enzyme is used in nick translation reaction for preparation of labelled DNA probe during hybridization. Modified version of this enzyme called *Klenow* is more commonly used especially in many DNA labeling reactions, such as random primer labelling or 5' end labelling. Klenow is a large fragment of DNA polymerase 1 obtained on cleavage by the protease, subtilin. It retains 5'-3' polymerase and 3'-5' exonuclease activity (Fig. 3.2). The small fragment has 5'-3' exonuclease activity.

Among other DNA polymerases used in recombinant DNA technology are *T4 DNA polymerase, Terminal deoxynucleotidyl transferase, Taq polymerase,* etc. T4 DNA polymerase has both 5'-3' and 3'-5' exonuclease activities. In absence of nucleotides, the exonuclease activity is very strong. The enzyme is used in 5' end labelling reaction and also during Sanger's plus-minus sequencing reactions. Terminal deoxynucleotidyl transferase is an un-usual polymerase, since it is template independent, unlike most other polymerases. It adds nucleotides randomly using 3' end of the DNA molecule as a primer. This enzyme is used for 3' end labelling of DNA.

Taq polymerase is an extremely useful enzyme during polymerase chain reaction (PCR) used for selective amplification of a desired gene (explained in details in chapter 4). This enzyme is thermostable and therefore, performs polymerization reactions at higher temperatures. One additional DNA polymerase important in molecular biology research is reverse transcritpase, which is an RNA dependent DNA polymerase and makes DNA copies using RNA as a template. These copies are called complementary DNAs (cDNAs). These enzymes naturally are involved in the replication of retroviruses. Similar to DNA polymerase, reverse transcriptase requires magnesium, dNTPs, an RNA template and a primer with a 3' OH terminus for polymerization reaction. RNase H activity is found to be associated with reverse transcriptase that can specifically degrade the RNA strand of an RNA-DNA hybrid.

RNA polymerases are used for synthesis of RNA from DNA template (Fig. 3.3a). T7 RNA polymerase is a commonly used polymerase for *in vitro* transcription reactions. This enzyme is structurally simple compared to multimeric prokaryotic and eukaryotic RNA polymerases and performs a highly specific promoter recognition and transcription in the absence of any additional protein factors.

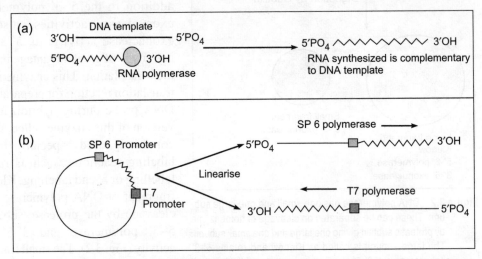

Figure 3.3   **(a)** RNA polymerase is DNA dependent RNA polymerase and synthesizes RNA complementary to DNA template in 5′-3′ direction. **(b)** Special vectors are used in recombinant DNA technology to prepare sense-and anti sense riboprobes. This is achieved by transcribing the cloned gene in the vector, once using T7 polymerase and later using SP6 polymerase.

In recombinant DNA technology, T7 RNA polymerase is used as a powerful tool in driving high-level expressions of cloned genes and in preparing *in vitro* a large amount of defined RNA transcripts. Like T7 RNA polymerase, SP6 RNA polymerase also displays a stringent specificity for its own promoter to start transcription. This feature is utilized in developing vectors for *in vitro* transcription. In these vectors, gene can be cloned in such a way that it is flanked by 2 promoters (namely T7 and SP6) on either side. Using polymerases specific for these promoters, namely T7 polymerase and SP6 polymerase, it is possible to synthesize sense- and anti-sense strands during *in vitro* transcription (Fig. 3.3b).

3.1.2 Ligases

DNA ligases catalyze the formation of a phosphodiester bond between 5′ phosphate and a 3′ hydroxyl end separated by a nick (Fig. 3.4a and b). Bacterial DNA ligases,

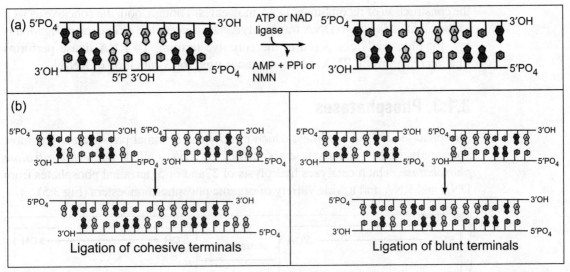

Figure 3.4 **(a)** Ligase joins the two DNA fragments by synthesizing a phosphodiester bond between 5'PO$_4$ and 3'OH termini separated by a nick. Ligase requires ATP (T4 DNA ligase) or NAD (*E. coli* DNA ligase) for making a bond. **(b)** Ligase can join both cohesive and blunt termini of the DNA fragments.

e.g., from *E. coli, B. subtilis*, use the hydrolysis of NAD as their energy source whereas ATP is the cofactor for DNA ligases from bacteriophages (e.g., T4 and T7) and eukaryotic cells. *In vivo,* the enzyme is absolutely essential for joining of Okazaki fragments during DNA replication. The enzyme is present ubiquitously in all living organisms. Most commonly used ligases in molecular biology research are *E. coli* DNA ligase and T4 DNA ligase.

The ligation reaction is carried out by a three step mechanism involving two relatively stable intermediates. The first step is the covalent transfer of the S-adenyl group of NAD or ATP to an epsilon-amino group of a lysine residue in the enzyme forming a ligase-adenylate intermediate with the concomitant release of NMN or ppi. This reaction occurs regardless of the absence or presence of DNA substrates. The AMP transfer results in the activation of 5' phosphate group by forming a new pyrophosphate linkage in the second intermediate, DNA-adenylate.

The DNA-AMP intermediate releases a stoichiometric amount of AMP on formation of a phosphodiester bond with a 3' OH group of the acceptor molecule. This final step is the nucleophilic displacement of the AMP by the attack of 3' OH at the activated 5' PO$_4$ group. The individual steps of the ligase reaction are reversible.

Many different factors including temperature, enzyme concentration, substrate DNA concentration, and the shape and flexibilities of the DNA molecules affect the rate of ligation and the nature of the end products. Therefore, the molar ratio of two DNA molecules to be ligated as well as the lengths of the DNA fragments affect the outcome of the ligation products. DNA ligases play the most important role in the construction of recombinant DNA molecules. Though, both the cohesive as well as the blunt ends of two DNA fragments can be ligated, the efficiency of ligation is low for blunt ends compared to the cohesive termini. T4 RNA ligase performs ligation reactions on RNA substrates and is used for 3′ labelling of RNA.

3.1.3 Phosphatases

Phosphatases are the enzymes, which remove the terminal phosphates. The most commonly used phosphatase is from *E. coli* called *bacterial alkaline phosphatase,* which catalyzes hydrolysis of 3′ and or 5′ terminal phosphates from DNA and RNA and a wide variety of organic phospho-monoesters (Fig 3.5).

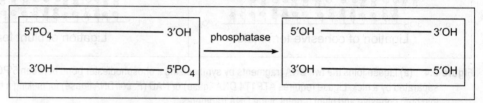

Figure 3.5 Alkaline phosphatase dephosphorylates the termini by removing the terminal phosphate. This process is referred to as baptisation.

The rate and efficiency of phosphatase reaction depends on pH, ionic strength, and temperature. The optimum pH is between 8–9.5 and the reaction has no detectable activity below pH 6. The optimum temperature for reaction is around 65°C. The activity increases up to 10 fold as the concentration of the salts, e.g., KCl or NaCl increases from 0.1–1 M.

Zn^{2+} is required for catalytic activity as well as for stability of the enzyme. The enzyme is mainly used for dephosphorylating the termini from DNA or RNA. During molecular cloning, dephosphorylation of 5′ PO_4 of the vector termini prevents self-ligation of vector, thereby increasing the yield of intermolecular ligation products.

3.1.4 Polynucleotide Kinases

Kinases transfer the phosphate group. They are an extremely diverse group of enzymes that catalyze the transfer of gamma phosphoryl group from NTP to an acceptor molecule. Depending on the nature of the phosphate acceptor, kinases are largely divided into protein kinases, carbohydrate kinases, and *polynucleotide kinases.*

The most commonly used polynucleotide kinase is T4 polynucleotide kinase, which transfers terminal phosphate group of ATP to the 5′ OH group of DNA,

RNA, and oligonucleotides. This reaction is fully reversible and thus, under appropriate conditions, polynucleotide kinase catalyzes both phosphorylation of dephosphorylated end termed as *forward reaction* or can exchange phosphate with phosphorylated end called *exchange reaction* (Fig 3.6). This enzyme is widely used to label the 5′ termini of nucleic acids.

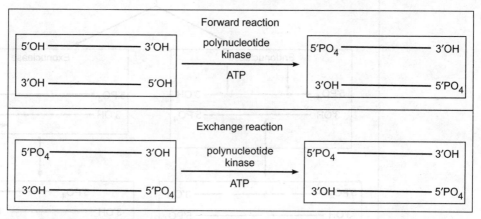

Figure 3.6 Polynucleotide kinase phosphorylates the termini by transferring the phosphate group from ATP, either in exchange reaction or forward reaction. Forward reaction is more efficient.

3.2 DNA DEGRADING ENZYMES

Nucleic acid degrading enzymes are referred to as *nucleases*. These can be either sequence-independent nucleases comprising of both exo- and endonucleases or sequence-dependent nucleases called *restriction endonucleases*.

3.2.1 Nucleases: Endo- and Exonucleases

They comprise large families of enzymes carrying out diverse reactions and with a range of substrate specificity. They can be DNAses, degrading DNA molecule or RNAses, degrading RNA molecule, whereas some can degrade either of them. Further, they can be single strand- or double strand specific. These enzymes can be endonucleases, which cleave DNA randomly along the length or exonucleases, which remove mononucleotides one by one starting from either 5′ or 3′ end (Fig. 3.7).

On cleavage of polynucleotide chain, the termini produced can be 3′ phosphate, 5′ hydroxy, or 5′ phosphate and 3′ hydroxy. Another class of promising new generation nucleases includes artificial and semi-artificial nucleases.

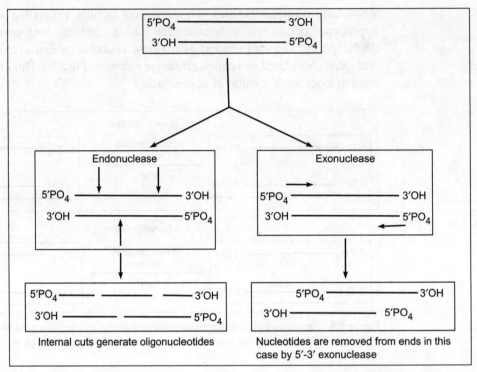

Figure 3.7 Nucleases are either endonucleases that cleave nucleic acids along the length or exonucleases which attack the molecules from one end.

Chemical nucleases are typical artificial nucleases that make use of redox active compounds (Phenanthronine-copper and Ferrous-EDTA) as the cutter and an attached oligonucleotide as the site specific recognition module.

Semi-artificial nucleases include chimeric nucleases composed of heterogenous structural parts brought together by the use of recombinant DNA techniques and hybrid nucleases derived from classical nucleases by incorporating a piece of synthetic oligonucleotide in the active site to harness sequence specificity.

Among the most commonly used DNases is *bovine pancreatic deoxyribonuclease* (DNase I) that cleaves both single stranded DNA (ssDNA) and double stranded DNA (dsDNA) to produce 5′ PO_4 dinucleotide and 5′ PO_4 oligonucleotides. The enzyme requires divalent cations, Ca^{2+} and Mg^{2+} as cofactors for cleavage.

DNase I is distinct from DNase II, a lysosomal acidic DNase found in various organs such as thymus, liver, and spleen. DNase I prefers a duplex region for cleavage, whereas DNase II prefers single stranded region for its activity. DNase I is commonly used as nonspecific nuclease to eliminate contaminating cellular DNAs in routine RNA preparations. It is also used to remove template DNAs after *in vitro* transcription, in random deletion mutagenesis, and in the reaction of nick translation for preparation of labelled probe for hybridization reactions.

Micrococcal nuclease is another commonly used nuclease with Ca^{2+} dependent phosphodiesterase activity, which cleaves both DNA and RNA to yield 3′-phosphomononucleotide and 3′-phosphooligonucleotide end products. This enzyme is used for degradation of chromatin in nucleosomal ladder, removing nonspecific DNA and RNA from cell free extracts used for *in vitro* translation.

Among the commonly used RNases is *ribonuclease from bovine pancreas* (RNase A). It is frequently used to degrade RNA (Fig. 3.8a) in plasmid and other DNA preparations. This enzyme has associated DNase activity, which needs to be inhibited before using the enzyme for degrading RNA. Since the enzyme cleaves RNA at pyrimidine bases (C and U), it is used for RNA sequencing.

Ribonulcease H (RNase H) is a ribo-endonuclease that specifically degrades RNA strand of DNA-RNA heteroduplex (Fig. 3.8b). The oligoriboucleotides formed have 5′ PO_4 and 3′ OH termini. This enzyme plays an important role in cDNA synthesis and RNA mapping.

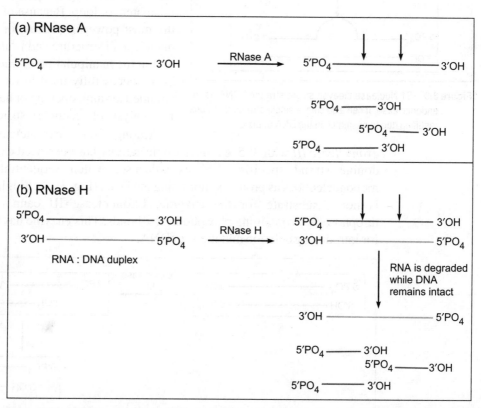

Figure 3.8 **(a)** RNase A degrades RNA along the length. **(b)** RNase H attacks RNA in RNA:DNA duplex and selectively cleaves RNA.

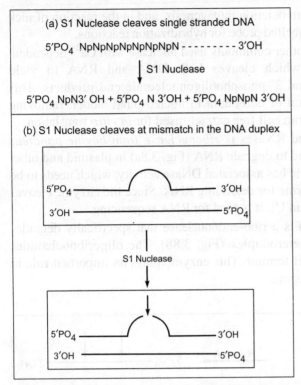

(a) S1 Nuclease cleaves single stranded DNA

5'PO$_4$ NpNpNpNpNpNpNpN - - - - - - - - 3'OH

↓ S1 Nuclease

5'PO$_4$ NpN3' OH + 5'PO$_4$ N 3'OH + 5'PO$_4$ NpNpN 3'OH

(b) S1 Nuclease cleaves at mismatch in the DNA duplex

5'PO$_4$ ———————— 3'OH

3'OH ———————— 5'PO$_4$

S1 Nuclease

↓

5'PO$_4$ ———————— 3'OH

3'OH ———————— 5'PO$_4$

Figure 3.9 S1 Nuclease cleaves single stranded DNA as an endonuclease. It can also cleave single stranded regions resulting due to mismatch in the DNA duplex.

Some endonucleases hydrolyze both DNA and RNA in a highly single strand specific manner. These nucleases degrade single stranded polynucleotides as well as single strand regions in double stranded polynucleotides. The ability to discriminate between double stranded and single stranded regions of polynucleotides makes these endonucleases invaluable in fine structure analysis of nucleic acids. Nuclease S1 from *Aspergillus oryzae* is highly suitable for single strand specific endonuclease, and is commonly used for manipulations of DNA and RNA in recombinant DNA work. The nuclease activity gives rise to 5'-mononucleotides (Fig. 3.9). This enzyme also cleaves double stranded DNA at the single stranded regions produced by nick, gap, mismatch, or loop. Because of this, it is one of the most powerful probes for the analysis of nucleic acid structure and is also an important tool in the manipulation of nucleic acid. It has been successfully used in cleavage of single stranded termini, opening of hairpin loops, and for analysis of secondary structure of RNA.

Among the exonucleases, *E. coli exonuclease III* with 3'-5' exonucleolytic activity has been used extensively. It is a double strand specific 3'-5' exonulcease, which sequentially releases 5' mononucleotides as products shortening the DNA (Fig. 3.10). Single stranded DNA is not a substrate for the enzyme. Exonuclease III cannot hydrolyze α-thiophosphate containing phosphodiester bonds. This enzyme has been used often for introducing deletion mutations in DNA.

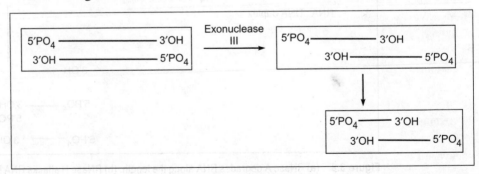

5'PO$_4$ ———————— 3'OH

3'OH ———————— 5'PO$_4$

Exonuclease III →

5'PO$_4$ ———————— 3'OH

3'OH ———————— 5'PO$_4$

↓

5'PO$_4$ ——— 3'OH

3'OH ———— 5'PO$_4$

Figure 3.10 Exonuclease III is an exonuclease in the direction 3'-5' and cleaves double stranded DNA shortening it sequentially.

Bal31 nuclease from *Alteromonas espejiana* is an extracellular nuclease with a wide range of substrate specificity. This enzyme catalyzes degradation of single stranded DNA, both endo- and exonucleolytically and of linear double stranded DNA exonucleolyticaly from 3′ termini generating blunt ends. This enzyme is used for progressive shortening of double stranded DNA for inducing deletion and mutation.

3.2.2 Restriction Endonucleases

Nucleases have been used as important tools in various recombinant DNA techniques, especially the class of nucleases referred to as *restriction endonucleases* is extensively used in generation of recombinant DNA molecules. Restriction endonucleases are DNases that recognize specific nucleotide sequence and cut double stranded DNA in a site-specific manner. They were discovered in 1950s as a part of restriction modification (R-M) system that bacteria operate for their protection against invading bacteriophages and foreign genetic elements. It is largely through this R-M system that bacteria destroy undesirable foreign DNAs and are considered equivalent of an immune system.

Till date, more than 2000 restriction enzymes have been documented, and they belong to different classes known as type I, type II, type III, and type IV (Table 3.1). The recommended nomenclature for R-M enzymes are based on the rules proposed by Smith and Nathans (1973) and updated by Szybalski *et al* (1988). These rules are as follows:

1. Among the first three letters, the first letter indicates the genus and second and third letters indicate the species name of the source organism, for example, Eco for *E. coli*, Hin for *H. influenzae*.

Table 3.1 Major differences between different classes of restriction endonucleases

Features	Type I	Type II	Type III	Type IV
R-M active structure	Single enzyme, three subunits, (RMS) complex Mg^{2+}	Separate enzymes, R: dimer, M: monomer	Single enzyme with two subunit complex	Separate monomeric enzymes
Cofactors	Restriction activity requires ATP hydrolysis	———	———	Restriction activity is stimulated by AdoMet
Recognition site	Asymmetric bipartite	Palindromic	Asymmetric	Asymmetric
Cleavage	Variable distance on either side	Same site	25–27 bp to 3′ side	14 bp to 3′ side

2. The first three letters are followed by strain or type identification in non-italic symbols or arabic numerals, e.g., Eco K for *E. coli* K strain. Hind for *H. influenzea* strain D.

3. This is followed without space by roman numerals to identify different enzymes produced by the same strain, e.g., Eco RI, Hind I, and Hind III.

R-M system generally consists of two enzymatic activities, a site specific restriction endonuclease responsible for digesting exogenous DNA, and a DNA modification methylase with identical sequence specificity. These enzymes are categorized based on their composition, cofactor requirement, recognition sequence symmetry, and cleavage characteristics.

Type II restriction enzymes are a large group of DNA endonucleases extensively used in recombinant DNA technology. They are distinguished from the others by two main characteristics:

1. The endonulcease activity is physically and functionally separate from the cognate modification methylase, and the nucleotide sequence specificities of the R and M enzymes are identical.

2. They recognizee specific nucleotide sequence along double stranded DNA and cut them at the recognition site.

Majority of them recognize tetra, penta or hexa nucleotides, whereas some can also recognize eight nucleotide sequences. The specific set of nucleotide sequences most readily cleaved is designated as a canonical sequence. Those sequences similar to the canonical sequence but are cleaved by enzymes at a moderately reduced rate are called non-cannonical sequences. The cleavage of DNA involves the hydrolysis of two phosphodiester bonds (1 per strand) either sequentially or simultaneously. The 5' PO_4 and 3' OH termini produced can be either blunt ends (flush) or 5' or 3' protruding (cohesive or sticky) ends with various numbers of base extensions (Fig. 3.11).

The recognition sequences of majority of type II restriction endonucleases consist of inverted complimentary sequences called *palindromic sequences*. A subgroup of type II enzyme called type IIs recognizes asymmetrical 4-7 bp sequences and cleaves DNA 1-20 bp away from the recognition sequence. These restriction enzymes are not only sequence specific but also structure sensitive, for example, restriction sites in a single stranded DNA or Z DNA are resistant to cleavage by restriction endonucleases.

Many times, one canonical sequence is recognized and cleaved by more than one enzyme; such groups of enzymes are referred to as *isoschizomers*. Isoschizomers can have different restriction patterns, for example, Sma I and Xma I both recognize the sequence 5'CCCGGG3', but Sma I cleaves between the middle C and G (5'CCCGGG3') to give blunt ended fragments whereas Xma I cleaves at 5'CCCCGGG3' to generate fragments with 5 base protruding 5' ends. Some isochizomers are differently sensitive to methylation. For instance, Mbo I, Sau 3AI,

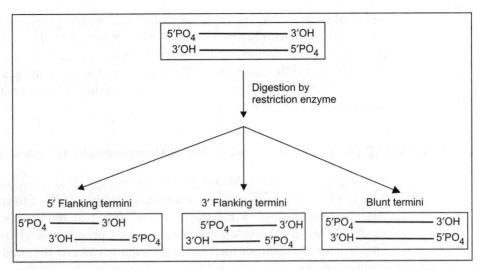

Figure 3.11   Restriction endonuclease cleaves the double stranded DNA to generate either 5′ flanking, 3′ flanking, or blunt termini.

and Dpn I, all recognize and cleave the sequence 5′GATC3′. Methylation of the A base renders the site resistant to Mbo I. Sau 3AI cannot cleave the sequence if C is methylated, whereas Dpn I requires methylation of A for cleavage.

Type II restriction enzymes exercise rigorous sequence discrimination between canonical site and other sequences, however under suboptimal conditions, some of them display moderate cleavage activities on slightly different recognition sequences. Sometimes, under suboptimal conditions, they show relaxed specific activity called *star activity*. With the star activity, the recognition specificity is lowered, for instance, from hexanucleotide to the core tetranucleotide, or from a core tetranucleotide to a dinucleotide. Nevertheless, the cleavage by the star activity occurs at the same site as the canonical site. Low ionic strength and high pH may also favour appearance of star activity.

Base modification or substitution in the recognition sequence seriously affects the action of several restriction enzymes. The restriction activity of most type II enzymes is blocked by methylation on the recognition site by the cognate methylases. Intercalation of ethidium bromide may inhibit the cleavage of DNA by restriction enzymes. Similarly, base modification and substitution on the recognition site seriously affects the action of several restriction enzymes. Sequence specificity of restriction enzymes can be differentiated by gel electrophoretic analysis of their characteristic cleavage patterns on some standard DNAs, such as phage lambda, SV40, ϕ X174, or plasmid pBR322.

Reaction conditions, such as quality of DNA, ionic strength, i.e., salt concentration, pH, and temperature are important determinants for restriction endonuclease reaction. Some enzymes have similar optimal reaction conditions, which allow their grouping in the same restriction buffer systems and becomes very convenient while dealing with more number of enzymes. The unit of restriction enzyme is usually defined as the amount of enzyme required to digest completely 1µg of standard DNA in one hour. These restriction enzymes are generally stored at –20°C in buffers containing 50% glycerol and adding BSA or gelatin can further prevent activity loss.

SUMMARY

Development of recombinant DNA technology was possible because of development of several tools to manipulate nucleic acids. Enzymes can either degrade nucleic acid called nucleases or they can modify the DNA. DNA modifying enzymes include polymerases, ligases, phosphatases, kinases, etc. These enzymes are stringent for their substrate requirement and the conditions required for manipulation.

Nucleases are either sequence independent or sequence dependent. Sequence independent nucleases include endonucleases, which cleave along the length of the DNA or exonucleases which digest DNA from either end. On the other hand, sequence dependent restriction endonucleases would only cleave the DNA if specific sequence is present with predicted set of DNA fragments, sizes of which can be determined. Single stranded DNA or RNA is not cleaved by restriction endonucleases. RNases would selectively degrade RNA, either endonucleolytically or exonucleolytically.

Among modifying enzymes, polymerases synthesize polynucleotide in a template dependent manner with the exception of terminal deoxynucleotidyl transferases, while ligase joins two DNA molecules with suitable terminal separated by a nick. Phosphatases remove terminal phosphates from DNA, RNA, or oligonucleotides, while polynucleotide kinase transfers phosphates to these molecules. Availability of these enzymes to make specific changes to DNA molecules in the test tube forms the basis of recombinant DNA technology.

SHORT ANSWER QUESTIONS

1. Explain why a primer is necessary for a polymerase to initiate DNA synthesis.
2. Why is Klenow preferred during many reactions of polynucleotide synthesis?
3. What will happen if 3′–5′ exonuclease activity of polymerase is non-functional?
4. Why are type II restriction endonucleases preferred for cleavage of DNA in recombinant DNA work?
5. Explain why the forward reaction catalyzed by polynucleotide kinase is used more frequently than the exchange reaction.
6. What are the most suitable termini for ligase reaction?

FURTHER READING

- Brown, T.A. (1998). *Molecular Biology Labfax. Vol.I: Recombinant DNA*, 2nd ed., Academic Press, London.
- Brown, T.A. (2006). *Gene cloning and DNA Analysis: An Introduction*, 5th ed., Blackwell Scientific Publishers, Oxford.
- Smith, H. O. and Nathans, D. (1973). A suggested nomenclature for bacterial host modification and restriction systems and their enzymes. *J. Mol. Biol.* **81**, pp 419–423.
- Szybalski, W., Blumenthal, R. M., Brooks, J. E., Hattman, S., and Raleigh, E. A. (1988). Nomenclature for baterial genes coding for class-II restriction endonucleases and modification methyltransferases. *Gene* **74**, pp 279–280.

Techniques in Molecular Biology

4

OVERVIEW

- Molecular biology learning and research starts with a clear concept on the requirement of a molecular biology lab including safety practices to be followed.

- Various techniques, either classical or modern, are integral part of molecular biology research. Some of them are not absolutely molecular biology techniques, yet are essential. These have been grossly classified as biophysical and biochemical techniques.

- The other class which is part and parcel of molecular biology is recombinant DNA techniques.

- It is imperative to understand the classical techniques up to their modern variations under each of these techniques. This understanding would be very valuable for choosing one or the other for a particular application.

4.1 REQUIREMENTS OF A MOLECULAR BIOLOGY LABORATORY

While preparing a Molecular biology laboratory, in which various molecular biology techniques including recombinant DNA can be carried out, the following four aspects are of concern: *construction, containment facilities, specific equipments*, and *appropriate laboratory practices*.

4.1.1 Construction

In general, a laboratory of 600 square feet size is adequate for accommodating 20 students of Master level, or 5–6 students of Ph.D. level. The dimension therefore would depend on the number of students and also the level of learning. The laboratory should be dirt-proof, preferably air-conditioned, and with a sink. The working tables should be sturdy in nature. The working tables with granite tops should have reagent racks, storage drawers/cabinets, and should be illuminated. The windows should have fly-screens; the doors need to have door closers.

4.1.2 Containment Facilities

This is particularly important for recombinant DNA and PCR work. A laminar air flow hood, and a PCR workstation are minimally required. In recombinant DNA work, one uses bacteria and such other organisms that need to be contained in the lab. Further, for aseptic requirement, a laminar air flow hood is necessary. PCR being a highly sensitive technique requires a sterile working station for avoiding contamination and spurious results.

4.1.3 Equipments

The following equipments are minimally required.

Minor equipments Vortex, magnetic stirrer, pH meters, oven, incubator, water bath, microfuge, platform shaker, microwave oven, pouch sealer, and refrigerator.

Electrophoresis equipments Horizontal DNA gel apparatus, vertical protein gel apparatus, electrophoresis power supply (constant current-voltage), isoelectric focusing/2-D gel apparatus, and Western blot transfer apparatus.

Major equipments High speed refrigerated centrifuge, UV/Vis spectrophotometer, PCR thermal cycler, gel documentation system, UV-transilluminator, -20°C deep freezer, -80°C freezer, Millipore distillation apparatus, phase contrast microscope, immunofluorescence microscope, ELISA reader, shaker-waterbath, ice flake machine, ultra centrifuge, lyophilizer, top-loading balance, and micro-balance.

4.1.4 Special Practices

For most of the molecular biology experiments, clean glassware and plasticware, and practice of aseptic techniques are mandatory. Thus, all the reagents, reagent containers (glassware and plasticware), pipetting devices and plasticware, filtration assemblies are to be sterilized, either through dry sterilization or steam-sterilization techniques, and by Millipore filtration as necessary.

The washing and sterilization should be performed in a separate adjacent room dedicated for this purpose. Other safety practices include: (a) proper handling and disposal of toxic and carcinogenic chemicals, (b) biosafety practices at the level of BL1 (P1) or BL2 (P2) depending on the usage of hazardous biological materials. The reader may refer any microbiology textbook for the details of P1 and P2 safety levels.

4.2 BIOPHYSICAL TECHNIQUES

Biophysical techniques are those used for obtaining information on the physical characteristics of biomolecules such as shape, size, structure, etc. These are also useful for their observations *in situ* in cells, and their separation from each other.

4.2.1 Microscopy

Microscopy, although a major tool in cell biology, is also useful for molecular biology studies of some specific nature. For *in situ* localization of macromolecules in the cells, particularly changes of their localization in response to any specific signal and mobility from one cellular compartment to the other, protein-protein interactions, etc., certain types of microscopy are very valuable. Further, structural details of macromolecular complexes both *in vivo* and *in vitro* are also better understood through high-resolution electron microscopy.

Various types of microscopy, all of which envisage to observe small invisible (through naked eye) structures/components, can be grouped under two broad categories based on the limit of resolution: *light microscopy* and *electron microscopy*. The resolutions of the light and electron microscopy are 0.2 μM and about 1-2 nM, respectively. Thus, electron microscopy can resolve objects hundred-fold better than light microscopy.

Light Microscopy

By using a contemporary compound light microscope (Fig. 4.1), one can obtain up to 1000-fold magnification of an object, with a proper resolution. The resolution of a microscope refers to the ability of the microscope to distinguish objects with small distances. Therefore, resolution is even more important than the magnification in the microscopy. The resolution in case of light microscopy is measured by the following two parameters: the wave length of visible light (λ) and the numerical aperture (NA) of the lens. The relationship is as follows:

$$\text{Resolution} = \frac{0.61\lambda}{\text{NA}}$$

The value of λ is fixed at ~0.5 μm (the wave length of visible light being 0.4 to 0.7 μm). The NA refers to the size of the cone of light that enters the lens of the microscope after it passes through the object or specimen (Fig. 4.1B). It is calculated by the equation,

$$\text{NA} = n \sin \lambda,$$

where n is the refractive index of the medium through which light travels between the specimen and the lens. The n for air is 1.0, whereas it is 1.4 in case of an oil immersion lens. The angle λ corresponds to half the width of the cone of light. The maximum value of λ is 90°, and thus the value of sin λ is 1. Therefore the maximum value of the numerical aperture NA = 1.4 × 1 = 1.4. Thus, theoretically, the maximum resolution of a light microscope is = (61 × 0.5)/1.4 = 0.22 μm.

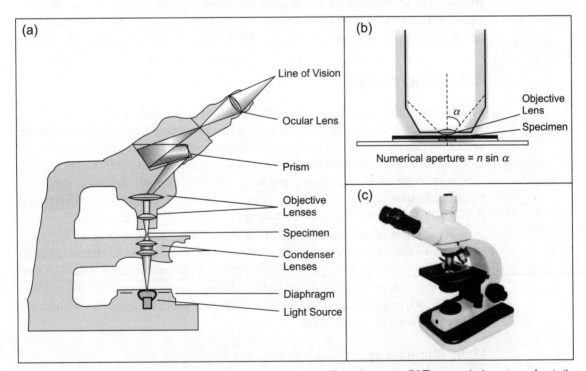

Figure 4.1 Compound light microscope. **(a)** Optical path of a compound light microscope. **(b)** The numerical aperture refers to the size of cone of light which enters through the specimen. **(c)** Picture of a compound light microscope.

Light microscopy which is generally used for observing cell structures can be of various types, namely bright field microscopy, phase contrast microscopy, differential interference contrast microscopy, fluorescence microscopy, confocal microscopy, and multi-photon excitation microscopy. For a basic concept, these are briefly described below. The students are advised to learn further details from cell biology textbooks.

Bright field microscopy

In this type of microscopy which is the simplest, light passes through the specimen directly, and the various components are distinguished by their differential absorbance of light. Often, to enhance the contrast, the objects are stained with coloured dyes that bind to different types of macromolecules, such as proteins and nucleic acids, prior to microscopic observations (Fig. 4.2).

The tissues, in particular, are processed in a specific manner for staining and better observations. They are treated with specific fixatives, such as alcohol, acids, acetone, and formaldehyde, for better preservation of structures. Further, the tissues are sliced into thin sections by the process of microtomy, for clarity up to cellular level.

For microtomy, the specimens after fixation are stained, dehydrated and embedded into paraffin blocks. These blocks are sectioned up to the desired thickness (~5 μm) using a microtome. Alternatively, the tissues can be sectioned while being frozen at –20°C with a cryotome, particularly when one wants to retain antigenecity of various target antigens for immunofluorescence or immunocytochemical studies.

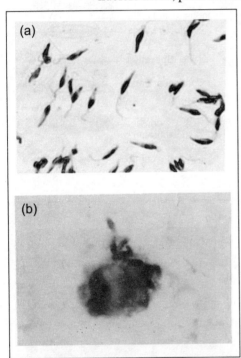

Figure 4.2 Cells stained with a coloured dye Giemsa for morphological observation through light microscopy. **(a)** *Leishmania donovani* promastigotes, **(b)** Promastigotes in the process of infecting a macrophage cell (P388D1). **(See Colour Plate 2)**

Phase contrast and DIC or Nomarski interference microscopy

By these methods, one can visualize living cells without modifying them as described above. These microscopes basically use optical variances of the light microscopes, the systems which convert variations in density or thickness between different parts of the cell to variations in contrast that are seen through these microscopes. In these systems, when light passes through the different structures like nucleus and cytoplasm, it slows down so that the phase in one structure is altered for the other. In this manner, these two types of microscopy convert these differences in phases to differences in contrast, resulting into improved and clearer images of the cells with clarity of internal compartments in particular (Fig. 4.3).

Light microscopy has been made more effective and powerful by the use of *video-imaging* and *computer-mediated image analysis systems*. With these aids, visualization of small objects that are otherwise undetectable has been possible. For example, visualization of movement of cell organelles along the microtubules

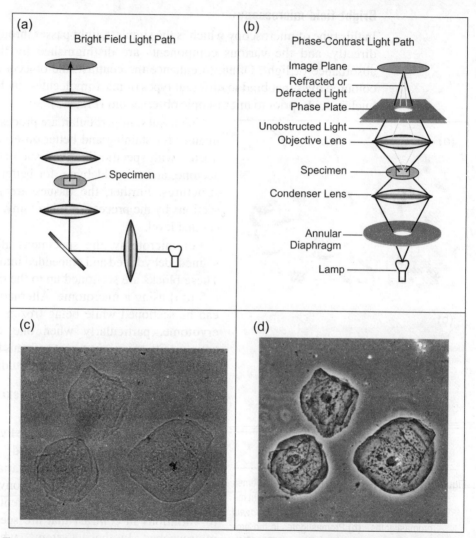

Figure 4.3   Differences between a bright field and a phase-contrast microscope. **(a)** Schematic representation of the optical path followed in a bright field microscope. **(b)** Schematic representation of the light path of a phase contrast microscope. **(c)** Bright field and **(d)** phase-contrast images of human squamous epithelial cells from buccal cavity (40X).

through video-enhanced time-lapse differential interference contrast (DIC) microscopy is now possible. Further, in case of DIC images of a thick structure like nucleus, details of the structure can be obtained by computer-aided reconstitution of a 3-dimensional structure from a series of thin optical sections or images.

Light microscopy has also evolved to the level of molecular analysis, in which macromolecule-specific labels are used. For example, presence of specific genes in the DNA, location of their expression (mRNA and proteins) in various tissues can be detected by using labeled nucleic acid probes and antibodies, respectively, and observation through light microscopy.

Fluorescence microscopy

It is one of the most powerful and versatile methods for localization of molecules in a cell or organism. A *fluorescent dye* is used to label the molecule of interest in a living or a fixed cell. The dye absorbs light at a particular wavelength (excitation wave length) and then emits light (fluorescence) at a specific and longer wavelength. The fluorescence-labeled object when viewed through a fluorescence microscope, only the specific fluorescence is observed due to the usage of a specific filter which can be imaged through a camera (Fig. 4.4).

The various applications of this method are: detection of endogenous proteins in a cell/organism by using protein-specific fluorescently-labeled antibodies, detection of expression of artificially introduced genes, like green fluorescent protein (GFP) that fluoresces naturally, and detection of small molecules like Ca^{2+} using ion-sensitive fluorescent dyes. Further details on these applications are described below.

Detection of specific proteins in fixed cells   The two reagents that are mainly required are: (a) antibody specific for the target protein, generally a monoclonal antibody and (b) a fluorochrome, such as rhodamine or Texas red that emits red fluorescence, fluorescein that emits green fluorescence, and

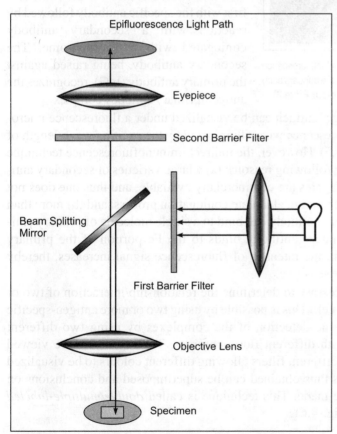

Figure 4.4   Schematic representation of the light path followed in an epifluorescence microscope.

Labels in figure: Epifluorescence Light Path; Eyepiece; Second Barrier Filter; Beam Splitting Mirror; First Barrier Filter; Objective Lens; Specimen

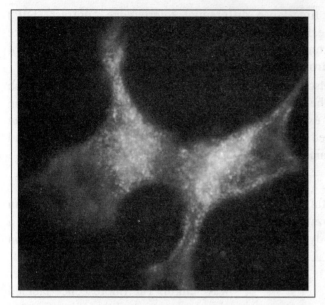

Figure 4.5 Image of melanotic melanoma cells by Immunoflourescence microscopy. Melanotic cells stained with DAPI (nuclei appears blue) and tyrosinase is depicted by red colour. **(See Colour Plate 2)**

Cy3 that emits orange fluorescence. Using these reagents, the target protein can be detected in a cell either by *direct immunofluorescence microscopy* or by *indirect immunofluorescence microscopy*. In either case, the cells/tissues are fixed with a fixative like paraformaldehyde, alcohol etc., permeabilized with a mild detergent like Triton X-100 prior to antibody reaction.

In the direct method, the target sample is reacted directly with the antibody conjugated with a fluorochrome, while in the indirect method, the sample is reacted first with the specific antibody followed by reaction with a secondary antibody conjugated with a fluorochrome. The secondary antibody, being raised against the primary antibody (IgG), recognizes the antigen-primary antibody complex.

In either case, the target antigen can be visualized under a fluorescence microscope in the form of fluorescence when illuminated by the exciting wave-length of the fluorochrome (Fig. 4.5). However, the indirect immunofluorescence technique is preferably used for the following reasons: (a) a large varieties of secondary antibody-fluorochrome conjugates are commercially available, and thus one does not need to make the reagent through elaborate conjugation process; and (b) more than one molecule of secondary antibody can bind to a single molecule of primary antibody, because the secondary antibody binds to the Fc portion of the primary antibody. For this reason, the intensity of fluorescence signal increases, thereby making this method more sensitive.

Many a time, it is necessary to determine the relationship/interaction of two or more antigens *in situ* in a cell. This is possible by using two or more antigen-specific antibodies followed by the detection of the complexes by using two different secondary antibodies with different fluorochromes. The results can be viewed separately by using two different filters allowing different colours to be visualized and imaged. The images thus obtained can be superimposed and conclusions on their relationship can be made. This technique is called *double/multiple-labeled immunofluorescence* (Fig. 4.6).

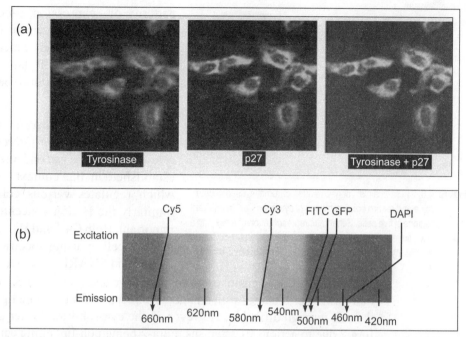

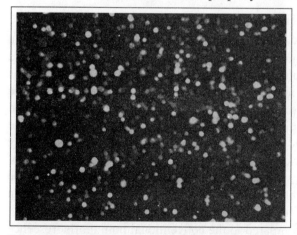

Figure 4.6 **(a)** Double labeled immunoflourescence. Melanotic melanoma cells stained with tyrosinase (red) and p27 (green) antibody. The third is the merged image of the two. **(b)** Schematic representation of the excitation and emission wavelength of some fluorochromes. **(See Colour Plate 2)**

Detection of heterologous expression of Green fluorescent protein in live cells/ tissues/organisms Green fluorescent protein (GFP) is a naturally fluorescent protein of about 28 kDa size. It occurs naturally in the jellyfish, *Aequorea victoria*. This fluorescence property of GFP makes it suitable for its use as a reporter gene product. Two of its most common uses are: (a) as a reporter gene for promoter analysis, (b) as a gene tag for understanding functions and localizations of unknown proteins. In the former, GFP gene is cloned in the vicinity of a putative promoter element. Such a construct is used for transfection of cells and studying the promoter activity through GFP expression under various conditions (Fig. 4.7). The GFP expression can be measured by fluorescence microscopy. In the latter, GFP gene is tagged with a gene of interest through recombinant DNA techniques. Such a construct can be used for transfection of cells and expression of the tagged protein as a chimeric protein. The

Figure 4.7 Fluorescence microscopy. K562 cells showing GFP expression under 20X objective. **(See Colour Plate 3)**

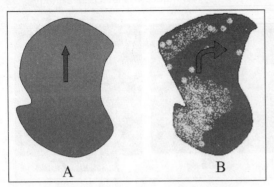

Figure 4.8 Fura-2, a Ca^{2+} sensitive fluorochrome, is used to determine and record concentrations of cytosolic Ca^{2+} in different regions of live cells. Schematic representation of a cell with a low concentration (a), and high concentration (b) of Ca^{2+} in synchrony with movement (arrow). **(See Colour Plate 3)**

pattern of expression, localization, and trafficking of the protein in question can be monitored by fluorescence measurement of the protein with the GFP tag. GFP does not interfere with the natural behaviour of the protein of interest.

Determination of intracellular Ca^{2+} and H^+ levels with ion-sensitive fluorescent dyes The role of Ca^{2+} as a second messenger is well established in the context of cell signaling which regulates various cellular processes. Similarly the H^+ ion concentration is equally important. Both Ca^{2+} and H^+ can be determined inside a cell by using specific fluorochromes, fura-2, and SNARF-1, respectively.

Fura-2 is a Ca^{2+}-sensitive fluorescence dye and it contains 5 caboxylate groups that form ester linkages with ethanol. This ester of fura-2 being lipophilic can enter the cell through diffusion. However, in the cytosol, fura-2 ester gets converted to fura-2 due to actions of esterases. Fura-2 being non-lipophilic cannot come out of the cell and it stays in the cytosol. One fura-2 molecule binds specifically to one molecule of Ca^{2+}, and if excited to a particular wavelength, it fluoresces (Fig. 4.8). Thus, the intensity of fluorescence is a measure of the concentration of cytosolic Ca^{2+} in the cell. In this way, one can measure and determine the dynamic pattern of changes of cytosolic calcium that can be related to specific cellular functions. Similarly, the concentration of H^+, i.e. cytosolic pH can be measured by fluorescent dyes such as SNARF-1.

Confocal and deconvolution microscopy

These techniques are improved variations of the conventional immunofluorescence techniques. In the latter techniques, often the fluorescence images of the objects are unclear, and sometimes blurred particularly in case of thick specimens (Fig. 4.9). This type of unclear images are due to emission of fluorescence by the molecules that are present above and below the plane of focus. In effect, the actual location of the target can be misleading, particularly in case of distribution of molecules that are cytoplasmic but close to the nuclear membrane. To partly overcome this problem for thick tissues, serial sections can be cut and used for immunofluorescence. Several such independent images thus obtained can be aligned to reconstruct the possible structures. However, a better solution has been provided by the development of the confocal and deconvolution microscopy.

In *confocal microscopy*, as compared to the conventional immunofluorescence microscopy, there is a pinhole or confocal aperture located in front of the detector, and it prevents lights that do not originate from the focal plane. In this way, the blurring effect caused due to fluorescence emitting from above and below the focal

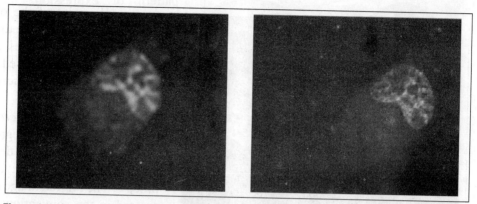

Figure 4.9 Mouse melanoma cells stained for a nuclear antigen (green) and tubulin (red) by their respective antibodies, as viewed under immunofluorescence microscope (a) and confocal microscope (b). The blurring effect in (a) is eliminated in (b). **(See Colour Plate 3)**

plane in the conventional fluorescence microscopy is eliminated (Fig. 4.10). Further, it combines fluorescence microscopy with electronic image analysis, thereby providing high-contrast images with clarity and accuracy. A laser beam is focused

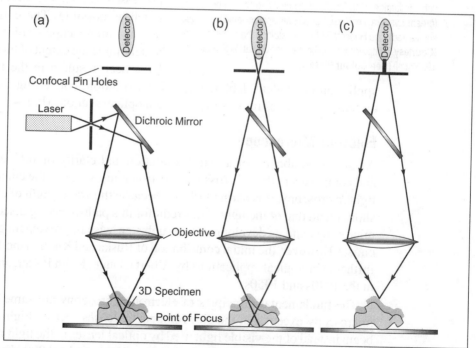

Figure 4.10 The confocal flourescence microscope. **(a)** Confocal microscope is different from normal fluorescence microscope only in having a laser, which is used to illuminate a pinhole whose image is focused at single point in the specimen. **(b)** Emitted fluorescence from this focal point in the specimen is focused at a second pinhole. **(c)** Emitted light from elsewhere is not focused here.

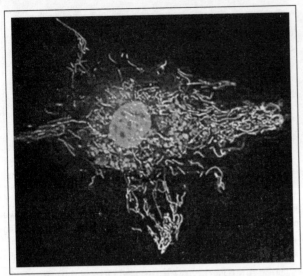

Figure 4.11   A deconvolved reconstructed image of live bovine pulmonary artery endothelial cells stained with LysoTracker Red DND-99, dihydrorhodamine 123 and Hoechst 33258. The oxidized product, rhodamine 123, localized primarily to the mitochondria (green), LysoTracker Red DND-99 stain accumulated in the lysosomes (red), and Hoechst 33258 dye stained the nucleus (blue). (**Courtesy**   http://probes.invitrogen.com/lit/catalog/3/sections/1925.html) (**See Colour Plate 3**)

on a single point of the object, and it is scanned across and down to build an image. A series of such images are stored in a computer and are processed to build a three-dimensional accurate fluoresecence image of the target.

In *deconvolution microscopy*, sharper images with a further degree of resolution can be obtained by deconvolution, a computationally intensive mathematical process. The blurred image of a point source created in the conventional microscopy is called its *point spread function*. For deconvolution, one first obtains a series of blurred images by focusing the microscope on a series of focal planes, eventually resulting in a blurred three-dimensional image. Several such images are then processed by computer to remove the blur as much as possible. The computer program uses the microscope's point spread function to determine the extent of blurring that would have been resulted in the image, and then applies an equivalent deblurring or deconvolution, converting the blurred three-dimensional images into a series of clean optical sections (Fig. 4. 11).

Electron Microscopy

As discussed above, for a better resolution and clarity of *sub-cellular* structures and complexes in the cells/tissues, electron microscopy is the choice as opposed to light microscopy. It is particularly so, because the wavelength of electrons is much shorter than that of the light. The credit for this path-breaking discovery of electron microscopy and its development through step-wise improvement goes to many scientists. However, the initial contribution of Ruska and Knoll in the 1930s were taken further to biological applications by Albert Claude, Keith Porter, and George Palade in the 1940s and 1950s.

The fundamental principles of electron microscopy are same as those of light microscopy except that the electromagnetic lenses focus a high-velocity electron beam instead of the visible light used by optical lenses in the light microscopy. In an electron microscope, with an accelerating voltage of 1,00,000 V, the wavelength of an electron is 0.004 nm. Thus, the theoretical resolution limit of an electron microscope is about 0.002 nm, which is 10,000 times that of a light microscope. However, due to uncorrectable aberrations of an electron lens, and considerably reduced effective numerical aperture, the practical resolving power of a modern electron microscope is about 0.1 nm. This limit gets increased to about 2 nm while in case of biological specimens due to additional factors like specimen preparation, contrast, etc. Nonetheless, the final resolution of an electron microscope is about 100 times better than that of a light microscope. The types of electron microscopy that are routinely used can be classified broadly under two categories: *transmission* and *scanning electron microscopy*. These are different with reference to applications and the details are described below.

Transmission electron microscopy (TEM)

In this type of microscope, electrons are emitted from a filament or cathode and accelerated in an electrical field. A condenser lens condenses the electrons and focuses on the samples in the form of an electron beam. An electromagnetic objective lens and a projector lens focus the electrons passing through the specimen onto a viewing screen or other detector. To prevent absorption of electrons by the atoms in air, the entire setup from the electron source to the detector is maintained in an ultra-high vacuum. The overall design of a TEM which is schematically presented in Fig. 4.12, is similar to that of a light microscope, except that it is larger and upside down.

Biological specimens require special preparations for TEM. Tissues are fixed with gluteraldehyde which crosslinks protein molecules to their neighbours, followed by staining with heavy metals such as lead and uranium. Further, to make the tissue penetrable by electrons, it requires to be cut into thin sections of about 50-100 nm thickness. For this, tissue or specimen is dehydrated and permeated by a resin that polymerizes to form a solid block of plastic. The block is sectioned thereafter with a diamond/glass knife on a special type of microtome. The sections are placed on a small circular metal grid and viewed in the microscope.

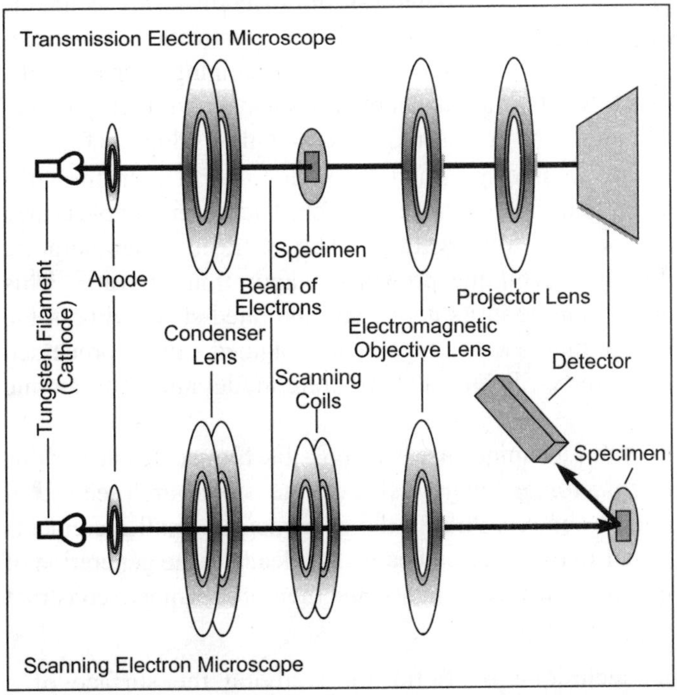

Metal staining is useful for a contrast in the image. Areas stained with metals appear dark on a micrograph because they scatter most of the incident electrons, and these scattered electrons are not focused by the electromagnetic

Figure 4.12 Transmission and scanning electron microscopes. Schematic representation of the optical path of scanning and transmission electron microscopes.

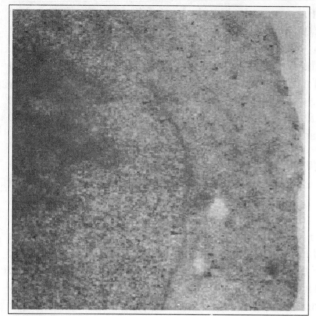

Figure 4.13 Electron micrograph showing localization of proteasomal subunit p28 (arrowheads) in sectioned chicken erythroleukamia cells. Cells were reacted with anti-p28 monoclonal antibody followed by detection with goat anti-mouse IgG conjugated to colloidal gold particles.

lenses and thus do not contribute to the image. Areas with less stain appear lighter creating a contrast in the image. Similarly, specific stains like osmium tetroxide stains certain cellular structures like membranes.

Specific macromolecules can be *in situ* localized in a cell by a variation of TEM called *immunogold electron microscopy.* The specimen is fixed and cryosectioned (for preserving antigenecity of the protein). Thin sections thus obtained are incubated with antigen-specific primary antibodies followed by electron-dense gold particles coated with protein A, a bacterial protein that binds to the Fc region of an antibody molecule. Detection can also be done with a secondary antibody conjugated with gold particles. Gold particles are electron-dense and thus appear as dark dots (Fig. 4.13).

Cryoelectron microscopy This technique is useful for direct visualization of particles without fixation or staining. For example, modeling of ribosome structure, other similar large proteins and viral caspids has been achieved by this technique. The processing of the samples is done as follows.

An aqueous suspension of a sample/specimen is applied to a grid in the form of a thin film, frozen in liquid nitrogen and maintained as such. The frozen sample is then placed in the microscope for observation. The ultra-low temperature (-196°C) that is used for this technique prevents evaporation of water. This facilitates observation of the sample in its native and hydrated state, without the need for staining etc. A large number of images thus obtained can be processed through computer to build a three-dimensional structure/model almost to atomic resolution.

A further improvement of this technique has been possible by an extension of this technique called *cryoelectron tomography.* A three-dimensional architecture of organelles can be generated by moving/tilting the specimen in small increments around the axis perpendicular to the electron beam. This leads to the generation of images of the object viewed from different angles, and their processing to construct a three-dimensional image of the object.

Metal shadowing This technique is useful for studying the surface of a specimen. In this, a thin film of a heavy metal such as platinum is evaporated onto the dried specimen. The metal is sprayed onto the specimen from an angle so that surfaces of the specimen that face the source of evaporated metal molecules are coated more heavily than others. In this manner, due to this differential coating, a shadow effect is created resulting in a three-dimensional image of the specimen.

In case of thick specimens, the organic material must be dissolved away after shadowing. This leads to the presence of a thin replica of the surface of the specimen, which is further contrasted with a film of carbon, placed on a grid for observing under electron microscope (Fig. 4.14).

Freeze-Fracture and freeze-itch electron microscopy Freeze-fracture electron microscopy helps in studying the interior of cell membranes. Specimens are

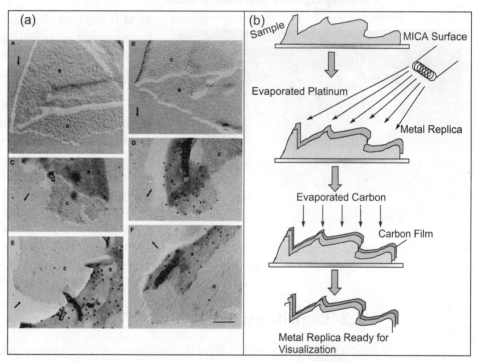

Figure 4.14 **(a)** Electron micrographs of isolated cornified Cell Envelopes prepared by freeze-drying and metal shadowing. The samples are from the epidermis of a newborn Loricin knock out (LKO) mouse (*A*) and a wild-type mouse (*B*), respectively. Similarly prepared samples of LKO mouse forestomach CE labeled with antiloricrin (*C*) and anti-SPRR1 (*E*) antibodies. Normal mouse forestomach CE labeled with antiloricrin (*D*) and anti-SPRR1 (*F*). Labeling was visualized by protein A complexed to 10 nm colloidal gold. c, cytoplasm. Arrows indicate the direction of the shadowing. Scale bar: 200 nm. (**Courtesy** Michal *et al.*, 2002, Journal of investigative dermatology118:102-9) **(b)** The schematic representation of the procedure of metal replica for sample preparation in electron microscopy.

frozen in liquid nitrogen (-196°C) followed by fracturing them with a knife blade. Most often, this process splits the lipid bilayer and reveals the interior of membrane. The material is then shadowed with platinum, the organic material is dissolved with acid, thus producing a metal replica of the surface of the specimen. When viewed through electron microscope, many surface bumps corresponding to large trans-membrane proteins that span the lipid bilayer are observed.

A similar method with slight variation called *Freeze-etch electron microscopy* can be used for visualization of the external surface of cell membranes in addition to the interior as above. The method is similar as above except that after fracturing with a knife blade, the ice level is lowered around the cells and to a lesser extent within the cells by the sublimation of ice in a vacuum, a process called *freeze-drying*.

Scanning electron microscopy (SEM)

Figure 4.15 Scanning electron microscope (SEM) image of frontal side of adult *Chironomus ramosus*. Compound eye showing individual ommatidial unit.

SEM is used for obtaining a three-dimensional image of cells (Fig. 4.15). Unlike TEM, in this method, electron beam does not pass through the specimen. The surface of the cell/specimen is thinly coated with a heavy metal, and the electrons are scattered or emitted from the metal-coated surface of the cell. In this way, as the electron beam moves across the cell, a three-dimensional image is generated and is displayed on a cathode ray tube like a television. The only limitation of the technique is its resolution limit which is only about 10 nm and much less than a TEM. Thus, it is used for studying whole cell structure than its *sub-cellular* structures.

4.2.2 Centrifugation

Biochemical analysis of sub-cellular structures and isolated macromolecules is important for the understanding of molecular biology of the cell. An important prerequisite for studying biochemical and physiological properties of organelles and biomolecules is preservation of their biological properties during separation of cellular components. Centrifugation is an indispensable key technique for separation of various elements of cellular origin. Analytical centrifugation is concerned with the study of purified macromolecules, whereas preparative centrifugation is devoted to the actual separation of cells, sub-cellular organelles, and other particles of biochemical functions.

When a particle revolves around an axis, a force is developed and acts away from the axis of rotation. This force is called the *centrifugal force*. A particle suspended in a medium takes some time to sediment, which depends upon the size or the molecular weight, the viscosity or the density of the medium under normal gravitational force. To speed up the process of sedimentation, the external speed of force is allowed to act on particle, which depends also on the distance of the particle from the axis of rotation in addition to the other two factors mentioned earlier (Fig. 4.16a).

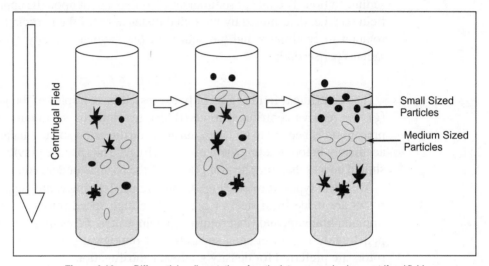

Figure 4.16a   Differential sedimentation of particulate suspension in a centrifugal field

Ultracentifugation

Ultracentifugation is the process of separation of particles whose molecular weights are very low and which do not get sedimented under fields of force applied in ordinary centrifugation. To achieve this, ultra speed, extra protection is offered to the motor by the application of refrigeration and vacuum. When designing a centrifugation protocol, it is important to remember the following:

(a) More the density of biological structure, faster is its sedimentation rate in a centrifugal field.
(b) More massive a biological particle, faster it moves in a centrifugal field.
(c) More the density of biological buffer system, slower will be the movement of the particle.
(d) Greater the frictional co-efficient, slower will be the movement of the particle.
(e) Greater the centrifugal force, faster the particles will move.
(f) The sedimentation rate of a given particle will be zero when the density of the particle equals the density of the medium.

Biological particles moving through a viscous medium experience a frictional drag whereby the frictional force acts in the opposite direction to sedimentation and equals the velocity of the particle multiplied by the frictional co-efficient. The frictional co-efficient depends on the size and shape of the biological particle. As the sample moves towards the bottom of the centrifuge tube, its velocity will be increased owing to the increase in radial distance.

At the same time, particles also encounter a frictional drag proportional to their velocity. The frictional force of a particle moving through a viscous pore is a product of its velocity and its frictional co-efficient and acts in the direction opposite of sedimentation. The rate of sedimentation is dependent upon the applied centrifugal field (G), i.e. determined by the radial distance (r) of the particle from the axis of rotation in centimeter and the square of the angular velocity (j) of the rotor in radians per second:

$$G = (j)^2 \times r$$

The centrifugal field is generally expressed in multiples of the gravitation field (g), the relative centrifugal force (RCF), revolutions per minute (rpm). A lot of precaution is needed to be taken during centrifugation such as balancing the tubes accurately before loading them into centrifuge. The specific gravity of the medium should be less than the specific gravity of the particle under sedimentation.

Different types of rotors (Fig. 4.16b) are used during centrifugation. Low speed rotors are made from steel or glass, while high speed rotors are made from aluminium, titanium, and fiber reinforced composites. *Fixed angle rotors* are used for pelleting during differential separation of biological particles whereas *swing out* rotors are preferred for density gradient centrifugation.

Density Gradient Centrifugation

This technique is widely used for separating proteins, macromolecules, organelles, etc. In the most common procedure, a continuous density gradient is first prepared

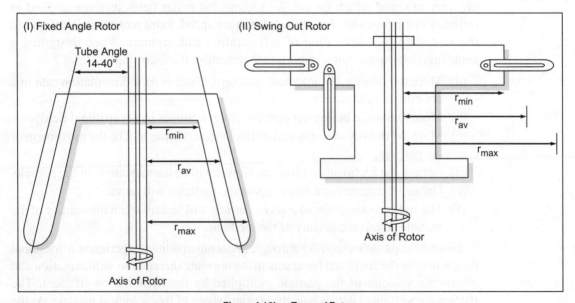

Figure 4.16b   Types of Rotors

from a suitable material such as sucrose, cellulose, glycerol, etc. using a gradient maker so that the density of the medium is greatest at the bottom of the tube (Fig. 4.17). The mixture of macromolecules to be resolved is laid on the top of the gradient. Centrifugation of the tube in a swing out rotor at a high speed causes each type of macromolecules to sediment down this density gradient at its own rate determined primarily by its particle weight and by density and shape in the form of separate bands. After the centrifugation is stopped, the contents of the tube are drained off in small volumes carefully through a pinhole in the bottom of the tube.

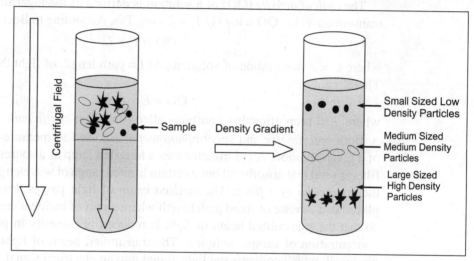

Figure 4.17 Separation of particles in a density gradient

4.2.3 Spectroscopy

Spectroscopic analysis provides a wealth of information about the structure, dynamics and function of biomolecules based on their interactions with electromagnetic radiations, namely X-rays, UV/visible and infra red light.

UV-Visible Spectroscopy

According to the electromagnetic theory of light, the light travels in the form of waves described by three attributes—wavelength, frequency, and energy. The visible portion of the electromagnetic spectrum extends from 380 to 900 nm and the ultraviolet (UV) from 200 to 380 nm. Each different molecular structure of chemical substance has a unique resonant frequency. When the frequency of an incident radiation on a substance coincides with this resonant frequency or energy level, its radiant energy is imparted to the substance. This phenomenon of energy transfer is referred to as *absorption*.

Liquids and other transparent materials display colour when they selectively absorb light in the visible region. Although this effect can be observed by the eye only in the visible region, similar absorption process takes place in the neighbouring UV and infrared regions.

The amount of radiant energy absorbed by a sample at a certain wavelength depends upon the concentration of sample. *Beer-Lambert's law* states that the optical density of a solution is directly proportional to the concentration and the optical path length through the solution. If I_0 is the intensity of radiation incident on a solution and I is the intensity of transmittant light, then *transmittance* (T) is, $T = I / I_0$ and percent transmittance (%T) = T × 100.

The *optical density* (OD) of a solution is defined as the logarithm of reciprocal transmittance, i.e. OD = log (1/ T) = 2 – log T%. According to Beer-Lambert's law,

$$OD \; \alpha \; C \times L,$$

where C = concentration of solution and L = path length of light through solution. Therefore,

$$OD = E \times C \times L,$$

where, E = proportionality constant called *extinction coefficient*.

Both *colorimeter* and *spectrophotometer* are used to measure the absorbance of a given solution. A colorimeter uses a tungsten lamp as a source of radiation. A filter is used that absorbs all but a certain limited range of wavelength referred to as the *bandwidth of a filter*. The isolated beam of light passes through the sample placed in a cuvette of fixed path length where a part of incident energy is absorbed so that the transmitted beam of light is reduced in intensity in proportion to the concentration of sample solution. The transmitted beam of light then falls on a photocell, which converts the light signal into an electrical signal, which is finally read out as transmittance or optical density on a suitable meter. The spectrophotometer uses a tungsten lamp for measurement in the visible region and a deuterium lamp for the UV region of the spectrophotometer.

A *monochromator* consisting of a grating or a prism or a combination of both is used to obtain a narrow band of wavelengths continuously through the exit slit of the monochromator (Fig. 4.18a). The beam of light transmitted by the sample is detected with the help of a suitable detector, either a phototube or a photo multiplier tube; and the optical density displayed on a suitable analog or digital readout. It is necessary to standardize or zero the instrument using the blank, i.e. without the sample. The best analytical procedure requires the zero to be reset before every measurement. It is also a good practice to start from the most dilute to the most concentrated solution, because even if the cuvette is rinsed between each measurement, the possibility of carry over should be minimized.

In both colorimetry and spectrophotometry, the most common practice is to prepare a set of standards and produce a concentration versus absorbance curve, which is linear. Absorbances of unknown compounds are measured and the concentration interpolated from the linear region of the plot. The cuvettes used, in both spectrophotometry and colorimetry, are an integral part of the system. These

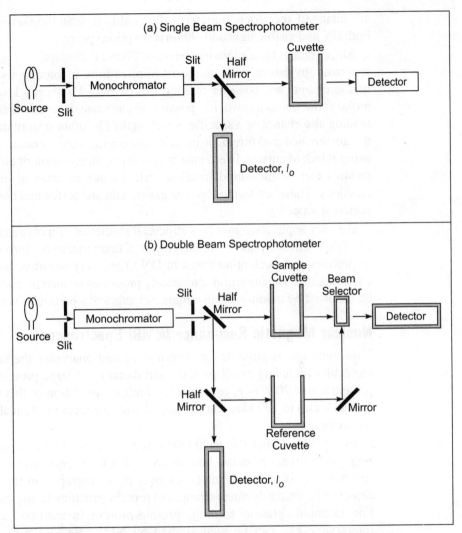

Figure 4.18 **(a)** Single beam spectrophotometer, **(b)** Double beam spectrophotometer

should be optically matched for the most precise and accurate work, wherein the optical faces should be parallel and the path lengths identical. In the UV range, quartz cuvettes must be used, whereas glass or plastic cuvettes are useful in the visible part of spectra. Because the cuvette may have its own absorption spectra, it is essential to determine the blank spectrum of cuvette and solvent and to subtract this from that of the solvent containing the sample to obtain a true spectrum of sample. This is achieved in a *dual beam spectrophotometer* with one beam passing through the blank and other passing through the analytical sample (Fig. 4.18b). Modern spectrophotometers are usually connected to a computer, which facilitates storage, presentation and analysis of the data. The major advantage of spectrophotometer is the facility to scan the wavelength range over both UV and visible light and obtain absorption spectra.

Since many biomolecules possess distinct absorption spectra, absorption spectroscopy has a wide range of applications in biochemistry. Quantitative measurements are possible making use of Beer-Lambert's law. If the value of molar extinction is known, it is possible to calculate the concentration directly from reading absorbance at a specific wavelength. The other quantitative application of the absorbance measurement includes measuring concentrations of biomolecules using standard curves. In enzyme assays, the concentration of either a substrate or product can be measured to allow calculations of rates of enzyme catalyzed reactions. These are termed as *rate assays* and are performed continuously over a period of time.

Another application involves structural studies of biopolymers such as proteins and DNA. Chromophores, which are part of these molecules (aromatic amino acids in proteins and nucleotide bases in DNA) are very sensitive to their immediate environments. Denaturations / assembly processes in such molecules can therefore be followed by monitoring absorbance changes at a particular wavelength.

Nuclear Magnetic Resonance (NMR) Spectroscopy

This technique is suitable for determining and analysing the structure of small molecules including small proteins and domains of large proteins. Although any protein above 20 kDa is generally the limit of resolution of this technique, larger molecules up to 100 kDa can be studied with the recent technical advancement of the technique.

In case of proteins, this technique is preferably used when a protein is insoluble (e.g., membrane protein) and hence cannot be crystallized. Further, NMR spectroscopy is often advantageous over crystallography in the sense that it can detect and monitor dynamic changes of protein structure during various conditions. For example, protein folding, protein-protein interaction, enzyme-substrate interaction, etc. can be well studied by NMR. Besides proteins, other small molecules of diverse types, such as RNA, side chains of carbohydrates, small ligands of receptors, etc. can be subjected to NMR for structural studies up to building 3-dimensional structures.

The principle of the technique is as follows. Study sample in the form of a concentrated solution is placed in a strong magnetic field. Certain atomic nuclei particularly those of hydrogen have a magnetic moment or spin (they have an intrinsic magnetization). The spin aligns along the strong magnetic field. However, this can be changed to misaligned excited state by applying radiofrequency (RF) pulses of electromagnetic radiation. Eventually, the excited nuclei return to their aligned state and they emit RF radiation. This radiation can be detected, measured, and displayed in the form of a spectrum. The nature of radiation depends on the environment of the hydrogen nucleus, and one excited nucleus influences the other neighbouring nuclei for their excitation and radiation. Thus, it is possible by 2-dimensional NMR to distinguish signals from hydrogen nuclei positioned at different amino acid residues in the protein molecule. In this way, it is also possible to identify and measure the small shifts in the signals which occur when the hydrogen nuclei are closely placed for interaction: eventually the size of such a shift correlates with the distance between the interacting pair of hydrogen atoms. Thus, using NMR, one can get information about the distances between the parts of the protein molecule. This information is essentially combined with the knowledge of the amino acid sequence, and through computation, a 3-dimensional structure of the protein is generated (Fig. 4.19).

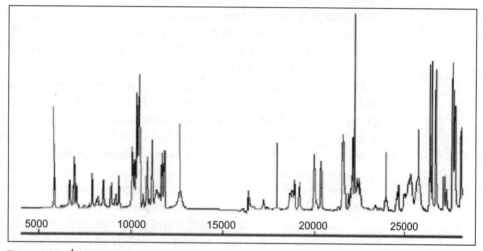

Figure 4.19 ^{1}H-NMR spectrum of a protein.

Circular Dichroism Spectroscopy

Circular dichroism (CD) is a form of spectroscopy based on the differential absorption of left- and right-handed circularly polarized light. It can be used to determine the structure of macromolecules including the secondary structure of proteins and the handedness of DNA.

Linearly polarized light is polarized in a certain direction, i.e. the magnitude of its electric field vector oscillates only in one plane, similar to a sine wave. While, in circularly polarized light, the electric field vector has a constant length, but rotates about its propagation direction. Hence, it forms a helix in space while propagating. If this is a left-handed helix, the light is referred to as left circularly polarized, and *vice versa* for a right-handed helix.

The electric field of a light beam causes a linear displacement of charge when interacting with a molecule, whereas the magnetic field of it causes a circulation of charge. When combined, these two motions result in a helical displacement when light is impinged on a molecule. Since circularly polarized light itself is 'chiral', it interacts differently with chiral molecules. That is, one of the two types of circularly polarized light is absorbed to different extents. In a CD experiment, equal amounts of left and right circularly polarized light are radiated into a (chiral) solution. One of the two types is absorbed more than the other, and this wavelength-dependent difference of absorption is measured, yielding the CD spectrum of the sample.

Applications of CD spectroscopy

So far as the applications of this technique are concerned, in general, this phenomenon will be exhibited in absorption bands of any optically active molecule. As a consequence, circular dichroism is exhibited by biological molecules, because of the dextrorotatory (e.g., some sugars) and levorotatory (e.g., some amino acids) molecules they contain. Even more important is that a secondary structure will also impart a distinct CD to its respective molecules. Therefore, the alpha helix of proteins and the double helix of nucleic acids have CD spectral signatures representative of their structures.

Proteins are chiral objects. Elements of local and regular structure, called secondary structure within proteins, have a distinct handedness. For example, helices are mostly right-handed. Proteins absorb ultraviolet light, exciting electrons into high-energy states. Plane-polarized light (such as that created by Polaroid sunglasses) is the one that propagates in only a single plane. It is possible to generate circularly polarized light. Left- and right-handed circularly polarized light may be considered as the two components of plane-polarized light. When plane-polarized light passes through a chiral medium, its plane of polarization is rotated. If the light excites an electronic transition, a chiral molecule will absorb the left- and right-handed circularly polarized components differently. This differential absorption is termed *circular dichroism*. It may be used to estimate quantitatively the secondary structure content of proteins, giving the fraction of residues in helices, sheets, turns, and in coil conformations. Thus, it is widely used to characterize proteins under equilibrium conditions and to measure the kinetics of protein folding and unfolding.

The far-ultraviolet (UV) CD spectrum of proteins can reveal important characteristics of their secondary structure. CD spectra can be readily used to estimate the fraction of a molecule that is in the alpha-helix conformation, the beta-sheet conformation, the beta-turn conformation, or some other (e.g., random coil) conformation. These fractional assignments place important constraints on the possible secondary conformations that the protein can be in.

CD cannot, in general, determine where the detected alpha helices are located within the molecule, nor can it completely predict their numbers. Despite this, CD is a valuable tool, especially for showing changes in conformation. It can, for instance, be used to study how the secondary structure of a molecule changes as a function of temperature or of the concentration of denaturing agents, e.g., *guanidinium hydrochloride* or *urea*. In this way, it can reveal important thermodynamic information (such as the enthalpy and Gibbs free energy of denaturation) about the molecule that cannot otherwise be easily obtained.

Anyone attempting to study a protein will find CD a valuable tool for verifying that the protein is in its native conformation before undertaking extensive and/or expensive experiments with it. Further, there are a number of other uses for CD spectroscopy in protein chemistry not related to alpha-helix fraction estimation. The near-UV CD spectrum (>250 nm) of proteins provides information on the tertiary structure. The signals obtained in the 250-300 nm region are due to the absorption, dipole orientation and the nature of the surrounding environment of the phenylalanine, tyrosine, cysteine (or S-S disulphide bridges) and tryptophan amino acids.

Visible CD spectroscopy is a very powerful technique to study metal-protein interactions and can resolve individual d-d electronic transitions as separate bands. CD spectra in the visible light region are only produced when a metal ion is in a chiral environment, thus free metal ions in solution are not detected. This has the advantage of only observing the protein-bound metal, so pH dependence and stoichiometries are readily obtained.

CD gives less specific structural information than X-ray crystallography and protein NMR spectroscopy, both of which give atomic resolution data. However, CD spectroscopy is a quick method that does not require large amounts of proteins or extensive data processing. Thus, CD can be used to survey a large number of solvent conditions, varying temperature, pH, salinity, and the presence of various cofactors. The readers are advised to visit a Protein Circular Dichroism Data Bank (PCDDB), a deposition and searchable data bank for validated circular dichroism spectra located at http://pcddb.cryst.bbk.ac.uk/Proteins 2006;62:1–3.

4.2.4 X-ray Crystallography

The constituent macromolecules of a cell act as both structural and functional components. In order to understand their functions better, the purified components are subjected to various structural studies, such as X-ray crystallography, nuclear magnetic resonance spectroscopy, and circular dichroism spectroscopy. These biophysical methods provide information to build the three-dimensional structure of macromolecules and their complexes, and thus help us understand structure-function relationship which can be extended to *in vivo* functions. In our discussion to follow, we will focus on proteins as the major macromolecules of cells.

X-ray crystallography is the main technique that has been used to obtain a three-dimensional structure of proteins at atomic resolution. The results of this technique depends principally on the generation of crystals of proteins suitable for X-ray diffraction analysis. X-rays are a form of electromagnetic radiations, and they have a much shorter wavelength (about 0.1 nm) as compared to light. When a sample of a pure protein is subjected to a narrow parallel beam of X-rays, most of the X-rays pass through it. However, a small fraction is scattered by the atoms in the sample. If the sample is a well-ordered crystal, the scattered waves reinforce one another at certain points and appear as diffraction spots when recorded by an appropriate detector (Fig. 4.20).

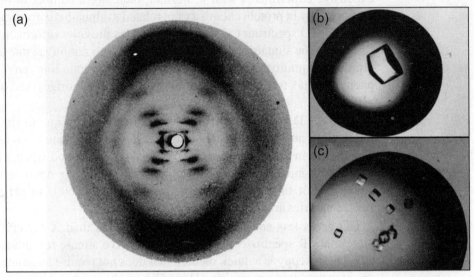

Figure 4.20 **(a)** X-ray diffraction structure of DNA structure B by Rosalind Franklin. **(b)** Single crystal of lysozyme in a drop of liquid. **(c)** Multiple crystals of lysozyme in a drop. (**Courtesy** Dr C.G. Suresh, NCL, Pune)

Although a hard task, the position and intensity of each spot in the diffraction pattern obtained as above, can be interpreted in terms of deducing a three-dimensional structure of the protein. In recent years, the process has been increasingly automated. Thus, the only difficult part that keeps limiting the use of this technique is generation of suitable crystals. The following are therefore the prerequisites for obtaining a good crystal. One requires highly pure and substantial quantity of the protein of interest. The protocol needs extensive and elaborate optimization demanding years of extensive trial and error experiments to look for the best crystallization conditions.

The diffraction pattern, obtained from good crystals, can be analysed to generate a three-dimensional *electron-density map*. Interpretation of this map that is translating its contour into a three-dimensional structure is highly complex and it requires information on the amino acid sequence of the protein. These two parameters, i.e. electron density map and sequence information are correlated through computer to generate the best possible fit. The reliability of this largely depends on the resolution of the crystallographic data. For example, 0.5 nm resolution might produce a low resolution map of the polypeptide backbone, whereas a resolution of 0.15 nm results in a more accurate atomic model.

As mentioned earlier, this technique has also been used to study macromolecular complexes. For example, recently the structure of ribosome, a large ribonucleoprotein complex, has been solved by using X-ray crystallographic technique.

4.3 BIOCHEMICAL TECHNIQUES

Biochemical techniques are used mostly for purification, identification and characterization of different biomolecules. Some of the most commonly used biochemical techniques often used in molecular biology research are described below.

4.3.1 Electrophoresis

Electrophoresis, a derivative of a Greek word, means carried by electricity. In this procedure, any charged particle can move towards its opposite charge in an electrical field. Based on this principle, among others, macromolecules, such as nucleic acids and proteins can be separated from each other based on their charges. Further, electrophoresis through a solid medium that imposes friction on the molecules, is very effective in separation based on both charge and molecular mass. Currently, the two most commonly used media are *polyacrylamide* and *agarose gels*. Although both are used for proteins as well as nucleic acids, agarose is preferred for nucleic acids and polyacrylamide gels for proteins. In general, electrophoresis is one of the most widely used techniques in biochemistry and molecular biology.

A few major types of electrophoresis used for analysis of nucleic acids and proteins are described here. For a basic concept as well as further details on electrophoresis, readers are suggested to refer to Lass (1989).

Gel Electrophoresis for Nucleic Acids

Gel electrophoresis is a procedure for separating a mixture of molecules through a stationary material (gel) in an electrical field. A gel is prepared which acts as a support for separation of the fragments of DNA. Gel electrophoresis of large DNA or RNA is usually done by agarose gel electrophoresis.

Agarose Gel Electrophoresis

Agarose is a less charged (purified) form of agar, which is a natural colloid extracted from sea weeds. It is a polysaccharide of a basic repeating unit agarobiose consisting of galactose and 3,6-anhydrogalactose. Agarose, upon melting in an aqueous buffer by boiling and subsequent cooling, forms a gel with certain porosity that the porosity can be changed to an extent depending on the concentration of agarose used. For example, a range of 0.8–1.5% agarose gel is used for various types of nucleic acid analysis; lower percent gel is suitable for separation of larger molecules, while higher percent is for smaller sizes (Fig. 4.21). Thus, by varying the concentration of agarose, fragments of DNA from about 200 to 50,000 bp can be separated using standard electrophoretic techniques. Further, a large number of different types of agarose that are commercially available (see Sigma catalogue),

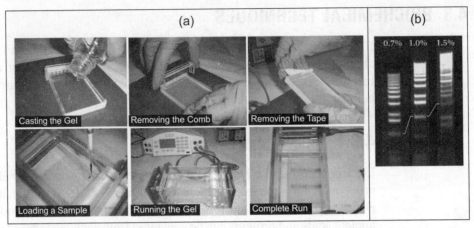

Figure 4.21 **(a)** Pictures depicting the procedure for casting agarose gel. **(b)** Picture depicting the difference in the DNA samples run in different concentrations of agarose gel. **(See Colour Plate 4)**

differ in their properties such as gelling temperature, electroendosmosis, etc. Thus, one has to choose the type suitable for a specific requirement and application.

For the details of the procedure for preparation of gels and other information regarding electrophoresis, readers can refer to the Maniatis protocol. In summary, the molten agarose is poured onto a gel plate containing a comb for forming wells, cooled to room temperature to form a gel. Thereafter, the comb is removed, the gel plate is transferred into a DNA submarine gel apparatus, the DNA samples are loaded into the wells along with a tracking dye (e.g., Bromophenol blue) and are electrophoresed in Tris-acetate-EDTA (TAE) or Tris-borate-EDTA (TBE) buffer at a constant voltage (10V/cm of the gel). DNA being negatively charged due to phosphate backbone, migrates towards anode. Electrophoresis is continued until the tracking dye which is mixed with the sample reaches about three-fourth of the gel towards the anode (Fig. 4.22). The gel is stained with 0.2–0.5 μg/ml of *ethidium bromide*, a nucleic acid-intercalating fluorescent dye, and is observed on an *UV-transilluminator*

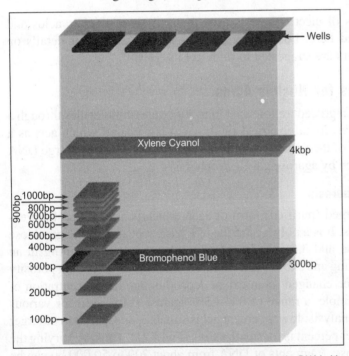

Figure 4.22 The schematic representation of the gel showing 100bp DNA ladder and the position of xylene cyanol and bromophenol blue after agarose gel electrophoresis. **(See Colour Plate 4)**

(260–300 nm). Ethidium bromide being a mutagen needs to be handled with gloves and to be deactivated prior to its disposal. The results can be documented (photographed) and analysed by a gel documentation instrument.

DNA molecules of mixed sizes can be separated by agarose gel electrophoresis based on their shape, topology, and molecular size. In case of linear DNA molecules, smaller molecules will migrate faster than the larger molecules. However, it is experienced further that the molecular sizes of any extracted or isolated nucleic acids can be better determined by the following relationship:

Log molecular size is inversely proportional to the mobility of molecules.

Thus, the gel can be calibrated with linear nucleic acid molecules of known sizes as reference for determining size of any unknown molecule (Fig. 4.23). Further, different forms of DNA, for example plasmid DNA of supercoiled, nicked-circular and linear froms, migrate differently in agarose gel (Fig. 4.24). Thus DNA agarose gel electrophoresis is very useful for various experiments with nucleic acids.

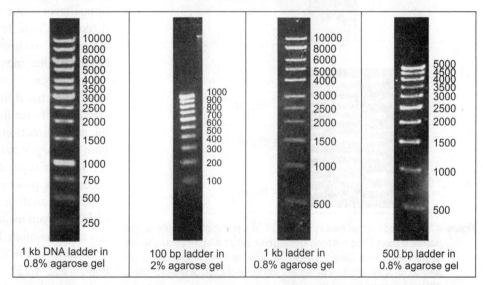

Figure 4.23 The pictures of agarose gel loaded with different DNA ladders available after electrophoresis.

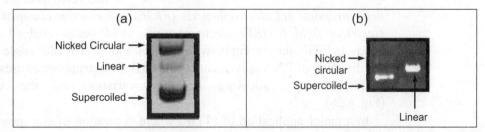

Figure 4.24 Three forms of plasmid DNA. **(a)** After plasmid extraction, depending on the handling different forms of plasmid can be visualized. The figure shows three forms of plasmid. **(b)** The first lane shows two forms of the plasmid. The second lane shows the linearised form of the plasmid after restriction digestion.

Pulsed-field gel electrophoresis

As mentioned above, DNA agarose gel electrophoresis can be used for separation of molecules up to 50 kbps. For resolution of larger size DNA, for example, DNA of sizes up to several million base pairs (mega bases), a modified form of agarose gel electrophoresis, called the *Pulsed-field gel electrophoresis* (PFGE) is used. The basic PFGE was discovered by Schwartz and Cantor (1984).

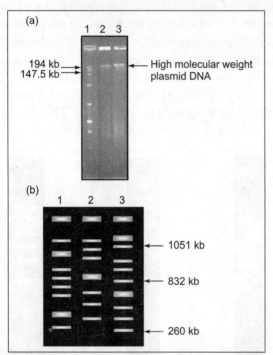

In this electrophoretic technique, the direction of current flow in the electrophoresis chamber is periodically altered. Further, instead of using constant current, it uses pulses of current, with relatively long pulses in the forward direction and shorter pulses in the opposite or sometimes sideways direction. This forces the molecules to reorient themselves before further movement through the gel. Because of this reorientation of electrical field, the smaller size DNA molecules will move in the new direction more quickly than the larger molecules, resulting in their separation from each other. This process therefore allows fractionation of large pieces of DNA, even as large as some of the chromosomes found in lower organisms in the form of discrete bands (Fig. 4.25).

Figure 4.25 Pulsed field gel electrophoresis (PFGE). **(a)** A high molecular weight plasmid of about 200kb has been resolved on it. Lane 1 is high molecular weight DNA marker. Lanes 2 and 3 are different amounts of high molecular weight plasmid. **(b)** Schematic representation of separation of chromosomes from three different organisms (lanes 1-3).

In the recent time, two modified versions of the PFGE, namely *field inversion gel electrophoresis (FIGE)* and *contour-clamped homogeneous electrical field (CHEF) electrophoresis* yield better-resolved straight DNA bands. In FIGE, the current is switched back and forth at 180° angle with respect to the direction of DNA movement; while CHEF electrophoresis uses multiple electrodes, thereby allowing accurate variation of the electric field (Fig. 4.26).

In a major application of PFGE, where separation of chromosome-size DNA molecules are envisaged, a special preparation is required for extraction and analysis of intact DNA. We know that large size DNA molecules (e.g., genomic

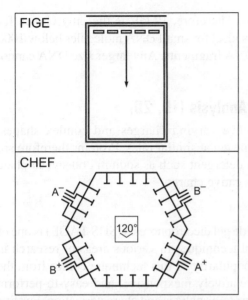

FIGE

CHEF

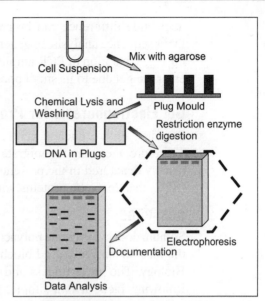

Figure 4.26 Two modified versions of pulsed field gel electrophoresis: Field inversion gel electrophoresis (FIGE) and Contour-clamped homogeneous electrical field (CHEF).

Figure 4.27 Flowchart showing the procedure followed during pulsed field gel electrophoresis. First the cell suspension is mixed with agarose, plugs are formed by chemical lysis and washing. Then the DNA is restriction digested and electrophoresed.

DNA) are prone to shearing while handling with pipette etc. In this preparation, the intact cells (e.g., bacteria, yeast) are embedded in agarose blocks/plugs, followed by extraction of DNA *in situ* by enzymes to degrade cell walls and proteins. This results in liberation of the intact naked DNA in the agarose. The agarose plugs are then trimmed, digested with restriction enzymes, if required, and loaded into the preformed wells in the agarose gel (Fig. 4.27).

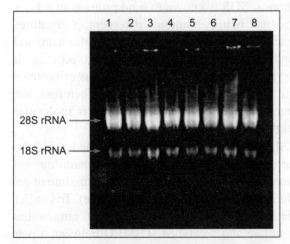

Figure 4.28 Formaldehyde-Agarose gel electrophoresis for RNA. 28S and 18S rRNAs are seen as intact bands, while the smear is of mRNA.

Formaldehyde-Agarose gel electrophoresis for RNA

For separation and analysis of RNAs, formaldehyde-agarose denaturing gel electrophoresis is performed (Fig. 4.28). The denaturing condition is required because of melting of the secondary structure of RNA, unlike DNA. For the detailed protocol, readers can refer to Maniatis protocol (Sambrook and Russell, 2001).

Polyacrylamide gel electrophoresis (PAGE) for nucleic acids

Although originally used for separation of proteins, PAGE is also used for separation of nucleic acids. Because of its high resolutive power, DNA molecules with even a single nucleotide difference can be resolved. Therefore, PAGE is the only choice for DNA sequence analysis. In general, it is ideal for small DNA molecules below 1000 bp, e.g. separation of PCR amplified DNA fragments. Any larger size DNA cannot enter the gel due to its small pore size.

Gel Electrophoresis for Protein Analysis (1D, 2D)

Proteins, unlike nucleic acids, can have varying charges and comlex shapes. Therefore, they may not migrate into the gel at similar rates. Proteins therefore are usually denatured in the presence of a detergent such as sodium dodecyl sulphate (SDS) that coats the proteins with a negative charge.

SDS–PAGE

Sodium dodecyl sulphate polyacrylamide gel electrophoresis (SDS-PAGE) is one of the most extensively used biochemical techniques in various areas of research in Biology. The powerfulness and thus popularity of this technique comes from the following facts. It is a simple, fast, relatively inexpensive, and easy-to-perform technique of high resolving power for protein analysis and characterization. Further, the porosity of the gel can be controlled by various combinations of concentrations and ratios of the acrylamide and the cross-linking agent, bis-acrylamide, as compared to the agarose gels.

Due to the use of SDS, an anionic detergent, this technique can be used for determination of molecular weight of proteins and also for determining subunits of proteins when used in conjunction with β-mercaptoethanol, a reducing agent. β-mercaptoethanol breaks disulfide bonds, and consequently the proteins dissociate into their subunits. SDS is known to bind proteins in a 1.4:1 ratio, and being an anionic detergent, it denatures proteins being bound to hydrophobic side chains and imparts uniformly negative charges to proteins. It also results in extended conformations of proteins with similar charge to mass ratio. Therefore, the proteins migrate only based on their molecular weights.

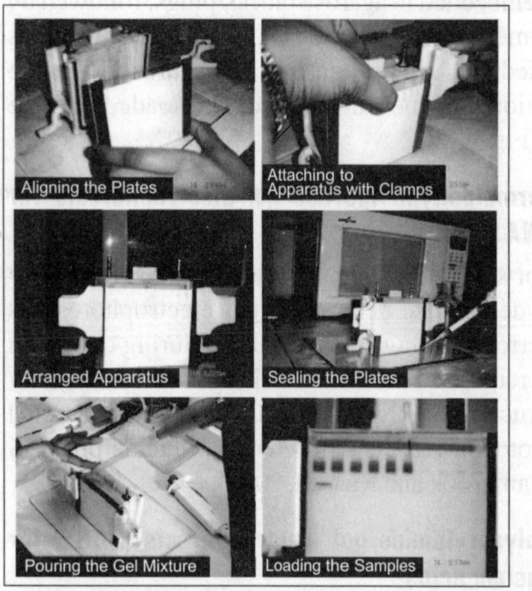

SDS-PAGE is most often performed using the Laemmli (1970) protocol. The separating gel (running gel) is prepared by pouring a mixture of gel solutions (acrylamide + bis-acrylamide), Tris buffer (pH 8.6), SDS, a polymerizing agent, ammonium persulfate and a catalyst, TEMED between a pair of glass plates sealed from three sides, or in a gel casting chamber (Fig. 4.29). This method of polymerization is called the *chemical polymerization*.

Figure 4.29 Pictures depicting the procedure followed while performing SDS-PAGE. First the plates are aligned, and then these are attached to apparatus using clamps. The plates are then sealed with agar, gel is casted and samples are loaded. **(See Colour Plate 4)**

In this, TEMED donates an electron to the persulfate generating a persulfate free-radical, which in turn generates acrylamide free-radical and induces polymerization of acrylamide with crosslinker bisacrylamide being inserted at an ordered manner (Fig. 4.30). In this way, a polyacrylamide gel of defined porosity (determined mainly by acrylamide-bisacrylamide concentrations and ratios) can be prepared.

The concentration of polyacrylamide in SDS-PAGE generally ranges from 7.5–12%. However, for very small proteins, such as histones, 15% gels are used. Further, a gel of a gradient of porosity can also be prepared by a gradient maker which mixes two solutions of high and low concentrations to generate a gradient of concentrations from heavy to light towards the top of gel. Such gradient gels are often useful for resolving many proteins in a mixture with a wide range of molecular size.

Polymerization of acrylamide gel can also be carried out by *photopolymerization* method using riboflavin and visible light. Upon polymerization of the resolving gel, a stacking gel of 4-5% is prepared on top with a comb being inserted for making wells for deposition of protein samples (Fig. 4.29). In the stacking gel, the concentration and pH of Tris buffer is different (pH 6.8). This pH is the same as that of the sample

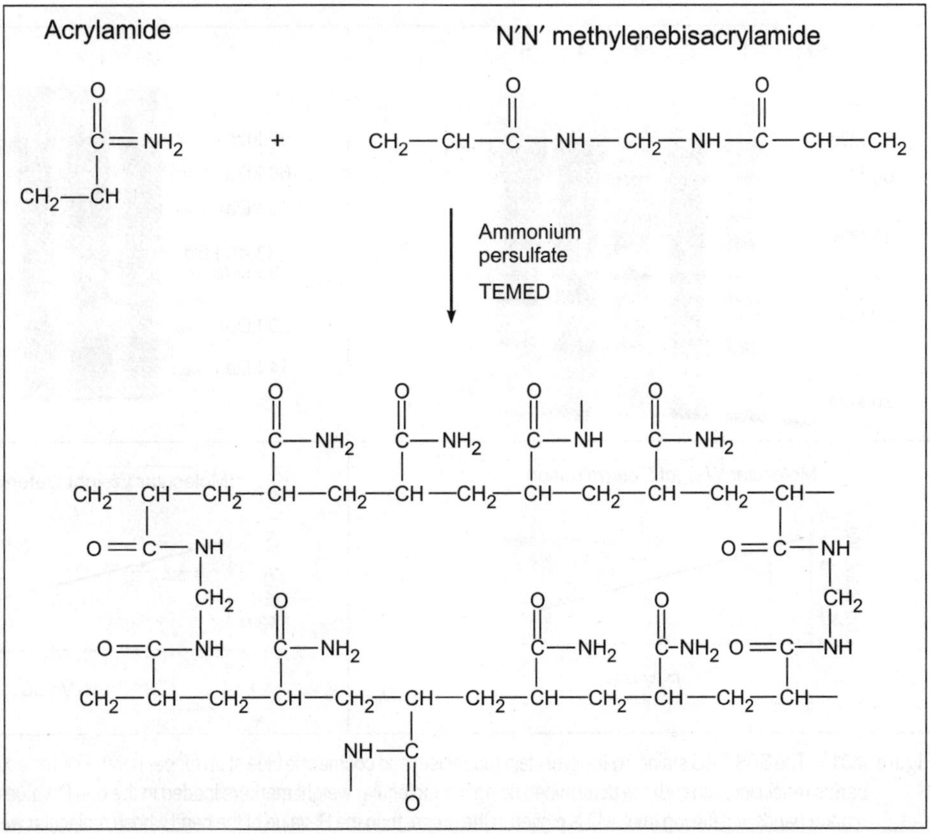

Figure 4.30 Chemical structures of acrylamide, bisacrylamide and cross-linked polyacrylamide after polymerization.

buffer which is used to prepare the protein samples. The protein samples are prepared by mixing with 2X Laemmli sample buffer (Tris-Cl, pH 6.8, SDS, β-mercaptoethanol, sucrose, bromophenol blue as a tracking dye), and boiling for 2-3 min in a boiling water bath.

The samples are loaded in the preformed wells in the stacking gel. During electrophoresis, in the stacking gel, the proteins get stacked between the leading ion Cl⁻ and the trailing ion glycine (comes from the electrophoresis buffer, Tris-glycine, pH 8.3), before they start getting separated as they enter the resolving gel. In the resolving gel, both the Cl⁻ and glycine ions move faster than the proteins which now migrate based on their sizes. Thus, the stacking gel helps in concentrating the proteins prior to their separation in the resolving gel. The readers are advised to refer to Janson Ryden (1989) for further details.

There is an inverse relationship between the mobility of the proteins and their log molecular weight. Thus, by running proteins of known molecular weights, the gel can be calibrated and thus by interpolations, the molecular sizes of protein subunits can be easily determined (Fig. 4.31). Proteins are detected on the gel by using

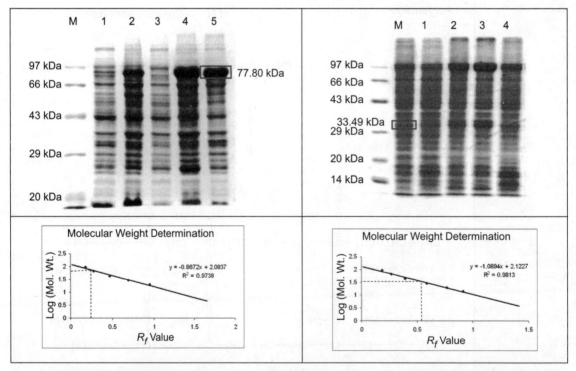

Figure 4.31 The SDS-PAG stained with silver stain (upper left) and coomassie blue stain (upper right). The molecular weights of particular bands (encircled with red) are determined using the molecular weight markers loaded in the gel. R_f value of the molecular weght marker bands and the log (mol. wt.) is plotted in the graph, then the R_f-value of the band whose molecular weight is to be determined is put in the equation and the log (molecular weight) is determined. **(See Colour Plate 5)**

protein-specific stains, such as coomassie brilliant blue, amido black, silver stain, gold stain, etc.

As described above, SDS-PAGE is a denaturing gel electrophoresis. Besides SDS, urea can also be used as another denaturing agent, either alone or in combination with SDS in specific requirements. SDS in combination with urea is useful for separation of proteins with molecular weight difference of as low as 1–2 kDa [Pal and Modak (1984)]. Since SDS solubilizes most of the proteins, SDS-PAGE is also useful for analysis of insoluble proteins, such as membrane proteins. Another major application of SDS-PAGE is immunocharacterization of proteins by Western blot analysis, which is described later in this chapter.

For separation and analysis of native proteins, non-denaturing PAGE is used. In such electrophoresis, the percentage of polyacrylamide is as low as 6–7.5%; due to the larger sizes of the native proteins, higher porosity is desirable. Further, non-denaturing PAGE is also useful for determining enzyme activity staining *in situ* in gel. In such cases, the activity can be observed, either through negative staining (e.g., amylase activity) when the substrate is impregnated in the gel while casting, or through positive staining by layering the substrate after the gel electrophoresis (e.g., LDH, tyrosinase, etc.).

Two-dimensional Gel Electrophoresis

One of the major limitations of the one-dimensional SDS-PAGE is its inability to separate proteins with identical or very little molecular weight difference. Indeed, in a cell with several thousands of proteins or polypeptides, there are many which will be similar in molecular size, albeit of different charges, due to differential charges contributed by the side chain amino acids. Thus, if one wants to resolve them from each other, with reference to analysis of such array of proteins particularly during various conditions in the same tissue or from different tissues or during *in vitro* manipulations or *in vivo* changes, a more resolutive method is highly desirable. Such a method called the *two-dimensional gel electrophoresis* has been very elegantly developed by O'Farrell in 1975, which is also popularly called the O'Farrell gel electrophoresis. In this method, the proteins are separated based on their charges by a method called *isoelectric focusing* in the first dimension followed by their separation by SDS-PAGE in the second dimension. The first dimension can also be NEPHGel instead of isoelectric focusing for some specific purpose. This modification was subsequently introduced in 1977 (O'Farrell *et al.*).

The cellular proteins are extracted and denatured with 8M urea. These are separated on a polyacrylamide gel with a pH gradient established by ampholines, a mixture of polyamino-polycarboxylic acids. Generally ampholines of a broad pH range 3.5 (+ve end) to 10 (−ve end) are used. Proteins migrate on the gel based on their charge, and get focused at a pH where their charge is neutralized, i.e. their net charge is zero (Fig. 4.32). This pH is their isoelectric point. Thus, this method is called isoelectric focusing (IEF). The gel strips containing proteins thus separated

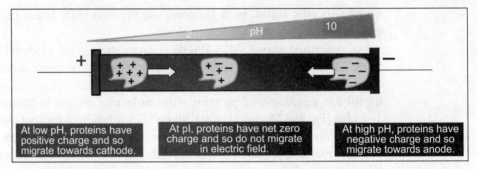

Figure 4.32 Principle of Isoelectric focusing. There is a pH gradient along the gel, proteins migrate in the electric field according to their charge and get focused at a pH where their charge is neutralized, i.e., their net charge is zero. **(See Colour Plate 5)**

based on their charge are treated with SDS sample buffer and are placed on top of a precast SDS polyacrylamide gel of appropriate concentration, and the proteins are subjected to SDS-PAGE. The proteins move out of the IEF gel strips, and get separated based on their molecular weight (Fig. 4.33). Thus, this method being highly resolutive, can resolve as many as a few thousands of polypeptides. Such analysis is often desirable in various experiments to understand the phenotypic variations of proteins in cells/tissues/organisms. This technique has been automated. There are various programs/software that are commercially available (Bio-Rad, Amersham, etc.) for analysing the variations as stated above.

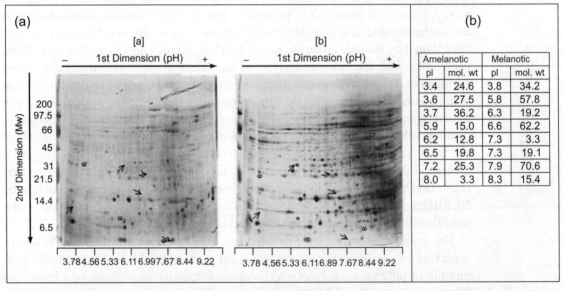

Amelanotic		Melanotic	
pI	mol. wt	pI	mol. wt
3.4	24.6	3.8	34.2
3.6	27.5	5.8	57.8
3.7	36.2	6.3	19.2
5.9	15.0	6.6	62.2
6.2	12.8	7.3	3.3
6.5	19.8	7.3	19.1
7.2	25.3	7.9	70.6
8.0	3.3	8.3	15.4

Figure 4.33 Two dimensional gel electrophoresis. **(a)** Differential protein profile in amelanotic and melanotic melanoma cells as seen by two-dimensional gel electrophoresis of B16 amelanotic melanoma cells [a] and B16 melanotic melanoma cells [b]. Proteins were separated in 1st dimension on immobilized pH gradient (3-10) and in 2nd dimension by SDS-PAGE. Unique protein bands are depicted by ω, and over expressed protein bands are labeled by $\rightarrow$. Isoelectric point (pI) and molecular weight of these protein spots are tabulated in **(b)**.

As described above, the two-dimensional profile of proteins/polypeptides are highly complex. Thus, in order to make a meaning of the results, these protein spots are to be identified. In recent times, a modified mass spectrometric method called *matrix-assisted laser desorption ionization time-of-flight* (MALDI-TOF) is in use for identification of polypeptides up to 50 amino acids in length. However, longer peptides can also be analysed after being cleaved into smaller fragments by a sequence-specific protease, such as trypsin, which cleaves proteins immediately after arginine or lysine residues, generating 5–75 amino acid long peptides. This is particularly important in the area of proteomics.

Peptides mixed with an organic acid are dried onto a metal or ceramic slide. The samples thus prepared are blasted with a laser which leads to ejection of the peptide in the form of an ionized gas in which each molecule carries one or more positive charges. Such ionized peptides are then accelerated in an electrical field and they fly toward a detector. The time of the flight of these peptides is determined by their mass and charge; larger the size of the peptide, slower it moves (Fig. 4.34). In this manner, one can determine the molecular mass of the peptide. Further, mass spectrometry can also be used for determining the sequence of amino acids of peptides. With such information, one can refer to genome databases to deduce the full amino acid sequence of the protein from the genome sequence, and thus characterize the protein or polypeptide. Readers are referred to Chapter 11 for further information.

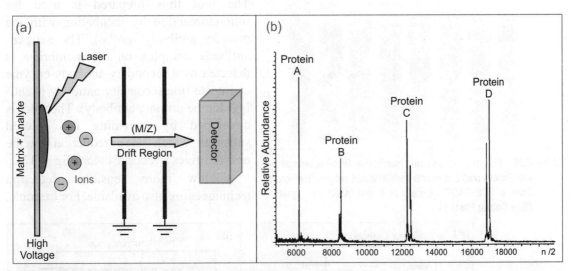

Figure 4.34 MALDI-TOF (Matrix Assisted Laser Desorption Ionization – Time of Flight). **(a)** The analyte is mixed with the matrix and due to the high voltage, ions are formed. These ions move towards the detector through the drift region in vacuum and the time taken is used to determine m/z ratio. **(b)** The image shows MALDI TOF mass spectrum of a mixture of 4 proteins: Proteins A-D; using 2,5-dihydroxybenzoic acid (DHB) as the matrix.

Western Blot Analysis

By this technique, one can identify, characterize, and quantify specific proteins separated on a one- or two-dimensional gel. The proteins from the gels are transferred onto membranes (nitrocellulose, PVDF etc.), and then the membranes are reacted with protein-specific antibodies as probes for detection of specific proteins (Towbin *et al.*, 1979).

Proteins are electrotransferred by using wet transfer apparatus or semi-dry transfer apparatus or vacuum-transfer apparatus (Fig. 4.35). For wet transfer in particular, one uses 20% methanol in the transfer buffer for removal of SDS from the gel and at the same time for preventing diffusion of the protein bands during the long period of transfer. The blot is stained with Ponceau Red S, a membrane-specific protein stain for detecting protein transfer (Fig. 4.36). The blot is then subjected to saturation of its unbound surface by a protein like BSA. The blot thus prepared is used for immunoreaction by incubating with the primary antibody (probe). The antigen-antibody complex on the membrane is detected by a secondary antibody-enzyme conjugate (the secondary antibody is anti-IgG for the primary antibody). The color is developed by providing the colored substrate which upon reduction by the enzyme gives a specific color (Fig. 4.37).

A few more sensitive detection techniques are also available. For example,

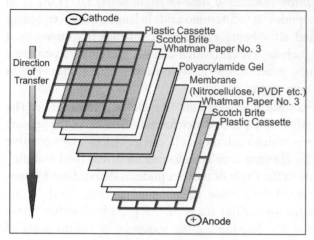

Figure 4.35 Western blot transfer assembly.

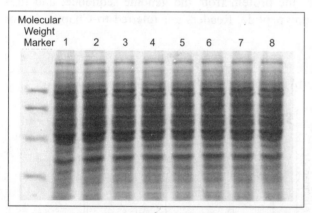

Figure 4.36 Ponceau red S stained membrane. Proteins are stained with Ponceau red S stain after transfer to nitrocellulose membrane from a SDS-PAGEl. Lanes 1- 8 are K562 cell lysates. **(See Colour Plate 5)**

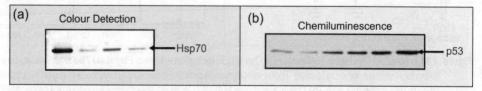

Figure 4.37 Protein detection by Western blot analysis. In colour detection, substrates which produce insoluble colour (e.g., BCIP/NBT) are used. While in chemiluminescence, substrate which produces chemiluminescence (e.g., Luminol) is used. The enzyme that is conjugated with the secondary antibody is horseradish peroxidase.

chemiluminescence, in which the substrate is a luminescent compound which upon reduction by the enzyme generates photons which can be detected by an X-ray film. In another approach, the secondary antibody can be radio-labeled, and thus can be detected by autoradiography on an X-ray film.

4.3.2 Chromatography

Chromatography is the most useful group of techniques available for the separation of closely related compounds in a mixture. It is a separation technique where separation is effected by differences in the equilibrium distribution between two immiscible phases, namely the stationary phase and the mobile phase. These differences in the equilibrium distribution are the result of the nature and degree of interaction of the compounds with these two phases. The *stationary phase* is a porous medium through which the sample mixture percolates under the influence of a *mobile phase*.

Chromatographic procedures are named after the main type of interaction between the sample and the stationary phase. Thus, there is *absorption, partition, ion exchange, molecular exclusion,* and *affinity chromatography*. Based on the mobile phase used, chromatographic procedures can be classified as liquid chromatography and gas chromatography. Based on the stationary phase used, chromatographic techniques are classified as paper chromatography, thin layer chromatography, column chromatography, gas-liquid chromatography, and high-pressure liquid chromatography.

Paper or Thin Layer Chromatography

It is the easiest to perform and requires simple apparatus. A narrow zone or spot of the sample mixture to be separated is placed at one end of the paper strip or TLC plate and allowed to dry. The strip or plate is then placed in this end dipping in the solvent mixture, taking care that the sample spot/zone is not inversed in the solvent (Fig. 4.38). As the solvent moves towards the other end of the strip, the test mixture separates into various components; this is known as the development. The strip or plate is removed after an optimal development time, dried and the spots/zones are detected using a suitable location reagent. The paper or silica gel acts as an inert support, which holds the more polar phase of the solvent mixture while the less polar phase acts as the mobile phase. Separation results from differences in partition equilibrium of the components in the mixture. An important characteristic used for identification of compounds is R_f value which is calculated as,

$$R_f = \frac{\text{Distance moved by the substance from the origin}}{\text{Distance moved by the solvent front from the origin}}$$

These chromatographies can be run in different ways such as ascending chromatography, descending chromatography, circular chromatography, and two-dimensional chromatography.

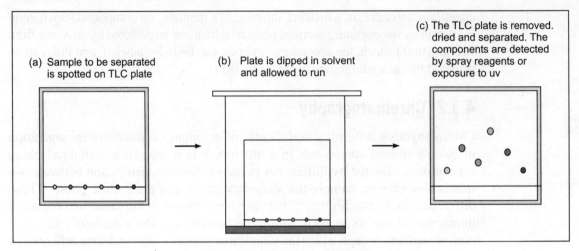

(a) Sample to be separated is spotted on TLC plate

(b) Plate is dipped in solvent and allowed to run

(c) The TLC plate is removed. dried and separated. The components are detected by spray reagents or exposure to uv

Figure 4.38   Thin layer chromatography

In *ascending chromatography* after the material is spotted onto a paper, it is folded in a cylindrical form and joined at three places. This paper cylinder is saturated with the solvent for one–two hours before dipping into the solvent system, which is kept in the petridish and enclosed by a jar. The unit is kept on a clean glass plate and sealed all around to make it airtight. The solvent is allowed to run overnight. Next day the paper is taken out, the solvent front marked with a pencil and is air-dried.

In *descending chromatography*, the solvent is kept at the top in a trough, and the end of the paper close to the spotting line is dipped in the solvent, and the solvent flows downward.

In *circular chromatography*, a circle of 12 inches diameter is cut out from a paper. An inner circle of 1.5–2 inches is drawn out by a pencil and the spots are applied on this circle. A small hole is made in a center of the paper and the paper wick is inserted into it. The paper after spotting is fitted in such a way that the wick dips into the solvent kept in a small petridish and the edges are supported on the cords.

In *two-dimensional chromatography*, a margin of 1.5–2 inches is made on both sides and the material under test is spotted at the junctions. The sample is allowed to move in the first solvent for 18 hours, air-dried and then run at right angle to the first run to the second solvent. The drawback of this technique is that only one sample can be separated in each run, however the advantage is of good resolution. After the chromatograms are developed, they are dried and the positions of the separated components are located by spraying a suitable indicator.

Column Chromatography

It is used extensively in biochemical studies. It is useful for both sample purification and quantitative analysis. A typical column chromatographic system using a gas or liquid mobile phase consists of a stationary phase, a column, a mobile phase and a delivery system, an injector system, a detector and recorder, and a fraction collector.

Successful chromatographic separations depend upon the correct choice of stationary phase and mobile phase so that the sample to be separated has different distribution co-efficient. Columns are filled with the stationary phase. The sample is applied at the top of the column and mobile phase is passed through the column normally under atmospheric pressure. Elution of the sample is performed either by passing the mobile phase in a fixed ratio or as a gradient depending upon the type of separation aimed at. The eluate is collected in fractions of small volumes for later analysis. Various types pf chromatographic media are available for column chromatography which can be used for adsorption, partition, ion exchange, exclusion, or affinity chromatography.

Gel Filtration Chromatography

This is one of the most useful and powerful tools of separating proteins from each

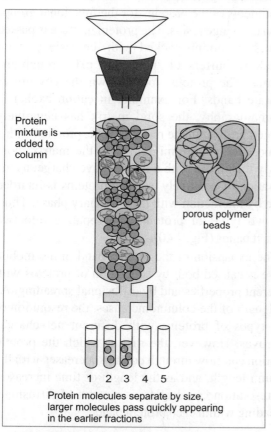

Protein mixture is added to column

porous polymer beads

1 2 3 4 5

Protein molecules separate by size, larger molecules pass quickly appearing in the earlier fractions

Figure 4.39 Gel filtration chromatography

other on the basis of size and is known as *molecular exclusion chromatography* or *gel filtration*. The mixture of proteins dissolved in a suitable buffer is allowed to flow by gravity down a column packed with beads of an inert, highly hydrated polymeric material, which has previously been washed, and equilibrated with the buffer (Fig. 4.39). Commonly used materials are Sephadex, Agarose, which are polysaccharide derivatives and Biogel, which is a polyacrylamide derivative.

All these materials can be prepared with different degrees of internal porosity and therefore each has a different molecular range over which molecules can be fractionated. For example, Sephadex G-75 will separate the proteins between the molecular range of 3000–80000. All the proteins above 80000 would be excluded from the gel. In the column, proteins of different molecular size would penetrate into the internal course of the beads to different degrees and thus travel down the column at different degrees. Very large protein molecules cannot enter the pores of the beads. They are excluded and remain in the exclusion volume (void volume) of the column defined as the volume of the aqueous phase outside the beads. Small proteins are retarded by the column while large proteins pass through rapidly. Proteins of intermediate size will be excluded by the beads to a degree that depends on their size and hence the term exclusion chromatography.

Molecular weight of a protein can also be determined by using the resolving power of molecular exclusion chromatography. Other kinds of macromolecules, e.g. viruses, ribosomes, nuclei, etc. can also be separated by molecular exclusion chromatography using gels with different degree of internal porosity.

Ion Exchange Chromatography

This chromatography technique separates solutes based on their charges and acid base properties. Ion exchange resins are insoluble in aqueous medium and contain functional groups that are capable of exchanging ions from the surrounding medium. Synthetic resins based on a polystyrene matrix are most commonly used.

The principle of separation depends on the differential affinities of various amino acid constituents of the proteins for the functional groups of ion exchange resins. The protein mixture is passed through a column packed with the resin and are eluted by buffers of increasing pH through the column. The proteins move down the column as separate bands. For example, in cation exchange chromatography, the solid matrix has negatively charged groups. In the mobile phase, proteins with a net positive charge migrate with the matrix more slowly than those with a net negative charge, as the migration of positively charged proteins is retarded more by interaction with the stationary phase. Thus, the two types of proteins can separate into two distinct bands (Fig. 4.40).

The expansion of the protein band in the mobile phase is caused both by separation of proteins with different properties and by diffusional spreading. As the length of the column increases, the resolution of two types of proteins with different net charges improves. However, the rate at which the protein solution can flow through column decreases with the column length, and as the length of time increases, the resolution can decline as a result of diffusional spreading within each protein band.

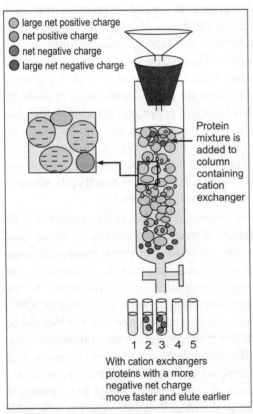

large net positive charge
net positive charge
net negative charge
large net negative charge

Protein mixture is added to column containing cation exchanger

1 2 3 4 5

With cation exchangers proteins with a more negative net charge move faster and elute earlier

Figure 4.40 Ion-exchange chromatography **(See Colour Plate 6)**

Affinity Chromatography

It is based on binding affinity of the proteins to the substrates, co-enzymes, inhibitors, activators, and other ligands specifically and reversibly. The column is packed with an insoluble matrix covalently attached to a ligand. Only the proteins, which have affinity for this ligand, would be retained on the column while others will pass through the column. For example, during antibody (IgG) purification from serum through a Protein A-Sepharose column, IgG will bind to its ligand Protein A. The specifically adsorbed protein is then subsequently removed by altering the mobile phase, such as change in pH, salt concentration, etc. to favour dissociation (Fig. 4.41).

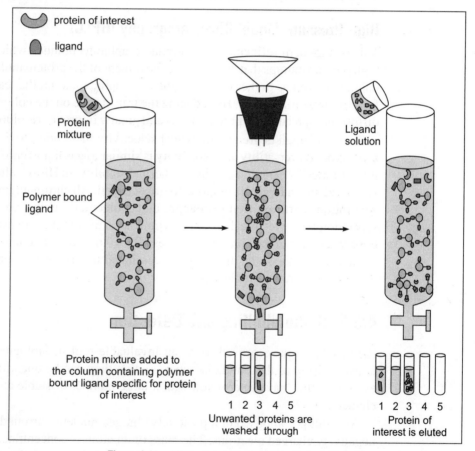

protein of interest

ligand

Protein mixture

Polymer bound ligand

Ligand solution

Protein mixture added to the column containing polymer bound ligand specific for protein of interest

1 2 3 4 5
Unwanted proteins are washed through

1 2 3 4 5
Protein of interest is eluted

Figure 4.41 Affinity chromatography **(See Colour Plate 6)**

Gas Chromatography

This is the method of choice for analysing volatile substances rapidly and accurately. The volatile material is injected into a column containing a liquid adsorbent supported on inert solid. The basis for separation of the component of volatile mixture is a difference in the partition co-efficient of the components as they are carried through the column by an inert gas, such as nitrogen, helium, argon, etc. The resolved compounds are carried over the detector, which presents an electrical signal to the recorder, which in turn displays the separated components of the mixture as a series of peaks. The area under these peaks is proportional to the amount of substances passing through the detector. Retention time is the time interval measured from the time of injection of the sample to the time where the peak appears. Unknown samples are identified based on the retention time of known standards. The method is best used for the estimation of fatty acids, bile acids, steroids, hormones, and alkaloids.

High Pressure Liquid Chromatography (HPLC)

This is a modern refinement in chromatographic techniques which makes use of high pressure pumps that speed up the movement of the protein molecules down the column as well as higher quality chromatographic material that can withstand the force of pressurized flow. By reducing the transit time on the column, HPLC limits diffusional spreading of protein bands and greatly improves resolution.

It offers a high degree of analytical selectivity in addition to the great speed of separation. Its versatility and sensitivity is similar to gas liquid chromatography and is not limited by sample volatility or thermal stability. In HPLC, after the mixture is separated, the individual components are detected. All chromatographic modes, i.e. adsorption, partition, ion-exchange, and molecular sieving are possible. High separation efficiency is achieved by optimization of column parameters, particularly the particle size of the column. Particles of 5–10 microns are commonly employed. Such a small size of the matrix has necessitated the use of high pressure to obtain adequate flow rate of mobile phase.

4.3.3 Radiolabelling and Detection

Radioisotopes have proved to be an invaluable tool in biological research. The unusual utility of these radioisotopes are that they can be quantitated and located even after they have intermixed with a large number of stable atoms of the same element.

An atom is composed of a positively charged nucleus surrounded by a cloud of negatively charged electrons. The mass of an atom is concentrated in the nucleus. Atomic nuclei are composed of two major particles, *protons* (P) and *neutrons* (N). Protons are positively charged particles with a mass several times greater than orbital electrons. The number of orbital electrons is equal to the number of protons, thus making atoms electrically neutral. The number of protons present is known as the *atomic number* (Z). Neutrons are uncharged particles with a mass approximately equal to that of a proton. Thus, sum of protons and neutrons in a given nucleus is the *mass number* (*A*). Thus, A = Z + N.

Since the number of neutrons in the nucleus is not related to the atomic number it does not affect the chemical property of the atoms.

Atoms of a given element may not necessarily contain the same number of neutrons. Atoms of the same element with different number of neutrons and therefore with different mass numbers are called *isotopes*. The number of isotopes of a given element varies. There are three isotopes of hydrogen, ^{1}H, ^{2}H, and ^{3}H; seven of carbon ^{10}C–^{16}C; and twenty or more of some of the elements of high atomic number.

The ratio of neutrons to protons in the nucleus determines whether an isotope of a given element is stable enough in nature. Elements of higher atomic number with neutron to proton ratio in excess of one are unstable. Unstable isotopes or *radioisotopes* as they are more commonly known are often produced artificially but may occur in nature. These radioisotopes, in an attempt to become stable, emit particles and/or electromagnetic radiation as a result of changes in the composition of atomic nucleus. This is referred to as *radioactive decay* for the production of stable isotope. There are many types of radioactive decays. The most relevant to biochemists are decay by *negatron emission,* decay by *positron emission*, decay by *alpha particle* emission, *electron capture* and decay by *emission of gamma rays*. Decay by negatron emission involves conversion of a neutron into a proton by the ejection of a negatively charged beta particle called a negatron.

$$\text{Neutron} \longrightarrow \text{proton} + \text{negatron}$$

As a result of this, nucleus loses a neutron but gains a proton. Thus, the atomic number increases by one, but atomic mass remains the same. An isotope ^{14}C, frequently used in biological work, decays by negatron emission.

$$^{14}\text{C} \longrightarrow {}^{14}_{7}\text{N} + \beta^-$$

Many of the commonly used radioisotopes in biology work, e.g. ^{3}H, ^{35}S, ^{32}P decay by this mechanism. Decay by positron emission involves emission of positively charged beta particles referred to as positron. This occurs when a proton is converted to a neutron. Positrons are extremely unstable. Once they have dissipated their energy, they interrupt with electron and are annihilated. As a result of positron emission, nucleus loses a proton and gains a neutron; therefore, the atomic number decreases by one and mass remains constant. An example of an isotope decaying by positron emission is $^{22}\text{Na}_{11}$.

$$^{22}\text{Na}_{11} \longrightarrow {}^{22}\text{Ne}_{10} + \beta^+$$

Decay by Alpha Particle Emission

Isotopes of elements with high atomic number frequently decay by emitting alpha particle. An alpha particle is a helium nucleus containing two protons and two neutrons ($^4He_2$). Emission of alpha particles results in decrease in atomic number by two and atomic mass by four. An example of a radioisotope decaying by alpha particle emission is Radium.

$$^{226}Ra_{88} \longrightarrow {}^4He_2 + {}^{222}Rn \longrightarrow {}^{206}Pb_{82}$$

In electron capture, a proton captures an electron orbiting in the innermost K shell. The proton becomes a neutron and electromagnetic radiation (X-rays) is given out, e.g.

$$^{125}I_{53} \longrightarrow {}^{125}Te_{52} + \text{X-ray}$$

In contrast to emissions of alpha and beta particles, gamma emission involves electromagnetic radiation similar to, but with a shorter wavelength, than that of X-rays. These gamma rays then result from a transformation from the nucleus of an atom and frequently accompany alpha and beta particle emission. Emission of gamma radiation in itself does not lead to any change in atomic number or mass. Gamma radiation has low ionizing power but high penetration.

$$^{131}I_{53} \longrightarrow {}^{131}Xe_{54} + \beta^- + \text{gamma rays}$$

The usual unit used in expressing energy associated with radioactive decay is the *electron volt*. Isotopes emitting alpha particles are more energetic in the range of 4.0–8.0 MeV, whereas beta and gamma particles have decay energies of less than 3.0 MeV. Radioactive decay is a spontaneous process and occurs at a definite rate characteristic of the source and follows an exponential law. Thus, the number of atoms disintegrating at any time is proportional to the number of atoms of the isotope present at that time.

The *half-life period* is a natural constant for each isotope and is defined as the time by which one half of the active atoms of the radioisotope have disintegrated. For example, tritium has a half-life of 12.26 years, whereas ^{14}C has 5670 years. Most commonly used ^{32}P and ^{35}S have relatively shorter half-lives of 14.20 days and 81.20 days, respectively.

Units of Radioactivity

The international system of units uses *becquerel* (Bq) as the unit of radioactivity, which is defined as one disintegration per second. However, the frequently used unit for radioactivity is *curie* (C_i). One curie is defined as the quantity of radioactive material in which the number of nuclear disintegrations is same as that in one gram of radium, namely, 3.7×10^{10} (or 37GBq). For biological purposes, this unit is too large, so microcurie (μC_i) and millicurie are used.

The unit curie refers to the number of disintegrations actually occurring in a sample per seconds (dps) and not to the disintegrations detected by the radiation counter which is generally only a proportion of the disintegrations occurring and are referred to as counts per seconds (cps). Normally, in experiments with radioisotopes, a carrier of the stable isotope of the element is added. It therefore becomes necessary to express the amount of radioisotope present per unit mass. This is the *specific activity*.

Interaction of Radioactivity with Matter

Alpha particles have considerable energy (3–8 MeV) and all the particles from a given isotope have the same amount of energy. They react with matter in two ways. First, they *cause excitation* wherein energy is transferred from the alpha particle to orbital electrons of neighbouring atoms. The alpha particles continue its path with less energy. The excited electron eventually falls back to its original orbital emitting energy as photons of light in the visible or nearly visible range. Secondly, alpha particles cause *ionization* of particles in their path, in which the target orbital electron is removed completely. The atom becomes ionized and forms the ion-pair consisting of positively charged ion and an electron. Despite their initial high energy, alpha particles are not very penetrating.

Negatrons are very small, rapidly moving particles that carry single negative charge. Due to their speed and size, they are *more penetrating* and *less ionizing* than alpha particles. Negatrons are emitted over a range of energy. The maximum energy level varies from one isotope to another ranging from 0.018 MeV for ^{3}H to 4.81 MeV for ^{38}Cl. The difference in E_{max} affects the penetration in radiation. For example, beta particles from ^{3}H can travel a few millimetre in air, whereas those from ^{32}P can penetrate over 1 metre of air.

Gamma rays and X-rays are electromagnetic radiations and therefore have no charge. They rarely collide with neighbouring atoms and travel great distances before dissipating all their energy. They are therefore highly penetrating.

Detection and Measurement of Radioactivity

Three commonly used methods for detecting and quantifying radioactivity are based on: ionization of gases, excitation of solids, or solutions and the ability of radioactivity to expose photographic emulsion (autoradiography). Any radiation measuring device has two components: a *detector* and a *scaler.* In ionization detectors, energy rich particles of radioactive radiation transform the molecules of the gas of the detector with which they collide into ions. These charged particles migrate in high-tension field creating a current flow of an intensity proportional to the number of gas ions. The widely used Geiger–Muller counters are based on this principle.

In *scintillation counters*, radiations interact with substances called flours, and a small flash of visible light is produced. This process constitutes the basis for the operation of all scintillation detectors. Two types of scintillation counters widely used in biological research are: (a) crystal counters, and (b) liquid scintillation counters. Crystal counters employ sodium iodide crystals as scintillators. Gamma rays are detected at high efficiency with the crystal counters than with Geiger–Muller counters. The beta rays are not detected by these crystal counters and are measured by liquid scintillation counters. In liquid scintillation counting, a solution of fluorescent substance dissolved in aromatic solvents replaces the crystal scintillator. This solution is called as *liquid scintillator*. The radioactive substances are dissolved directly or suspended in liquid scintillator, beta particles emitted on their decay interact with the liquid scintillator to produce very small flash of light, which is then detected by photomultiplier tube. The most commonly used fluorescent substance and solvent system consists of PPO (2,5-diphenyl oxazole), POPOP (2,2′-paraphenylene bis 5-phenyl oxazole), and toluene. Most commonly used radioisotopes in biology are counted using liquid scintillation counters. The main advantage of this counter is the facility for simultaneous determination of two isotopes in a mixture.

Though the emission of radiation is isotropic, the radiation measuring instrument may or may not record all the disintegration of a radioisotope. The disintegrations recorded by a scaler are known as counts per minute (cpm). The efficiency of the detector is obtained by the following formula:

$$\% \text{ efficiency} = \text{cpm/dpm} \times 100$$

Autoradiography

This is a very sensitive method and has been used in a wide variety of biological experiments. This allows locating the distribution of radioactivity in biological specimens of different types. In general, weak beta emitting isotopes are more suitable for autoradiography for cell and tissue localization experiments. This is because the ionizing track of the isotope is short due to the low energy of the negatrons, and this gives a discrete image. Conversely, for location of DNA bands in electrophoretic gel, ^{32}P is useful, as the more energetic negatron will leave the gel and produce a strong image.

For an autoradiograph, a radiation source emanating from within the material to be imaged is required along with a sensitive emulsion. The emulsion consists of a large number of silver halide crystals embedded in a solid phase such as gelatin. As energy from the radioactive material is dissipated in the emulsion, the silver halide becomes negatively charged and is reduced to metallic silver, thus forming a latent image. Photographic developers are designed to show these silver grains as blackening of the film and fixers remove any remaining silver halide. Thus, a permanent image of the location of the original radioactive event remains.

Safe Handling of Radioisotopes

Radioisotopes are the most noxious substance, as they inflict health hazards even when used in a trace amount. Therefore, it is absolutely necessary to use radioisotopes in the specified area and avoid any exposure to radioisotopes. It is advised to segregate the radioactive work, use minimum quantity of radioactivity needed, maximize the distance between the user and the source, minimize the time of exposure, and maintain shielding all the time. It is also necessary to dispose the radioactive waste very carefully and use personal monitoring devices to check accidental contamination due to radioisotopes.

4.3.4 Sequencing Techniques

The sequencing technique for both DNA as well as protein allows determination of nucleotides sequence and amino acid sequence, respectively, as they occur in these polymers.

DNA Sequencing

DNA sequencing can be primarily performed by either Sanger's dideoxy chain termination sequencing method or by Maxam-Gilbert sequencing method.

Sanger's dideoxy chain termination sequencing method

This method is also known as sequencing by *enzyme copying*. In this method, single stranded DNA is hybridized to an oligonucleotide primer which is around 20 bases long and the primer is extended using the *Klenow fragment* of DNA polymerase that produces DNA complementary to the template. Sanger's method (Fig. 4.42) involves carrying out DNA synthesis reaction in four separate tubes and to include in each tube a different *chain terminator*. The chain terminator is a dideoxy nucleotide like dideoxy ATP (ddATP). This nucleotide cannot form a phosphodiester bond because it lacks the necessary 3'OH group. Thus, wherever dideoxy nucleotide is incorporated into a growing DNA chain, DNA synthesis stops.

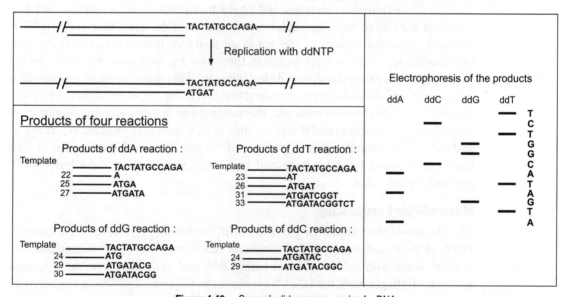

Figure 4.42 Sanger's dideoxy sequencing for DNA

The dideoxy nucleotides by themselves do not permit any DNA synthesis at all, so an excess of normal deoxy nucleotides are used with just enough dideoxy nucleotide to stop DNA strand extension at random. Each tube contains a different dideoxy nucleotide, e.g. ddATP in tube 1, so the chain termination will occur wherever A has to be incorporated, ddCTP in tube 2, ddGTP in tube 3, and ddTTP in tube 4. One radioactive deoxynucleotide is also incorporated into the tube, so the DNA products become radioactive.

At the end of the reaction, a series of DNA fragments of different length are produced in each tube. In tube 1 all the fragments end with A, in tube 2 all end with C, in tube 3 all end with G and in tube 4 all end with T. All four reaction mixtures are electrophoresed in parallel lanes in high resolution polyacrylamide gel under denaturing conditions, so all DNA fragments electrophoresed as single strand. Finally, autoradiography is performed to visualize DNA fragments, which appears as horizontal bands on an X-ray film.

To begin reading the sequence, the autoradiogram is read from the bottom successively. This manual sequencing technique is very sensitive but is still relatively slow. If a large DNA has to be sequenced, rapid automated sequencing methods are required. Indeed, very efficient *automated sequencers* which function based on Sanger's chain termination method have been developed, and these are extensively used. This procedure uses dideoxy nucleotides as in the manual method with an important exception: the primers used in each of the four reactions are tagged with a different fluorescent molecule, so the products from each tube will emit different colour fluorescence when excited by light.

After the extension reactions and chain terminations are complete, all four reactions are mixed and electrophoresed together in the same lane on a capillary tube gel. Near the bottom of the gel is an analyzer that excites the fluorescent oligonucleotides with a laser beam as they pass by, and then the colour of the fluorescent light emitted is detected electronically. This information then is passed to a computer, which has been programmed to convert the colour information to a base sequence. If it sees blue for example, this might mean that this oligonucleotide came from dideoxy C reaction and therefore ends in C. Green may indicate A, orange G, and red T. The computer gives a printout of the profile of each passing fluorescent band color coded for each base and stores the sequence of these bases in its memory for later use.

Maxam-Gilbert sequencing

The Maxam-Gilbert sequencing follows a different strategy. Instead of synthesizing DNA *in vitro* and stopping the synthesis reactions with chain terminator, this method starts with full-length end-label DNA and cleaves it with *base specific reagents*. To begin with, the DNA fragment to be sequenced is labelled either at the 5′ or 3′ end and divided equally into four tubes (Fig. 4.43).

In each tube, the chemical reaction is formed to modify one base at a time, for example, dimethyl sulphate (DMS) is used to methylate guanine. The reaction is performed under mild conditions, which will randomly modify some guanines per DNA strand. In the next reaction, another reagent called *piperdine* is used which causes loss of methylated base and then breaks the DNA backbone at the site of the

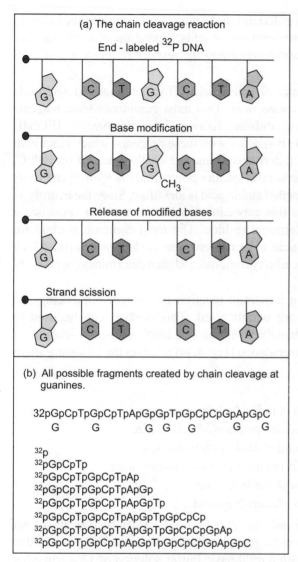

(a) The chain cleavage reaction

End - labeled ^{32}P DNA

Base modification

Release of modified bases

Strand scission

(b) All possible fragments created by chain cleavage at guanines.

32pGpCpTpGpCpTpApGpGpTpGpCpCpGpApGpC
G   G   G G G   G   G

^{32}p
^{32}pGpCpTp
^{32}pGpCpTpGpCpTpAp
^{32}pGpCpTpGpCpTpApGp
^{32}pGpCpTpGpCpTpApGpTp
^{32}pGpCpTpGpCpTpApGpTpGpCpCp
^{32}pGpCpTpGpCpTpApGpTpGpCpCpGpAp
^{32}pGpCpTpGpCpTpApGpTpGpCpCpGpApGpC

Figure 4.43 Maxam-Gilbert sequencing

lost base (apurinic site). Another reaction is performed by using acid that weakens the glycosidic bonds of both adenine and guanine followed by reaction with piperdine, which will break the DNA strand.

Similarly, *hydrazine* opens both thymine and cytosine rings and piperdine can remove these bases at the resulting apyrimidic site. In presence of 1 M NaCl, hydrazine is specific for cytosine only. After the reactions are over, the products are electrophoresed in a dentaurating gel of high resolution and bands are detected by autoradiography just as in the chain termination method.

Pyrosequencing

Pyrosequencing is a newly developed method of sequencing DNA in which the template to be sequenced is copied adding deoxynucleotides. As the new strand is being made, the order in which the deoxynucleotides are incorporated is detected, so the sequence can be read as the reaction proceeds. The addition of deoxynucleotide releases pyrophosphate, which can be converted by the enzyme sulfurylase into a flash of chemiluminescence.

Each deoxynucleotide is added separately, one after the other, with a nucleotidase enzyme in the reaction mixture so that if a deoxynucleotide is not incorporated in the polynucleotide, then it is rapidly degraded before the next one is added. Thus, this procedure makes it possible to follow the order in which the deoxynucleotides are incorporated in the growing chain and it is also automated.

Sequencing of proteins

The two major direct methods of protein sequencing are *mass spectrometry* and the *Edman degradation* reaction. It is also possible to generate the amino acid sequence from the DNA or mRNA sequence encoding the protein if this is known.

It is often desirable to know the unordered composition of amino acids in a protein, prior to finding the ordered sequence, as this knowledge can be used to find the errors, if any, in the sequencing process or to distinguish between ambiguous results. A generalized method for determining amino acid composition involves hydrolysis of proteins into its constituent amino acids, separation of amino acids by ion-exchange chromatography, and determining the respective quantities of each amino acid.

Before determining the sequence of amino acid, efforts are made to determine *the N-terminal and C-terminal amino acid.* Two most commonly used reagents for determining N-terminal amino acids are *1-fluoro-2,4-dinitrobenzene* (FDNB) and *Dansyl chloride.* The peptide is reacted with these reagents, which react with amino terminal residue and yield derivatives that are easily detected by HPLC. After the amino terminal residue is labelled, the polypeptide is hydrolyzed to its constituent amino acid and the labelled amino acid is identified. Since the hydrolysis stage destroys the polypeptide, this procedure cannot be used to sequence a polypeptide beyond its amino terminal residue. The most common method for determining C-terminal amino acid is to use enzyme *carboxypeptidase*, which removes single amino acid from carboxy terminus and then determines the released amino acid.

The Edman degradation is an important reaction for protein sequencing and automated Edman sequencers are widely used. This method can be used to sequence peptides up to approximately 50 amino acid long. A reaction scheme for sequencing a protein by Edman degradation (Fig. 4.44) involves the following steps:

1. Breaking disulphide bridges, if any, in the protein by oxidizing with performic acid.
2. Separation and purification of individual chains of protein.
3. Determining the amino acid composition of each chain.
4. Determining the terminal amino acid of each chain.
5. Breaking each chain into fragments of 50 amino acids long.
6. Separation and purification of the fragments.
7. Determining the sequence of each fragment.

The peptide to be sequenced is adsorbed on the solid surface such as glass fiber coated with polybrene. The Edman reagent phenylisothiocyanite (PTC) is added to the adsorbed peptide together with a mild basic buffer solution of 12% trimethyl amine. This reacts with amine group of N-terminal amino acid generating PTC derivative. The derivative can be detached by addition of anhydrous acid and identified by chromatography. This cycle can be repeated. The efficiency of each step is around 98% and allows around 50 amino acids to be reliably determined.

The Edman degradation is carried out on a machine called *sequenator*, that mixes reagents in proper proportions, separates the products, identifies them, and records the results. The other major direct method for determining sequence of amino acid is *mass spectrometry*. The protein to be sequenced is digested by an endoprotease and then the resulting solution is passed through a HPLC column. At the end of this column, the solution is spread out of a narrow nozzle charged to high

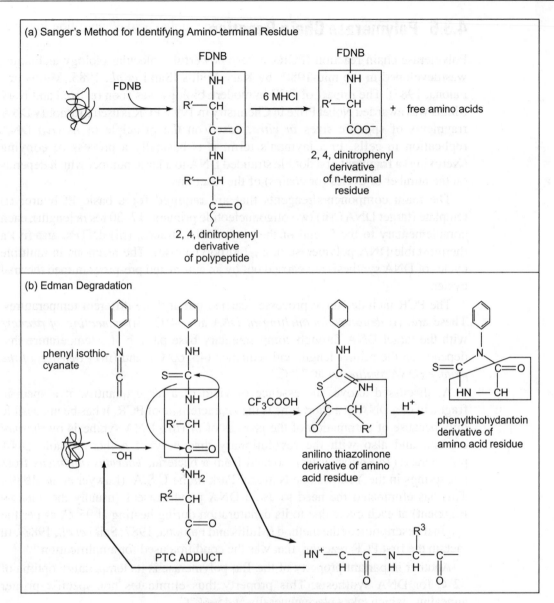

Figure 4.44   Sequencing of proteins

positive potential into the mass spectrometer. The charge on the droplet causes them to fragment until only single ion remains. The peptides are then fragmented and mass-charge ratios are measured. The mass spectrum is analysed by a computer and compared against a database of previously sequenced proteins in order to determine the sequences of fragments.

4.3.5 Polymerase Chain Reaction

Polymerase chain reaction (PCR), a very powerful molecular biology technique, was developed in the mid-1980s by Kary Mullis (Saiki *et al.*, 1985; Mullis and Falona, 1987). The impact of PCR in modern biology was soon realized and Kary Mullis was awarded Nobel Prize in Chemistry in 1993. PCR is used to amplify DNA fragments of specific sizes *in vitro*, based on the principle of *in vivo* DNA replication in cells. In a layman's term, it is basically a process of copying (Xeroxing) a fragment of a double stranded DNA to a large number which depends on the number of cycles (or chains) of the reactions.

The main components/reagents that are required for a basic PCR are: (i) template (target DNA), (ii) two oligonucleotide primers (17–30 nts in length), each complementary to the 5′ end of the target DNA strands, (iii) dNTPs, and (iv) a thermostable DNA polymerase (e.g., Taq polymerase). The reactions in multiple cycles of DNA synthesis are carried out by an automated pre-programmed thermal cycler.

The PCR includes three processes carried out at three different temperatures. These are: (i) *denaturation/melting of DNA* at ~94°C, (ii) *annealing of primers* with the target DNA through complementary base pairing at a temperature that depends on the primer length and sequence (45–60°C), and (iii) *extension of the primer (DNA synthesis)* at 72°C.

As discussed above, the essence of yield of a large quantity of a specific fragment of a DNA in a short time is the characteristic of PCR. It has been possible only because of automation of the process of *in vitro* DNA synthesis by *thermal cyclers*, and also with the revolutionary discovery of a thermostable DNA polymerase (*Taq polymerase*) isolated from a bacteria, *Thermus aquaticus* from hot springs in the Yellowstone National Park in the U.S.A. (Lawyer et al., 1989). This has eliminated the need to use a DNA polymerase 1 (usually the Klenow fragment) at each cycle due to its denaturation during heating at 95°C, as per the original description of the method (Mullis and Faloona, 1987; Saiki *et al.*, 1988). In fact, in the first PCR reaction, that was the condition used for amplification.

Another important property of the Taq polymerase is its temperature optima of 72°C for DNA synthesis. This property thus eliminates non-specific primer annealing, which takes place generally at 45–60°C.

The precise size of the DNA fragment that is yielded at the end of the PCR is dictated by the sequence of the primers annealing at the 5′ends of the double stranded DNA. However, in the initial two cycles, the 3′ ends of synthesized DNA are larger than the final size. *The actual size that is expected is only generated at the end of 3rd cycle.* Thereafter, the quantity of actual size DNA increases as a function of cycle number, and eventually what is noticed in a gel during its analysis is the expected size of the DNA; the low quantity of larger size DNA generated is not seen due to its proportionately very low amount (Fig. 4.45). Theoretically, the

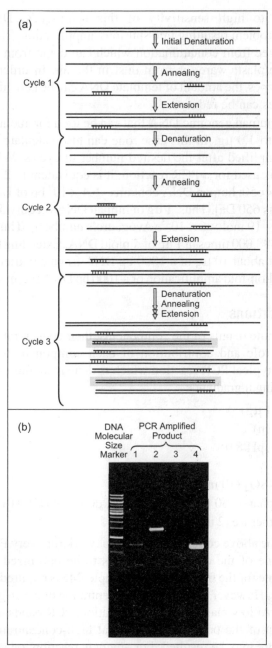

(a)

Cycle 1
- Initial Denaturation
- Annealing
- Extension

Cycle 2
- Denaturation
- Annealing
- Extension

Cycle 3
- Denaturation
- Annealing
- Extension

(b)

DNA Molecular Size Marker 1   PCR Amplified Product 2 3 4

Figure 4.45 Polymerase Chain Reaction (PCR). **(a)** The schematic representation of the polymerase chain reaction. In cycle 3, there is formation of the ds DNA products of desired length (shown with yellow background). **(b)** The agarose gel electrophoresis of the fragments obtained after PCR using specific primers. Lane 1 to 4 shows DNA products of 717bp, 1395bp, 1254bp and 612bp, respectively. **(See Colour Plate 7)**

total number of DNA molecules amplified during a PCR reaction, which depends on the number of cycles, is $2n$ (n = number of cycles). On an average, 25–30 cycles of PCR reactions are carried out, and during this, about 30 million-fold of amplification is expected.

PCR, although an easy technique, requires optimization of several parameters in order to get satisfactory amplification of the desired DNA fragment. The most essential parameters are:

(i) the quality of the template DNA,

(ii) the overall reaction conditions,

(iii) the appropriate thermostable DNA polymerase, and

(iv) the ideal oligonucleotide primers.

Further, due to its high sensitivity, utmost precautions of various kinds, such as use of dedicated autoclavable pipettes, reagents, glassware and plasticware specifically for PCR, and an aseptic PCR workstation, are necessary for optimum and desirable results. We describe some of these aspects below, however, the readers may refer Innis et al. (1999), McPherson and Moller (2000), Reece (2004) and Twyman et al. (2007) for further details.

The Quality of Template DNA

The template DNA can be of any source (e.g., Genomic DNA, cDNA, plasmid DNA), and any form in case of plasmid DNA (e.g., supercoiled, linear, etc.). However, the most important requirement is the intact primer binding sites as well as the intactness of the internal sequence between these sites. If these two criteria are taken care of, even the ancient DNA from fossils, mummies, etc. can be amplified by PCR.

As described earlier, due to high sensitivity of this techniques, DNA contamination can be a major problem resulting into spurious amplification. Thus, the template DNA should be free from contamination which can come from the researcher, reagents, glassware/plasticware, or from dust in the air. In order to reduce contamination of these types, the amount of template DNA can be kept high, while the number of PCR cycles can be reduced.

For example, the amount of human genomic DNA that can be used for about 25 cycles of PCR is between 0.1 to 1.0 µg. In this case, one can also calculate the amount of DNA that can be amplified after the desired number of cycles. When 1 µg of human genomic DNA is used for a PCR reaction, it is equivalent to $2.4 \times 10-19$ mol [$1 \times 10^{-6}/6.4 \times 10^9 \times 650$; human DNA contains ~$6.4 \times 10^9$ bp of DNA and average MW of a base pair is 650 Da]. Thus, 1 µg of human DNA is equivalent to 144,000 molecules [$2.4 \times 10-19$ mol $\times 6 \times 10^{23}$ (Avogadro's number)]. Thus, a single gene will be represented 288,000 times in 1 µg of diploid DNA. Extending this information further, one can get about 10 µg of a 1000 bp DNA fragment after 25 cycles of PCR reaction (an 8 million-fold amplification of a 1000 bp DNA fragment).

The Overall Reaction Conditions

It is imperative to know the minute details of the components, their concentration, sequence of adding, and the role and contribution of each component for a successful PCR experiment. A typical PCR reaction which is done in a volume of 25 to 100 µl contains the following ingradients:

(a) Template DNA (0.01 – 0.1 µg)
(b) Two primers (20 pmol each)
(c) Buffer, Tris-HCl (20 mM, pH 8.0)
(d) $MgCl_2$ (2 mM)
(e) KCl (25 mM) and $(NH_4)_2SO_4$ (10 mM)
(f) Deoxynucleotide triphosphates (50 µM each of dATP, dCTP, dGTP, dTTP)
(g) Thermostable DNA polymerase (2 units)

It is to be kept in mind that the above composition may not work for every PCR reaction. Concentration of some of the components needs to be optimized for practically any PCR reaction done for the first time. For example, Mg is required for the activity of DNA polymerase. However, an optimum concentration only can give rise to a specific PCR product. At low magnesium concentration, PCR bands may not be obtained due to inaction of the polymerase, while at high concentration, multiple bands of spurious nature may be resulted. For any new reaction, one can try a range of 1–5 mM of Mg. Similarly, lower potassium concentration may sometimes yield better result by allowing the polymerase to remain on the DNA template for a longer time (Foord and Rose, 1994).

So far as the sequence of addition is concerned, one needs to remember to add the polymerase at the last, because of the high cost of the enzyme. If it is added

Colour Plate I

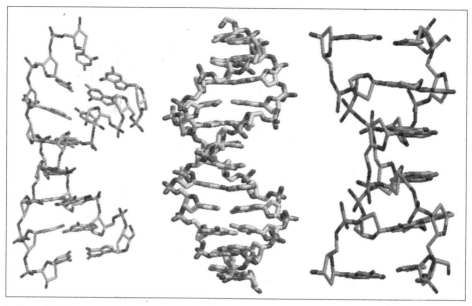

A, B, and Z forms of DNA (from left to right). Under physiological conditions, B is the most stable and favourable form of DNA. (Chapter 2, p. 15)

Tertiary structure of lysozyme C from *Gallus gallus*. (Chapter 2, p. 22)

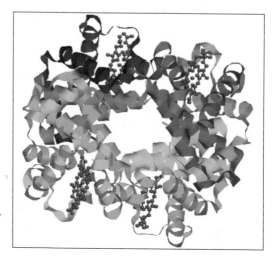

Quaternary structure of haemoglobin. It consists of two α-chains and two β-chains. (Chapter 2, p. 22)

Colour Plate 2

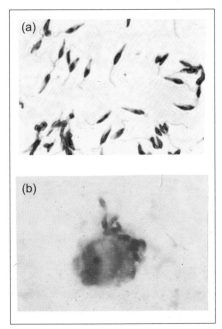

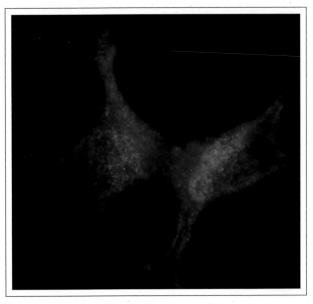

Cells stained with a coloured dye Giemsá for morphological observation through light microscopy. (a) *Leishmania donovani* promastigotes, (b) Promastigotes in the process of infecting a macrophage cell (P388D1). (Chapter 4, p. 47)

Image of melanotic melanoma cells by Immunofluorescence microscopy. Melanotic cells stained with DAPI (nuclei appears blue) and tyrosinase is depicted by red colour. (Chapter 4, p.50)

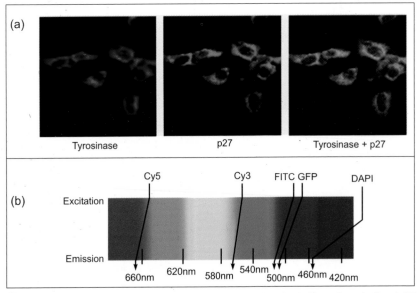

(a) Double labeled immunofluorescence. Melanotic melanoma cells stained with tyrosinase (red) and p27 (green) antibody. The third is the merged image of the two. (b) Schematic representation of the excitation and emission wavelength of some fluorochromes. (Chapter 4, p. 51)

Colour Plate 3

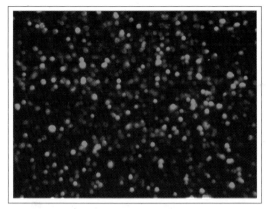

Fluorescence microscopy. K562 cells showing GFP expression under 20X objective. (Chapter 4, p. 51)

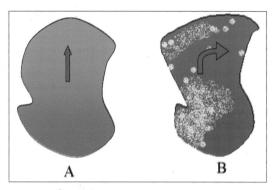

Fura-2, a Ca^{2+} sensitive fluorochrome, can be used to monitor the relative concentrations of cytosolic Ca^{2+} in different regions of live cells. (Chapter 4, p. 52)

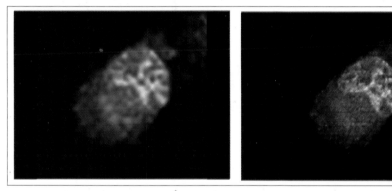

Fluorescence microscope image of mitotic cells. The wide-field microscope image is blurred, while the confocal image is clear. In this, the blurring effect caused due to fluorescence emitting from above and below the focal plane in the conventional fluorescence microscopy is eliminated. (Chapter 4, p. 53)

A deconvolved reconstructed image of live bovine pulmonary artery endothelial cells stained with LysoTracker Red DND-99, dihydrorhodamine 123 and Hoechst 33258. The oxidized product, rhodamine 123, localized primarily to the mitochondria (green), LysoTracker Red DND-99 stain accumulated in the lysosomes (red), and Hoechst 33258 dye stained the nucleus (blue). (Chapter 4, p. 54)

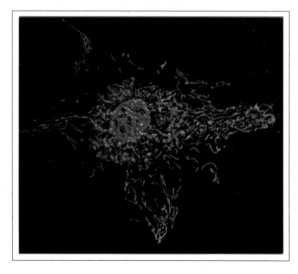

Colour Plate 4

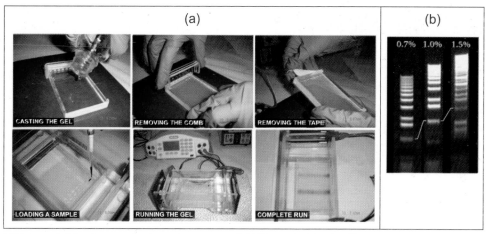

(a) Pictures depicting the procedure for casting agarose gel. (b) Picture depicting the difference in the DNA samples run in different concentrations of agarose gel. (Chapter 4, p. 70)

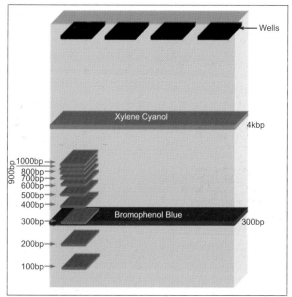

The schematic representation of the gel showing 100bp DNA ladder and the position of xylene cyanol and bromophenol blue after agarose gel electrophoresis. (Chapter 4, p. 70)

Pictures depicting the procedure followed while performing SDS-PAGE. First the plates are aligned, and then these are attached to apparatus using clamps. The plates are then sealed with agar, gel is casted and samples are loaded. (Chapter 4, p. 74)

Colour Plate 5

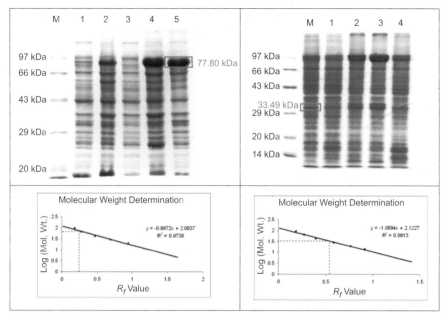

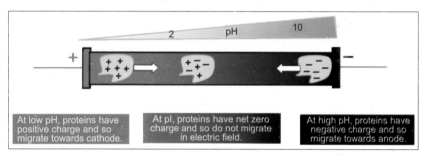

The SDS-PAG stained with silver stain (upper left) and coomassie blue stain (upper right). The molecular weights of particular bands (encircled with red) are determined using the molecular weight markers loaded in the gel. R_f value of the molecular weight marker bands and the log (mol. wt.) is plotted in the graph, then the R_f-value of the band whose molecular weight is to be determined is put in the equation and the log (molecular weight) is determined. (Chapter 4, p. 76)

Principle of Isoelectric focusing. There is a pH gradient along the gel, proteins migrate in the electric field according to their charge and get focused at a pH where there charge is neutralized, i.e., their net charge is zero. (Chapter 4, p. 78)

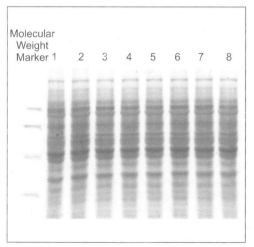

Ponceau red S stained membrane. Proteins are stained with Ponceau red S stain after transfer to nitrocellulose membrane from a SDS PAGel. Lanes 1-8 are K562 cell lysates. (Chapter 4, p. 80)

Colour Plate 6

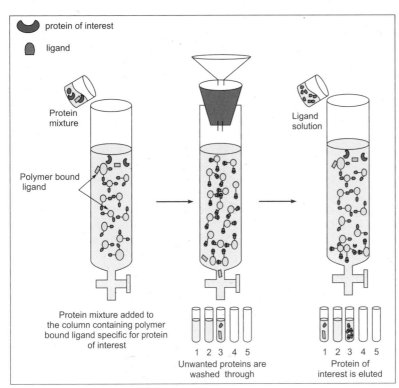

protein of interest

ligand

Protein mixture

Polymer bound ligand

Protein mixture added to the column containing polymer bound ligand specific for protein of interest

1 2 3 4 5
Unwanted proteins are washed through

Ligand solution

1 2 3 4 5
Protein of interest is eluted

Affinity chromatography. (Chapter 4, p. 85)

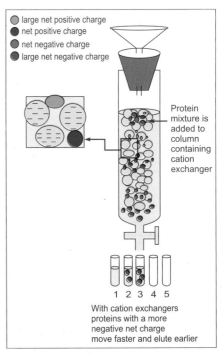

large net positive charge
net positive charge
net negative charge
large net negative charge

Protein mixture is added to column containing cation exchanger

1 2 3 4 5

With cation exchangers proteins with a more negative net charge move faster and elute earlier

Ion-exchange chromatography. (Chapter 4, p. 84)

Colour Plate 7

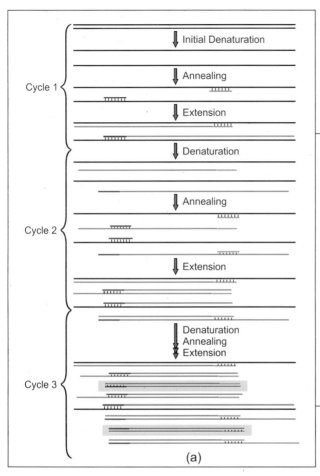

(a)

Polymerase Chain Reaction (PCR). **(a)** The schematic representation of the polymerase chain reaction. In cycle 3, there is formation of the ds DNA products of desired length (shown with yellow background). **(b)** The agarose gel electrophoresis of the fragments obtained after PCR using specific primers. Lane 1 to 4 shows DNA products of 717bp, 1395bp, 1254bp and 612bp, respectively. (Chapter 4, p. 97)

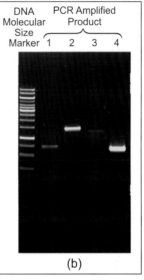

(b)

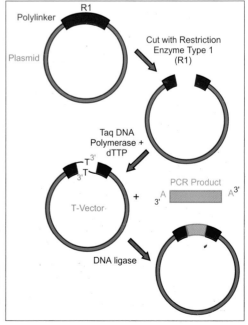

TA cloning. The plasmid vector is digested with restriction enzyme type 1, such that blunt ends are formed. Then Taq DNA polymerase with dTTP is added and thus results in 3′ T overhangs. PCR product has 3′ A overhangs. Therefore, these get easily ligated. (Chapter 4, p. 102)

Colour Plate 8

Chromatin immunoprecipitation (ChIP). In the live cells, the proteins are crosslinked to the DNA. Then cells are lysed and DNA is sheared. To this primary antibody of interest is added. After this, protein A/G agarose beads are added which binds to Fc portion of the antibody. This is then immunoprecipitated to enrich the protein of interest bound fragments. The protein DNA interaction is reverse crosslinked and the proteins are digested with proteinase K. The protected DNA is detected by PCR or hybridization with proper positive and negative controls. (Chapter 4, p. 108)

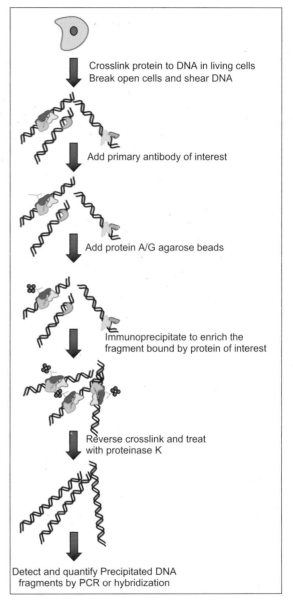

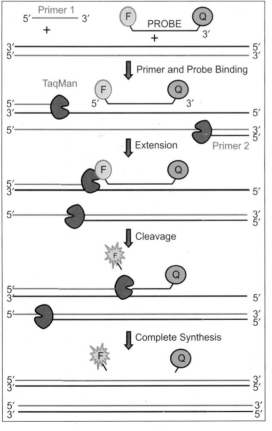

Real time PCR. The probe binds in the region which has to be amplified. It has fluorescent dye at one end and quencher at other end. When the region is amplified by the TaqMan, fluorescent dye is cleaved from the probe and thus quencher cannot quench its fluorescence and it is read by the machine. (Chapter 4, p. 104)

before and one finds an error in other components, the reaction mixture has to be thrown. Further, to avoid accidental omission of this enzyme, many products are available with colour, to ensure the addition.

Before placing the PCR reaction tube in the thermal cycler, a layer of mineral oil is generally added on top of the PCR mix to avoid evaporation of the sample. However, in case of a cycler with a heated lid (hot bonnet), it is not required. Now, the thermal cycler needs to be programmed with reference to the time it spends at each of the reaction temperature in a cycle. Generally, the following cycling conditions are used:

(a) Denaturation—94°C for 30 sec
(b) Annealing—60°C for 30 sec
(c) Extension—72°C for 1 min

However, this needs to be standardized depending upon the template, primers, sizes of the expected product, etc.

The number of cycles required for a PCR reaction also depends on the initial amount of the DNA template, the total amount of final product required, and the purpose of usage of the amplified DNA. In general, a range of 17–25 cycles are performed. The number of cycles is generally kept within this range to avoid replication errors. At the end, a final extension step of ~5 min. is kept to ensure replication of all DNA molecules in the double stranded form. Most of the pilot experiments for optimization are now done in a gradient thermal cycler in which many parameters can be standardized at a single time.

The Appropriate Thermostable DNA Polymerases

We have mentioned earlier about the first thermostable DNA polymerase, the Taq polymerase. This thermostable enzyme is a monomer of 94 kDa. Although it synthesizes DNA at 74°C, it remains stable even at 95°C. It has a 5′– to 3′ DNA polymerase activity, and a 5′– to 3′ exonuclease activity. However, unlike bacterial DNA polymerase 1, it lacks a 3′– to 5′ exonuclease (proof-reading) activity. Because of this, it introduces errors (mutations) in the amplified DNA at the rate of 1×10^{-4} per base pair of replicated DNA. However, the error rate may vary from one experiment to the other. Thus, if the DNA is to be used for cloning and subsequent functional studies in particular, it may affect such experimentation. In any case, for such purpose, the amplified DNA needs to be sequenced for its further use for such purpose. In case, the purpose is only for identification of a specific DNA product, the error may not affect the PCR experiment.

Another feature of this polymerase to remember is that it has a tendency to add a deoxynucleotide (usually an adenosine) at the 3′end of the newly synthesized DNA in a template-independent manner. Consequently, the PCR product will not have a blunt end, but will have a 3′ A overhang. This property is exploited for cloning experiments, which is described later.

There are several other thermostable DNA polymerases, discovered subsequently, which are sometimes preferably used over Taq polymerase. These are:

(a) Pfu, isolated from *Pyrococcus furiosus*,

(b) Pwo from *P. woesi*,

(c) Deep Vent™ from *Pyrococcus sp.*,

(d) Tfl from *Thermus flavus*,

(e) Tli from *Thermococcus litoralis*, and

(f) Tgo isolated from *Thermococcus gorgonarius*.

Their properties are summarized in Table 4.1. As seen in the table, each of the polymerases has some positive characteristic as well as some negative characteristic with reference to the PCR product. Therefore, an approach of blending of polymerases appears more useful and beneficial for obtaining authentic product with high yield.

Table 4.1 Comparative profile of different thermostable DNA Polymerases

DNA polymerases	Taq	Tfl	Tli	Tgo	Pfu
Source organism	*Thermus aquaticus*	*Thermus flavus*	*Thermococcus litoralis*	*Thermococcus gorgonarius*	*Pyrococcus furiosus*
Properties:					
5′-3′ exonuclease activity	+	+	–	–	–
3′-5′ exonuclease activity	–	–	+	+	+
Extension time/kb	1 min	1 min	2 min	2 min	2 min
Gross error rate (error/base)	5.0×10^{-4}	5.0×10^{-5}	2.8×10^{-6}	2.0×10^{-6}	1.3×10^{-6}
Ends of PCR product	3′-A	3′-A	Blunt	Blunt	Blunt

Generally all the components of the PCR mix are added together at 4°C. At this temperature which is below the hybridization temperature of the primers (45–60°C), mismatched primers will form and may be extended to an extent by the polymerase. In this manner, the mismatched primer may get stabilized at the undesired position, and with multiple cycles, a spurious product will be resulted. Thus, to increase the specificity and the yield of the PCR product, two very intelligent approaches are in use. In one approach, which is called *hot start method*, the Taq polymerase is added to the reaction mix which is already at the denaturing temperature (94°C). This strategy not only increases the specificity of the reaction but also the yield of the PCR product. In another approach, the Taq polymerase is mixed with a specific antibody (Kellogg et al., 1994) that binds to the enzyme and inactivates its activity. Thus, at low temperature, there is no replication of DNA, while at the high temperature, due to dissociation of the enzyme-antibody complex, the polymerase replicates the DNA. AmpliTaq Gold™ (Taylor and Logan, 1995) is an example of such commercially available modified polymerases.

To increase specificity of the PCR product further, under some circumstances like longer size of the product, one can employ *nested primers* for a second PCR reaction of the initially amplified products with the usual primers. Since these primers are located internally with respect to the primers for the first PCR, they are called the nested primers. The chances that the actual authentic product get amplified are thus enhanced using this approach.

The Ideal Oligonucleotide Primers

One of the most crucial components of a successful PCR is oligonucleotide primers. Therefore, one needs to know the important characteristics of an ideal set of primers for a PCR reaction. The two primers that recognize the anti-sense and the sense strands of the target DNA to be amplified are called the *forward* and the *reverse primers*, respectively. Each PCR reaction is unique with reference to primers, and one needs to design an ideal set of primers using certain essential criteria/characteristics which are as follows.

- *Size* Primers should be of 17 to 30 nucleotides in length, for specific annealing to a single unique sequence in the target DNA template.
- *GC content* In general, primers should contain 50% GC.
- *Annealing temperature* The temperature at which the primers anneal to the target sequence in the template is generally 5°C below the melting temperature. Ideally both the primers should anneal at the same temperature. Annealing temperature is determined by using the equation 2(AT) + 4(GT).
- *Sequence composition* As seen above, the composition of the primers, i.e. the types of nucleotides is very important.
- Further, long runs of a single nucleotide should be avoided to prevent nonspecific binding of the primers to repetitive sequences in the target DNA.
- Individual primers should not have complementary sequences within. For example, if the sequence is palindromic, it can fold back on itself and form a secondary structure which will impede the amplification of the target DNA.
- Two primers should not have any complimentarity. In such a case, there could be primer-dimer formation which will be replicated during the first cycles of the PCR reaction. As a result, amplification of target sequence will be hampered.

Thus, one can design ideal primers based on the above criteria. However, now there are several freely available software packages, some of which are also available on the web. These can help in the primer designing process (Rozen and Skaletsky, 1998).

Applications of PCR

As described earlier, PCR is a breakthrough in the field of life science research. In the area of fundamental research, other than its strength of reducing the need for various cloning techniques, it is used extensively for determining and quantifying the level of gene expression by reverse transcription- and real-time quantitative PCR, respectively. It is used for studies on genomics, evolution, genetics, mutations, and many other aspects. For example, for cloning experiments, PCR has reduced the number of elaborate steps used in the conventional cloning techniques. Further, due to the automation, sensitivity, and ease of performing, it has become a popular and efficient technology for various human applications, particularly in the molecular diagnosis of diseases, agriculture, forensics, and archeology. Some of the above techniques are described in detail of hereunder.

PCR cloning

As mentioned above, instead of obtaining DNA fragments from genome or cDNA from cells by conventional methods, equivalent PCR-amplified products, generated either by DNA PCR or RT-PCR, can be used directly for cloning experiments. This approach is particularly useful for cloning cDNAs of low abundant mRNAs in cells.

Further, while doing PCR, restriction sites can be added at the 5′-ends in the primers so as to overcome the less efficient blunt-end ligation of the fragment with the vector sequence.

In another approach, PCR fragments can directly be used for cloning experiments using the advantage that Taq polymerase most often adds a template-independent A residue at the 3′-end of amplified DNA. Such fragments with 3′-A overhangs can easily be ligated to linearized PCR cloning vectors with 5′-T overhangs, called the *T vectors* (Fig. 4.46). This cloning procedure is thus called the *TA cloning*, and it allows high-efficiency cloning of PCR products.

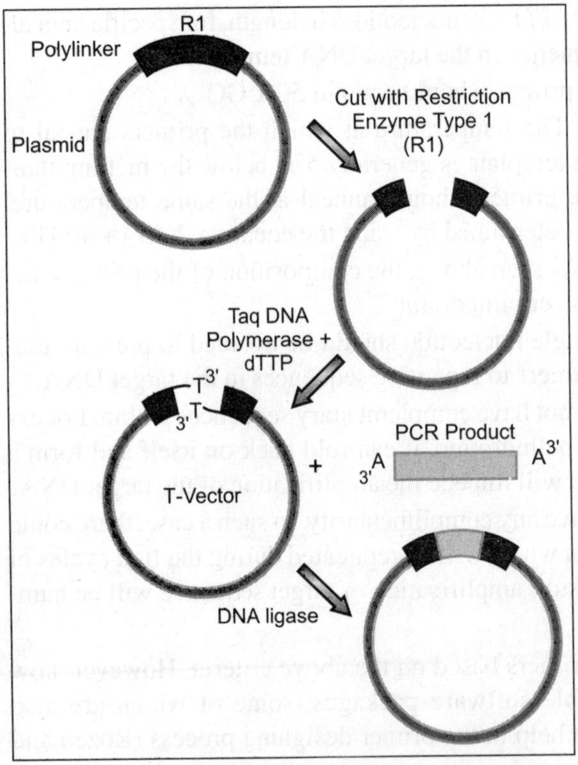

Figure 4.46 TA cloning. The plasmid vector is digested with restriction enzyme type 1, such that blunt ends are formed. Then Taq DNA polymerase with dTTP is added and thus results in 3′ T overhangs. PCR product has 3′ A overhangs. Therefore, these get easily ligated. **(See Colour Plate 7)**

RT-PCR

Reverse transcription-polymerase chain reaction (RT-PCR) is a method of RNA amplification and quantitation through an intermediate cDNA. Thus, there are two steps in this method. In the first step, total mRNAs (or total RNA) from the cell source are converted to cDNAs, and in the second step, a specific cDNA for the chosen mRNA is PCR amplified through the usual reaction process using gene-specific primers (Fig. 4.47).

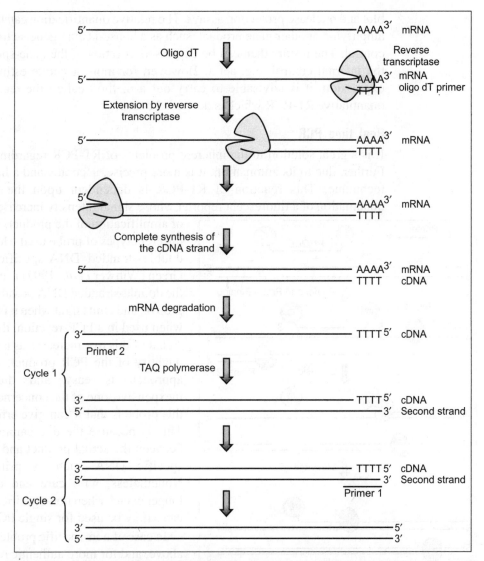

Figure 4.47 Reverse transcriptase PCR. mRNA is first reverse transcribed by reverse transcriptase using oligo dT primer. Then RNA is degraded and the first strand of cDNA is used in the PCR with the gene specific primers.

As seen in the figure, a single stranded cDNA is generated by using RNA/mRNA template, a primer (random or oligo dT), dNTPs, a buffer and a reverse transcriptase enzyme. The single-stranded cDNA is then subjected to PCR, and in the first cycle of PCR, it gets converted to a double-stranded cDNA by using a specific complementary primer by the Taq polymerase. The rest of the PCR is similar to the DNA PCR resulting amplified product.

Although it can be used for quantifying the level of gene expression, it is a semiquantitative technique, unlike other RNA estimation techniques like Northern blot and nuclease protection assays. The relative quantification can thus be done by amplifying another gene product, such as a house-keeing gene, actin as an internal control. The results then can be expressed as ratios of the gene-specific signal to the internal control, e.g. actin. However, for more accurate estimation of gene expression, it is advisable to carry out a method called the real time PCR or quantitative RT-PCR which is described below.

Real time PCR

It is a great solution to the inherent problems of RT-PCR regarding quantitation. Further, due to its automation, it is more precise, accurate, and a high throughput technique. This reaction of RT-PCR is dependent upon the detection and quantitation of a fluorescent reporter whose signal intensity increases as a function of amplification of the product. There could be different types of probe used. However, often a double-stranded DNA-specific dye SYBR® Green (Wittwer et al., 1997) is used. It binds to the double-stranded DNA possibly in the minor groove, and emits light when it is excited. Thus, when used in a PCR reaction, the fluorescence signal of this dye increases as a function of the quantity of the PCR product. Although, this approach is easy and the reagent is inexpensive, one of the concerned limitation of this probe is that it can give erroneous values. This is because the dye cannot discriminate between the actual product and the other non-specific DNA such as primer dimmers. Nonetheless, with care and elimination of longer cycles when non-specificity increases, it can safely be used for single PCR product.

In case of non-specific problem as discussed above, and for more authentic results, an alternative method for PCR quantification is TaqMan®. This quantification method is based on fluorescence resonance energy transfer (FRET) of hybridization probes. TaqMan probes are oligonucleotides that contain a fluorescence dye attached to the 5′ base, and a quenching dye to the 3′ base. The probe for the PCR reaction is synthesized with a sequence such that it binds to an internal region of the product. Initially, the reporter dye does not fluoresce due to the transfer of energy to the

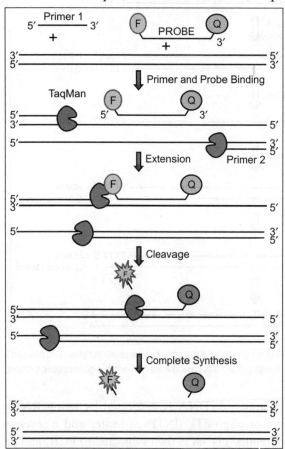

Figure 4.48 Real time PCR. The probe binds in the region which has to be amplified. It has fluorescent dye at one end and quencher at other end. When the region is amplified by the TaqMan, fluorescent dye is cleaved from the probe and thus quencher cannot quench its fluorescence and it is read by the machine. **(See Colour Plate 8)**

nearby quenching dye molecule. However, during PCR, when the polymerase replicate a DNA template on which a probe is bound, the 5′-3′ exonuclease activity of the polymerase cleaves the probe (Holland et al., 1991). This results in separation of the fluorescent and the quenching dyes resulting lack of FRET (Fig. 4.48). Thus, fluorescence increases in each PCR cycle in proportion to the rate of probe cleavage, which is measured in a modified specialized thermal cycler. Although most appropriate for quantification, this method is expensive due to high cost of the reagents and the equipment.

4.3.6 Methods for Determining DNA-protein Interactions

Study of DNA-protein interactions is extremely important in molecular biology, particularly in the context of regulation of transcription, and chromatin modulation in general. There are several methods to determine the DNA bases which interact with the protein(s), and also to identify and characterize the DNA-binding proteins (transcription factors). A few of these are discussed hereunder.

Filter Binding Assay

Nitrocellulose membranes (filters) can bind DNA under certain conditions. Single stranded DNA binds readily to nitrocellulose filters but double stranded DNA by itself does not. On the other hand, proteins bind to the filter easily. Thus, if a protein is bound to a double stranded DNA, then such a protein–DNA complex will bind to the filter. This is the basis of filter binding assay.

Labelled double stranded DNA is poured through the nitrocellulose filter. The amount of label in the filtrate or in the filter bound material is measured. As seen in Fig. 4.49a, when a labelled DNA alone is passed through the filter, almost all the labels are found in the filtrate confirming that the double stranded DNA does not bind to nitrocellulose. Whereas when a labelled protein is filtered, most of the label remains filter bound indicating that the protein is bound to the filter (Fig. 4.49b). However, when a labelled double stranded DNA is mixed with a protein, if the protein binds to the DNA, then the protein-DNA complex is retained on the filter and the label is recovered on the filter rather than in the filtrate (Fig. 4.49c).

Electrophoretic Mobility Shift Assay (EMSA)

Another method for detecting DNA-protein interaction is *electrophoretic mobility shift assay* or *gel retardation assay* which relies on the fact that a small DNA molecule has a much higher mobility in gel electrophoresis than the same DNA will have when it binds to a protein. A short double stranded DNA fragment is labelled, mixed with protein and the complex is electrophoresed. Then the gel is dried and subjected to autoradiography to detect the labelled species.

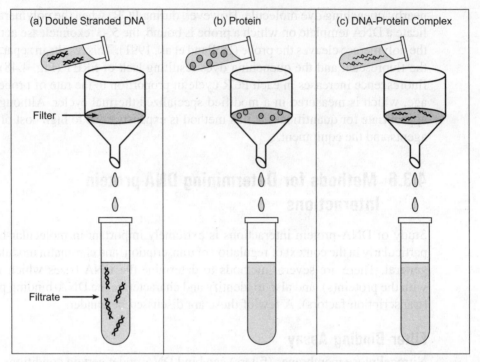

Figure 4.49   Nitrocellulose filter binding assay: **(a)** End labeled double stranded DNA when passed through a filter is recovered in the filtrate as it does not bind to the filter. **(b)** Labeled protein when passed through the filter binds tightly to it and therefore does not pass through the membrane. **(c)** When labeled double stranded DNA is mixed with a protein, protein-DNA complex is formed and is retained on the filter.

As shown in Fig. 4.50, the electrophoretic mobility of DNA is very high, and therefore it runs faster during electrophoresis than the DNA bound to the protein. This is because of the increase in the size of the complex, and it moves slowly during

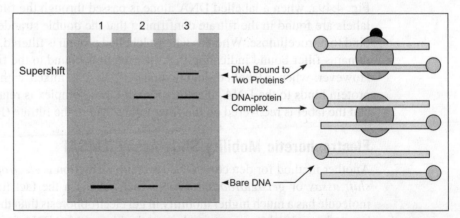

Figure 4.50   Gel mobility shift assay. Pure labelled DNA or protein-DNA complex is electrophoresed through agarose gel and then autoradiographed. Lane 1 shows the pure DNA which has the highest mobility. Lane 2 shows the mobility shift resulting due to binding of a protein. Lane 3 shows the supershift caused by binding of more than one protein (as shown as a sketch in the right hand side of the figure).

electrophoresis and its mobility is greatly reduced. Therefore, the technique is referred to as electrophoretic mobility shift assay or gel mobility shift assay. If the DNA is bound to more than one protein, then the mobility of the complex is reduced further because of its greater mass, and this is called as *supershift*. In this case, the protein could be another DNA binding protein, or a second protein that binds to the first one. A supershift is also seen if the complex is mixed with the antibody specific to the DNA-binding protein.

DNase Footprinting Analysis

Footprinting is a method for detecting DNA-protein interaction and also locating the bases in the DNA that are involved in this interaction. Different methods of footprinting are available, such as DNase footprinting, dimethyl sulphate footprinting, and hydroxyl radical footprinting. DNase footprinting relies on the fact that a protein when bound to DNA covers the binding site and protects it from attack by a DNase.

The first step in a footprinting experiment is to end label anyone of the strands of the double stranded DNA. Then, the protein is allowed to bind to the DNA and the DNA-protein complex is treated with DNase I under mild conditions so that when DNase I randomly cleaves the DNA; on an average one cut occurs per DNA molecule. Then, the protein is removed from the DNA, the DNA strands are separated, and the resulting fragments are electrophoresed on a high-resolution polyacrylamide gel along with the molecular weight markers (Fig. 4.51). Although the fragments will arise from the other end of the DNA, they will not be detected because they are unlabelled. The region of the DNA complexed with protein is not accessible for DNase, and therefore the bands corresponding to that region are not seen in the electropherogram. This protected area is called the *footprint*.

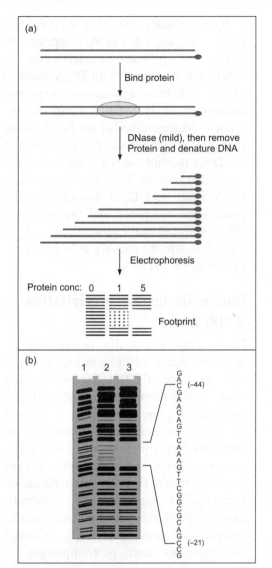

Figure 4.51 DNase footprinting. **(a)** Double stranded DNA is end-labeled and bound to the protein. DNA-protein complex is then digested with DNase I which cuts DNA randomly along the length, except where the protein is bound to the DNA. The DNA is then electrophoresed and autoradiography is performed to detect the DNA fragments. **(b)** By sequencing the DNA exact footprint can be obtained. Lane 1 is the DNA not bound to protein and lane 2 and 3 is the DNA bound to 2 different concentrations of protein. The DNA is sequenced by Sanger's dideoxy sequencing method. Exact sequence of the DNA fragment protected by protein can be obtained.

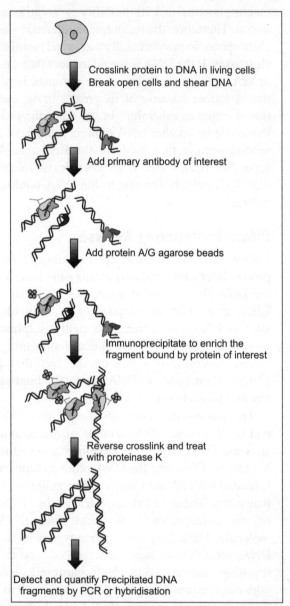

Crosslink protein to DNA in living cells
Break open cells and shear DNA

Add primary antibody of interest

Add protein A/G agarose beads

Immunoprecipitate to enrich the
fragment bound by protein of interest

Reverse crosslink and treat
with proteinase K

Detect and quantify Precipitated DNA
fragments by PCR or hybridisation

Figure 4.52 Chromatin immunoprecipitation (ChIP). In the live cells, the proteins are crosslinked to the DNA. Then cells are lysed and DNA is sheared. To this primary antibody of interest is added. After this, protein A/G agarose beads are added which binds to Fc portion of the antibody to enrich the protein-bound fragments. The protein DNA interaction is reverse crosslinked and the proteins are digested with proteinase K. The protected DNA is detected by PCR or hybridization with proper positive and negative controls. **(See Colour Plate 8)**

By sequencing the DNA used for this experiment, the bases in the footprint with which the protein interacts can be identified. A control with DNA alone is always included and more than one protein concentration is usually used, so the gradual disappearance of the bands in the foot print region reveals that protection of the DNA depends on the concentration of the added protein. Both the footprints of DNA alone and DNA complexed with protein are compared and the region of DNA protected by the protein can easily be identified which precisely tells where the protein binds to DNA.

Chromatin Immunoprecipitation (ChIP) Assay

ChIP assay essentially determines the macromolecular complexes interacting *in vivo* with specific DNA sequences in a gene, such as promoters and enhancers, for modulating gene expression at various conditions in a cell. As compared to the other two techniques described above, this technique helps one to determine a more global picture of chromatin modulation for transcription regulation of eukaryotic genes, in particular. These macromolecular complexes are assembled through modifications and protein-protein or protein-DNA interactions. They may be composed of basic transcription factors, promoter/enhancer-specific transcription factors, transcription co-activators and histone modifiers, such as histone acetyl transferases (HATs) and histone deacetylases (HDACs).

The principle underlining this assay is that DNA-bound proteins in living cells can be cross-linked to the specific region in the chromatin with the help of formaldehyde.

Formaldehyde reacts with primary amines located on amino acids and the bases on DNA molecules, forming a covalent cross-link between the specific protein and DNA to which they are bound. Following the cross-linking, the cells are lysed and the crude cell extracts are sonicated to shear the DNA to smaller size ranging from 100–1000 bp. The protein DNA complex is then immunoprecipitated using an antibody against the protein of interest. The DNA-protein cross-links are reversed by heating and subsequently the proteins are removed by treatment with Proteinase K. The DNA portion of the complex is then purified and identified by PCR using specific primers to the suspected binding region (Fig. 4.52).

4.4 RECOMBINANT DNA TECHNIQUES

Recombinant DNA technology deals with creation of new combinations of genetic material of different organisms. First of all, the DNA to be cloned and the suitable vehicle for transmission called *vector* are joined forming a recombinant molecule. Such recombinant molecules are then introduced into appropriate host organisms where they can be multiplied and selected for. The selection of appropriate methods and experimental strategies designed will depend on the biological problem to be addressed. If the interest is to clone the entire gene from an organism, it is advisable to use genomic DNA for cloning. On the other hand, if the interest is in expressing the gene, then cloning of cDNA prepared from mRNA is advisable.

4.4.1 Molecular Cloning

Here, we define cloning as the procedure of isolating a defined DNA sequence and obtaining multiple copies of the molecule *in vivo*. Cloning is widely used in several biological experiment and more importantly, in large scale production of biologically important molecules.

Constructing Gene Libraries

Following the isolation and purification of genomic DNA, it is fragmented with restriction endonucleases. These restriction endonucleases as explained earlier (Chapter 3) recognize specific sequences in DNA and cut the DNA to give cohesive or blunt ends. Many restriction endonucleases are known and the choice of the enzyme to be used depends on a number of factors. For example, the recognition sequence of 6 bp will occur every 4096 (4^6) bases assuming a random sequence of each of the 4 bases. This means that digesting genomic DNA with Eco RI recognizing sequence 5′-GAATTC-3′ will produce fragments, each of which is of an average of 4 kilobase; whereas enzymes with 8 bp recognition sequence will produce much longer fragments.

It is also possible to produce fragments of DNA by physical shearing, although the ends of fragments need to be repaired before ligation. The fragments made by either way need to be ligated to a suitable vector before transferring it to a host cell (Fig. 4.53). If enough clones are produced, there is a very high chance that any particular DNA fragment will be represented in at least one of the clones. It is possible to calculate the number of clones that must be present in a gene library to give a probability of obtaining a particular sequence. The formula is,

$$N = \frac{\ln (1 - p)}{\ln (1 - f)}$$

where N is the number of recombinants, p is the probability, and f is the fraction of genome in one insert.

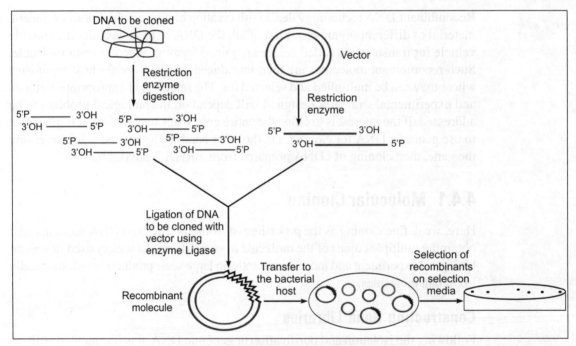

Figure 4.53   Strategy for genomic cloning

cDNA Libraries

Several thousand proteins are produced in a cell at any one time, all of which are translated from corresponding mRNA molecules. Libraries that represent mRNA from a particular cell or tissue are termed cDNA libraries. mRNA cannot be used directly in cloning since it is too unstable, instead cDNA molecules are synthesized to all the mRNAs, and this cDNA is ligated to the vector and cloned. The cDNA is synthesized using the enzyme, reverse transcriptase. Reverse transcriptase is a RNA dependent DNA polymerase and synthesizes a first strand DNA complementary to mRNA template using a mixture of four dNTPs. Since an eukaryotic mRNA contains a poly A tail at the 3′ end, complementary oligo dT primer can be used. Such primers provide a free 3′OH end which is used by reverse transcriptase to synthesize cDNA. Following the synthesis of the first strand, a second strand is synthesized to obtain a double stranded cDNA.

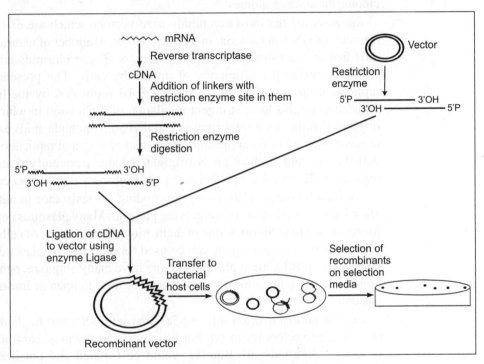

Figure 4.54 cDNA cloning

Since the ligation of blunt-ended DNA fragments is not efficient, the termini of cDNA molecules need to be polished. One of the ways to do this is to attach linkers with internal restriction site to the cDNA, and then generate cohesive termini by digesting with the corresponding restriction enzymes before using it for ligation (Fig. 4.54). It is also possible to generate the DNA fragments to be cloned by PCR as explained earlier.

4.4.2 Cloning Vectors

For cloning of any molecule of DNA, it is necessary that it is ligated to a proper cloning vector. The vectors used for cloning vary in their complexity, ease of manipulation, selection and the size of DNA that they can accommodate. For each vector, a suitable host also needs to be used. Vectors in general have been developed from naturally occurring molecules such as *bacterial plasmids, bacteriophages, or combinations of the elements that make them up such as cosmids.* For gene library constructions, there is a choice and trade-off between various vector types usually related to ease of manipulations needed to construct a library and the maximum size of foreign DNA insert of the vector. Vectors with the advantage of large insert capacity are difficult to manipulate, but are very useful for cloning large size genomes.

Plasmids are the most commonly used vectors, which are extra-chromosomal elements of DNA in bacteria. In the early 1970s, a number of natural plasmids were modified and constructed as cloning vectors. These plasmids are small in size, which increases the efficiency of uptake by cells. The presence of origin of replication ensures that the plasmid will be replicated by the host cell. Some replication origins have stringent regulation of replication in which replication is originated at the same frequency as cell division. Such plasmids exist in low copy number, while most other plasmids have relaxed origin of replication, which means that the plasmid replication is originated more frequently than chromosomal replication. Hence, these plasmids exist as high copy number per cell.

All these plasmids also have genes coding for resistance to antibiotics, which allows easy selection of cell containing plasmid. Many plasmids contain two genes for resistance to antibiotics, one of them allows the selection of cells containing the plasmid, while the other gene can be used for detection of plasmid containing the insert gene. Lastly, these plasmid vectors have many single recognition sites for a number of restriction enzymes, which can be used to open or linearize the circular plasmid.

Insertional inactivation (Fig. 4.55a) is a useful selection method for identifying recombinant vectors for insert. For example, a fragment of chromosomal DNA to be cloned digested with BamH1 would be isolated and purified. The plasmid pBR322 would be digested at a single site using BamH1. Since BamH1 generates

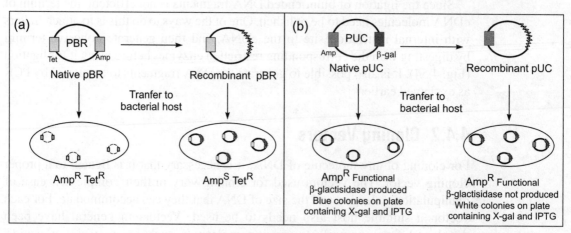

Figure 4.55 Insertional inactivation

sticky ends, it is possible to ligate plasmid and digested chromosomal DNA in the presence of enzyme ligase. The product of this ligation will include recombinant plasmid containing chromosomal DNA fragment as an insert along with some unwanted product such as recircularized plasmid without an insert, dimers of plasmid, etc. Most of these unwanted molecules could be eliminated during subsequent steps.

If the restriction digested plasmid is treated with the enzyme alkaline phosphatase prior to ligation, recircularization of plasmid is prevented since the enzyme removes the 5′ P group essential for ligation. The ligated DNA is then used to transform *E. coli*, which are made competent by prior treatment with calcium. After a brief incubation to allow expression of antibiotic resistance genes, the cells are plated onto a medium containing antibiotic, e.g. ampicillin. Colonies that grow on these cells are derived from cells that contain plasmid, since this carries the gene for antibiotic resistance. At this stage, it is not possible to distinguish these colonies containing plasmids with insert and those that contain simply recircularized plasmid. To do this, colonies are replica plated using a sterile velvet pad onto a plate containing tetracycline in the medium. Since the BamH1 site lies within the tetracycline resistance gene, this gene will be inactivated by the presence of insert that will remain intact in those plasmids that have merely circularized. Thus, those colonies that grow on ampicillin and not on tetracycline must contain plasmids with inserts. These colonies can then be recovered from the master plate for further growth.

Another very popularly used plasmid is pUC, which contains antibiotic resistance gene for ampicillin. In addition, most of the restriction sites which could be used for cloning are concentrated into a region termed *multiple cloning site* or *MCS*. MCS is part of a gene coding for a portion of polypeptide called β-galactosidase. When *E. coli* is transformed with pUC plasmid, β–galactosidase gene can be switched on by adding the inducer isopropyl β-D–thiogalacto pyranoside. The enzyme induced is able to hydrolyze its substrate called X-gal (5-bromo-4-chloro-3–indolyl–β–galactopyranoside) into a blue insoluble material. However, if the gene is disrupted by the insertion of a foreign fragment of DNA, a non-functional enzyme results which is unable to carry out hydrolysis of X-gal. Thus, a recombinant pUC plasmid can be easily detected, since *E.coli* containing this plasmid grow as white colonies in presence of X-gal, whereas those with intact pUC plasmid grow as blue colonies. This system termed *blue/white selection* (Fig. 4.55b) allows initial identification of recombinants very quickly and has been included in a number of vector systems.

Vectors Other than Plasmids

Though the plasmids are convenient vectors for cloning, they have limitation on the size of insert DNA that they can accept. Therefore, in order to propagate large fragments of DNA, a series of vectors are derived from *bacteriophage lambda,* which can easily accept insert size *up to 21kb*. Moreover, lambda vectors can efficiently enter into *E.coli* cells. Two types of lambda phage vectors have been developed, *lambda insertion vectors* and *lambda replacement vectors*. Insertion vectors accept less DNA than the replacement vectors, common example being lambda gt10 and lambda charon16A. In a replacement vector, a central region of DNA not essential for lytic cycle is removed and replaced by inserting a foreign DNA. The most notable examples of lambda replacement vectors are lambda EMBL and λZap.

Vectors have also been derived from single stranded bacteriophage such as M13. These vectors are very useful, as they produce DNA cloned in them in the single stranded form, which can be directly used for sequencing reactions. Yet another type of vector called *cosmid* was developed which is specially useful for cloning very large size DNA fragments. Cosmid vectors have been constructed to incorporate the *cos sites* form the lambda phage which are required for correct packaging of lambda DNA into lambda heads and also essential features of a plasmid such as origin of replication, a gene for drug resistance, and several unique restriction sites for insertion of the DNA to be cloned. The cosmid is linearized by restriction digestion and ligated to DNA for cloning and then packaged *in vitro* in the phage assembly mixture containing phage head precursors, tails and packaging proteins. The only requirement for DNA length to be packaged into viral capsids is that it should contain two cos sites separated from each other by 37–52 kb. Since the cosmid itself is very small, inserts of about 40 kb in length is readily packaged. Once inside the cell, the DNA recircularizes through its cos sites and from then on, behaves exactly like a plasmid.

Recent developments have allowed production of large insert capacity vectors based on bacterial artificial chromosomes (BACS), yeast artificial chromosomes (YACs), and virus p1 artificial chromosomes (PACs). Among these, most commonly used are YACs, which are linear molecules, composed of a centromere, a telomere, and a replication origin termed ARS (autonomous replicating sequence). The main advantage of YAC based vectors is the ability to form very large fragments of DNA, which can be stably maintained and replicated in yeast cells. Foreign DNA fragments up to 2000 kb have been cloned in YAC and these are the main vector of choice in various genome mapping and sequencing projects.

Expression Vectors

Specialized series of vectors are developed called expression vectors. In these vectors, transcriptional and translational signals needed for regulation of gene expression are included. Additionally, these vectors also may include a transcription termination sequence and a ribosome-binding site to enhance translation of gene of interest. Most vectors are inducible which avoids premature overproduction of a foreign product. Expression vectors frequently produce *fusion proteins* with one part of the protein coming from coding sequences in the vector, and another part from the sequences in the cloned gene itself. These fusion proteins have great advantage of being simple to be isolated by affinity chromatography. Different expression vectors are constructed for different hosts. Eukaryotic expression systems have the advantage that the protein products are usually soluble and are modified like in eukaryotic cells.

Though in most of the cloning procedures *E. coli* is used as a host, cloning in different eukaryotic cells is possible and many vectors have been specifically developed for this purpose. Vectors used for cloning in eukaryotes require an origin of replication and a marker gene that will be expressed by eukaryotic cells. Additionally, shuttle vectors have also been constructed which contain two origins of replication and two marker genes, one for yeast cells and another for bacterial cells. This means that the vector can survive in both, the hosts *E. coli* as well as yeast, and hence called a *shuttle vector.*

4.4.3 Transfer of Recombinant DNA to the Host Cells

In case *E.coli* is the host of choice, the recombinant DNA is transferred by the process of *transformation*. In this, *E. coli* cells are made *competent* by treating with calcium. During the state of competence, these cells efficiently take up the DNA, whose entry is facilitated by a brief heat-shock. In case the vectors used are bacteriophage based, the DNA can be transferred to *E. coli* by infection. On the other hand, if the host is an animal cell, the recombinant DNA is transferred by *transfection*, the efficiency of which can be increased by making the membrane permeable with divalent cations, high molecular weight polymers such as polyethylene glycol (PEG).

Alternatively, DNA can be introduced into both bacterial and animal cells by *electroporation*. In this process, the cells are subjected to high voltage pulse gradient causing many of them to take up DNA from the surrounding solution. Yet another way of transferring DNA to animal cells is by *lipofection*, wherein the recombinant DNA is encapsulated by lipid coated bilayers, which fuse with the lipid membrane of cells and release the DNA into the cell. *Microinjection* of DNA into the cell nuclei has also been performed successfully in many mammalian cells.

Once the DNA is transferred to appropriate host cells, the next job is to select the cells containing the desired recombinant molecule termed as *clones*. The first selection is always based on the marker gene of the vector used. Thereafter, more confirmatory tests are required to select the appropriate clone. These are described below.

4.4.4 Hybridization Techniques

Colony hybridization is the oldest method used to identify an appropriate clone. A large number of clones are grown in an agar plate and are replica plated on to a nylon membrane. The colonies are then lysed *in situ* and liberated DNA is then denatured being bound to the membrane. This is the target molecule in hydbridization. The membranes are then incubated with prehybridization mix containing non-specific DNA or yeast tRNA to block non-specific sites on the membrane. Following this, denatured gene probe is added which is prepared by different methods of labeling.

Under hybridization conditions, the probe will bind only to the cloned fragments containing at least part of its corresponding gene. The membrane is then washed to remove any unbound probe and the binding is detected by autoradiography. If non-radioactive labels are being used, the alternative method of detection must be employed. By comparing the patterns on the autoradiograph with the original plates of colonies, those that contain the desired gene can be identified and isolated for further analysis. Then, similar procedure is used to identify the desired genes cloned in bacteriophage vector and the process is termed as *plaque hybridization* and DNA from phage particle is immobilized on to the nylon membrane. The hybridization of nucleic acid on membrane is a widely used technique in recombinant DNA technology. Unlike solution hybridization, membrane hybridization does not go to completion since some of the bound nucleic acids remain embedded in the membrane and therefore inaccessible to the probe hybridization.

Southern in 1975 developed the method of Southern blotting/hybridization (Fig. 4.56) for detecting DNA fragments complementary to a probe. In Southern

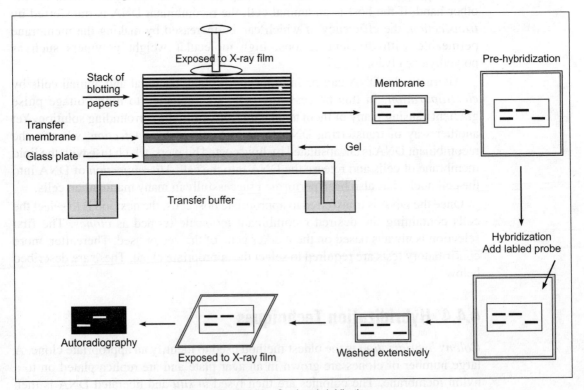

Figure 4.56 Southern hybridization

blotting, agarose gel is kept on a filter paper wick, which dips into a reservoir containing transfer buffer. Nylon membrane is then placed on top of the gel, and on top of that, a stack of paper towel is placed which can draw the transfer buffer through the capillary action. The DNA molecule is carried out by a flow and remains immobilized on the membrane. For efficient transfer of DNA, the gels are treated with alkaline buffer, which denatures the fragments prior to transfer. The gels are then equilibrated in neutralizing solution prior to blotting.

Alternatively, positively charged nylon membrane is used for transfer wherein DNA is transferred in native form and then alkali denatured *in situ* on the membrane. After the transfer, nucleic acid is to be fixed to the membrane, which is done by exposing the membrane to UV radiation. After the fixation step, the membrane is exposed to prehybridization mixture containing non-specific DNA. Following this, denatured labelled probe is added. Conditions of hybridization are chosen such that the rate of hybridization is maximized and non-specific binding of probe is avoided. This is referred to as *stringency conditions*. At high stringency, completely complementary sequences are bound whereas low stringency conditions allow hybridization to partially matched sequence. Stringency is most commonly controlled by concentration of salt and temperature during hybridization and also during post-hybridization washes. Higher stringency is achieved at high temperature and low salt concentration.

Northern hybridization involves transfer of RNA molecules on the gel to the nylon membranes. In both Southern and Northern hybridization, DNA and RNA molecules are first electrophoresed, resolved, and then transferred onto the membrane. However, these can be directly applied to slots/dots of a specific apparatus containing the nylon membrane. This is termed as *slot/dot blot* and can provide information about the abundance of specific DNA/RNA molecules. However, it does not provide information regarding the size of the fragments.

The Western blot refers to a procedure that involves transfer of electrophoresed protein bands from a polyacrylamide gel onto a nylon membrane. Specific antibodies are used to detect the presence of the desired proteins. Further details have already been discussed earlier.

4.4.5 Molecular Probes

In all hybridization techniques, a labelled probe is an essential component. The label of the probe allows any complementary sequence that the probe binds to be identified. These probes can be labelled using radioactivity or by using non-radioactive labels. Probes can also be labelled at one end or along the length. Most popularly used non-radioactive label is *digoxygenin*. If the probe is radioactively labeled, the detection procedure after hybridization is *autoradiography*; whereas if the probe is non-radioactively labelled, detection can be performed by colour reaction or by chemiluminescence. If the probe is labelled by digoxygenin, then the antibody to digoxygenin is first used to detect the presence of hybrid and then the secondary antibody tagged to horse radish peroxidase/alkaline phosphatase is added which will bind to the first complex and then the colour is developed using appropriate substrate.

End labelling reactions of probe are quite simple. *5′ end labelling* involves removal of 5′ phosphate by alkaline phosphatase followed by transfer of labelled phosphate by the enzyme polynucleotide kinase. On the other hand *3′ end labelling* can be done by using the enzyme terminal transferase. In both the end labelling reactions, the labels are transferred only at the end of DNA compared to the method that labels the DNA along the length.

Nick translation and *random primer labelling* (Fig. 4.57) are the most commonly used labelling techniques. In random primer labelling, the DNA to be labelled is first denatured and then annealed to mixture of random primers, which are hexamers. The Klenow fragment of DNA polymerase is then used for synthesis of short stretches of labelled DNA by extending primers annealed to the template. Similar to random primer labeling, polymerase chain reaction can also be used to incorporate radioactive and non-radioactive labels.

In nick translation, the reaction is started by adding low concentration of DNase I, which makes single stranded nicks in double stranded DNA. DNA polymerase holoenzyme is then used to fill the nicks using the appropriate dNTPs simultaneously making a new nick at 3′ site of the previous site. Thus, the nick is translated along

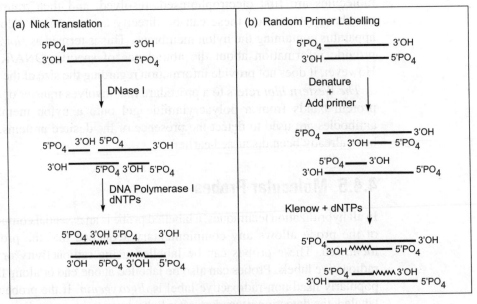

Figure 4.57 Preparation of probe

the DNA. After the probes are synthesized by any of the methods, they are quantitated and denatured before hybridization.

In both Northern and Southern hybridization, probes can be either DNA or RNA, whereas for Western blot the probe used is always antibody against the desired protein. If the antibody reacts with the protein, the complex formed is visualized by using secondary antibody tagged to a suitable enzyme, e.g. alkaline phosphatase or horseradish peroxidase and appropriate substrate reaction. Alternatively, it can be visualized by chemiluminescence. Once the positive clone is identified by these hybridization techniques, further characterization can be done.

SUMMARY

Requirements of a Molecular Biology Laboratory

Many reagents and biological materials, particularly for recombinant DNA work, that are used in a molecular biology laboratory necessitate certain specific requirements. These are, a) sturdy and dust-proof lab with appropriate furniture and working tables, b) containment facilities, such as a laminar airflow hood, 3) appropriate infrastructure including instruments, and 4) special microbiological practices up to P2/BL2 level.

Biophysical Techniques

Microscopy

As per broad classification, based on resolution, there are two types of microscopic techniques, namely, light microscopy and electron microscopy. By resolution, an electron microscope is about 100-fold more sensitive than a light microscope. There are two major types of electron microscopy, namely, transmission and scanning electron microscopy. On the other hand, there are a number of variations under light microscopy. These are, bright field, phase contrast, differential interference contrast (DIC) or Nomarski interference microscopy. Further, in the bright field type, there are variations such as, fluorescence, confocal and deconvolution microscopy. Light microscopy, in general, has been made more effective and powerful by the use of video-imaging and computer-mediated image analysis systems. With these aids, visualization of small objects, and movements of structures and organelles that are otherwise undetectable has been possible.

Centrifugation

Centrifugation is the process that involves use of centripetal force for separation of mixtures. Heavy components of a mixture tend to sediment faster and pellet down, whereas lighter substance tend to remain in the supernatant, depending on the centrifugal force applied. The rate of centrifugation is specified by the acceleration applied to the sample, typically measured in r.p.m or g. Analytical centrifugation is useful in studying the physical properties of the substances. Ultracentrifugation is a process of separation of particles whose molecular weight is very low. Additionally, density gradient centrifugations are performed to separate and purify macromolecules based on their mass.

UV-Visible spectroscopy

Various spectroscopic techniques are used for qualitative and quantitative analysis of biological molecules. UV-visible spectroscopy is most commonly employed technique in routine analytical work and biological research. If the molar extinction coefficient is known concentration of a biological molecules can be accurately calculated.

NMR spectroscopy

NMR is suitable for determining and analyzing the structure (up to 3-dimensional level) of small molecules including small proteins up to 20 kDa and domains of large proteins. It is particularly useful for Insoluble proteins (e.g., Membrane protein) which cannot be crystallized. Further, NMR spectroscopy is often advantageous over crystallography for detection and monitoring dynamic changes of protein structure under various conditions, e.g., protein folding, protein-protein interaction, enzyme-substrate interaction. Thus using NMR, one can get information about the distances between the parts of the protein molecule. This information is essentially combined with the knowledge of the amino acid sequence, and through computation, a 3-dimensional structure of the protein can be generated.

CD spectroscopy

CD is a form of spectroscopy based on the differential absorption of left- and right-handed circularly polarized light. In general, this phenomenon will be exhibited in absorption bands of any optically active molecule. As a consequence, circular dichroism is exhibited by biological molecules, because of the dextrorotary (e.g. some sugars) and levorotary (e.g. some amino acids) molecules they contain. Even more important is that a secondary structure will also impart a distinct CD to its respective molecules. Therefore, the alpha helix of proteins and the double helix of nucleic acids have CD spectral signatures representative of their structures. Circular dichroism spectroscopy may be used to estimate quantitatively the secondary structure content of proteins, giving the fraction of residues in helices, sheets, turns, and in coil conformations. Thus, it is widely used to characterize proteins under equilibrium conditions and to measure the kinetics of protein folding and unfolding.

X-ray crystallography

In order to understand structure-function relationship of macromolecules in a cell/organism, purified molecules are subjected to various structural studies. X-ray crystallography is one of them, and it provides information to build the three-dimensional structure of macromolecules and their complexes, which can be extended to their *in vivo* functions. For proteins, the main requirement is to generate protein crystals suitable for x-ray diffraction analysis. When a sample of a pure protein is subjected to X-rays, most of the X-rays pass through it. However, a small fraction is scattered by the atoms in the sample. If the sample is a well ordered crystal, the scattered waves reinforce one another at certain points and appear as diffraction spots when recorded by an appropriate detector. The diffraction pattern, obtained from good crystals, can be analysed to generate a three-dimensional electron-density map. The electron density map and sequence information of the protein are correlated through computer to generate an atomic model of the protein. The reliability of this largely depends on the resolution of the crystallographic data.

Biochemical Techniques

Electrophoresis

In electrophoresis, any charged particle can move towards its opposite charge in an electrical field. Based on this principle, among others, macro molecules, such as nucleic acids and proteins can be separated from each other based on their charge and mass when electrophoresed through a solid medium, namely, polyacrylamide and agarose gels. Although both are used for proteins as well as nucleic acids, agarose is preferred for nucleic acids and polyacrylamide gels for proteins. In general, electrophoresis is one of the most widely used techniques in biochemistry and molecular biology.

By varying the concentration of agarose, fragments of DNA from about 200 to 50,000 bp can be separated using standard electrophoretic techniques. The migration of nucleic acid molecules in an agarose gel is inversely proportional to the log molecular weight/size. Thus, by this technique one can detect and determine the size of a DNA fragment. However, different forms of DNA, for example plasmid DNA of supercoiled, nicked-circular and linear forms, migrate differently in an agarose gel. Thus the conformation of the molecule also contributes to its mobility. DNA in the gel is detected by ethidium bromide or similar DNA-intercalating stain through a UV transilluminator. There are a number of variations of agarose gel electrophoresis for various applications. For example, formaldehyde-agarose gel electrophoresis for analysis of RNA, pulse-field gel electrophoresis (PFGE) for larger size DNA up to several million base pairs (mega bases).

For analysis of proteins, Polyacrylamide gel electrophoresis (PAGE) of different types are used. They are called, native PAGE, SDS PAGE, Urea-PAGE etc. Among these, SDS PAGE is widely used, particularly for molecular weight determination, Western blot analysis etc. The migration of proteins on SDS PAGE is inversely proportional to the log MW of proteins. Proteins are detected on the gel by using protein-specific stains, such as Coomassie brilliant blue, amido black, silver stain, gold stain etc. Non-denaturing PAGE is useful for determining the size of a native protein, and also for measuring enzyme activity *in situ*. Another major application of SDS PAGE is analysis of specific proteins by Western blot analysis. In this, proteins separated on a SDS PAgel are electrophoretically transferred or blotted onto membranes (Nitrocellulose/PVDF), followed by immunoreaction of the blot with protein-specific antibodies to detect and quantify the proteins.

To overcome the limitation of resolution of one-dimensional PAGE, two-dimensional gel electrophoresis called O'Farrell gel electrophoresis are used. In this method, the proteins are separated based on their charges by isoelectric focusing in the first dimension followed by their separation by SDS PAGE in the second dimension. Thus, this method being highly resolutive, can resolve as many as few thousands of polypeptides on a gel. Such analysis is often desirable in various experiments to understand the phenotypic variations of proteins in cells/tissues/organisms. For identifying the protein/polypeptide spots in the 2-D gels, an advanced mass spectrometric method known as MALDI-TOF is used, based on their charge-to-mass ratio and deduced amino acid sequences.

Chromatography

Chromatography is the most useful technique used for separation of compounds in a mixture. It involves passing a mixture dissolved in a mobile phase through a stationary phase, which separates the compound from other molecules in a mixture and allows its isolation. Chromatography may be preparative or analytical. Preparative chromatography separates the components of mixture for further use, whereas analytical chromatography operates with smaller amounts of material and measures the relative proportions of compounds in a mixture. Chromatographies are named after the main type of interaction between the sample and the stationary phase, thus there is absorption, ion- exchange, molecular exclusion and affinity chromatography. Based on the mobile phase used it can be liquid chromatography or gas chromatography, whereas based on the stationary phase used, it could be paper chromatography, thin layer chromatography, gas layer chromatography or high pressure chromatography.

Radiolabelling and detection

Radioisotopes have been extensively used in biological research. Atoms of a given element with different mass numbers are called isotopes. These are unstable and emit particles and/ or electromagnetic radiations in an attempt to become a stable isotope. This process is known as radioactive decay and energy level associated with this decay are measured in terms of electron volt. Radioactive decay is a spontaneous process and occurs at a definite rate characteristic of the source. Isotopes emitting alpha particles are most energetic, whereas beta and gamma emitters have decay energies of less than 3 MeV. The commonly used unit to express radioactivity is curie, which is defined as the quantity of radioactivity material in which the number of radioactive disintegration is same as that in one gram of radium. For detection and quantitation of radioactivity methods used are based on ionization of gases, excitation of solids or solutions. Most commonly used is the liquid scintillation counting in biological experiments. Autoradiography is also commonly used for detection of radioisotopes. Radioisotopes being hazardous to health needs to be used very carefully and in restricted working space. Radioisotopes also needs to be disposed very carefully.

Sequencing techniques

Both DNA and proteins are sequenced to find out the sequence of nucleotides and amino acids, respectively in these polymers. Two main sequencing reactions used for DNA sequencing are Sanger'sdideoxy sequencing method and Maxam Gilberts chemical sequencing method. Automated sequencers function based on Sanger's chain termination method and are extensively used for sequencing of large DNA molecules. Recently, a new sequencing method calle pyrosequencing is developed. For protein sequencing two methods used are mass spectroscopy and Edman's degradation method. Automated sequencers are based on Edman's degradation.The C-terminal and amino- terminal amino acids can also be detected by using specific reagents.

Polymerase chain reaction

PCR is a technique for *in vitro* amplification of DNA fragments of specific sizes, based on the principle of *in vivo* DNA replication in cells. it is basically a process of copying a fragment of a double stranded DNA to a large number which depends on the number of cycles (or chains) of the reactions. The main components/reagents which are required for a basic PCR are, (1) template (target DNA), (2) two oligonucleotide primers (17–30 nts in length), each complementary to the 5′ end of the target DNA strands, (3) dNTPs, and (4) a thermostable DNA polymerase (e.g., Taq polymerase). The reactions in multiple cycles of DNA synthesis are carried out by an automated pre-programmed/programmable thermal cycler.

The PCR includes three processes, namely, denaturation/melting of DNA at ~94°C, annealing of primers with the target DNA through complementary base pairing at around 45–60°C, and extension of the primer (DNA synthesis) at 72°C. The yield of a large quantity of a specific fragment of a DNA in a short time is the characteristic of PCR, and it is due to (1) automation of the process of *in vitro* DNA synthesis by thermal cyclers, and (2) the discovery of a thermostable DNA polymerase (Taq polymerase).

There are a large number of applications of this technique. To name a few, PCR is used for diagnosis of various diseases, particularly pre-natal diagnosis, DNA finger print analysis (forensics), cloning, gene expression analysis etc.

Methods for determining DNA-protein interactions

Determining DNA-protein interactions are often required in understanding regulation of DNA functions. There are the following 4 major methods that are used for this purpose. Filter binding assay is a means of measuring DNA-protein interactions, based on the fact that double stranded DNA will not bind by itself to the nitrocellulose filter but the DNA-protein complex will. Electrophoretic mobility shift assay detects interaction between a protein and DNA by the reduction of the electrophoretic mobility of a small DNA that occurs on binding to a protein. DNAse footprinting is performed to find out the region on the DNA where protein binds. When the DNA-protein complex is subjected to DNAse and the resulting fragments are eletrophoresed, the protein binding sites shows up as gap or footprint in the pattern where the protein protected the DNA from degradation. Chromatin imunoprecipitation assay is particularly useful for determining in vivo DNA-protein complexes in a cell during modulation of gene expression.

Recombinant DNA techniques

For cloning a gene of interest a suitable vector is absolutely essential which will carry the gene into a host cell and ensure that it will replicate there. Variety of vectors have been developed and used for cloning. For all these vector molecules the physical map is known that allows easy manipulation of the DNA. They all have markers based on which their presence can be easily identified. Expression vectors are designed to yield the protein product of a cloned gene in the greatest amount. Once the recombinant DNA is produced by joining the vector and the foreign DNA, this is transferred to appropriate host. Initial screening of the desired clones is based on the marker gene of the vector. Later on the presence of foreign DNA needs to be confirmed in these clones by different techniques. Most of the techniques used are based on hybridization between a target and a labeled probe molecule. Colony hybridization allows identification of clones positive for the presence of cloned DNA. Southern hybridization is performed between DNA from clones electrophoresed and then immobilized on the membrane and a labeled probe, which can be DNA or RNA. Northern blot is similar to Southern blot except that it contains electrophoretically separated RNA instead of DNA. The intensities of the bands reveal the relative amounts of specific RNA and position of band indicates the length of the respective RNA. Western blot on the other hand involves hybridization between electrophoresed proteins transferred to the membrane and corresponding antibodies. Thus, positive signal in Southern blot indicates the presence of desired DNA in the clone whereas Northern blot confirms the presence of transcript and Western blot of proteins. In case the probe used in all these hybridizations is radioactively labeled the detection is done by autoradiography, and in case of non-radioactive probe it is done by either colour development or chemiluminescence.

Recombinant DNA technology has vast applications in various fields including biotechnology, human health and diseases and agriculture.

EXERCISES

Short Answer Questions

Microscopy

1. What makes an electron microscope more resolutive than a light microscope?
2. How does one calculate resolution of a microscope?
3. What is the principle of the phase-contrast microscopy?
4. Distinguish between confocal and deconvo-lution microscopy.
5. Why is an indirect immunofluorescence method more sensitive than a direct immunofluorescence method?

Centrifugation

1. When should one use fixed angle rotor and swing out rotor and why?
2. What is the advantage of using density gradient for centrifugation?

Spectroscopy

1. What is Beer-Lambert's law?
2. What components in proteins and nucleic acids absorb lights?
3. What is the advantage of using a double beam spectrophotometer as compared to a single beam spectrophotometer?
4. Why is it possible to calculate concentration of a compound in a solution if its molar extinction coefficient is known?

NMR, CD and X-ray crystallography

1. Among NMR, CD and X-ray crystallography, which is suitable under what situation regarding determining the structure of a protein?
2. What is the major factor that yields reliable data on crystallography?
3. How does one constitute a 3-dimensional structure of a protein from the electron density map?

Electrophoresis

1. What is the basis of separation of DNA in an agarose gel?
2. Mobility of which form of the plasmid DNA is considered for calculation of plasmid size?
3. Why is a pulsed field gel electrophoresis suitable for separation of chromo-some-size DNAs?
4. What is the principle of SDS PAGE and why is it used for MW determination of proteins?
5. Why is methanol used for transfer of proteins in Western blot analysis?
6. Why is the two-dimension gel electrophoresis more resolutive than SDS PAGE alone?

7. What is the relationship between the time-of-flight during mass spectrometry and the mass of a peptide?

Chromatography

1. What is a mobile phase and stationary phase in column chromatography?
2. How are proteins separated by ion-exchange chromatography?
3. Why is affinity chromatography a useful method for purification of certain components?
4. What is the principle of gel filtration?

Radiolabelling and detection

1. What is the unit of radioactivity?
2. Which radioisotopes are commonly used in biological experiments?
3. How does alpha particle interact with matter?
4. Why are gamma rays more penetrating than alpha particles?
5. What is the principle of scintillation counting?

Sequencing techniques

1. What is the function of dideoxynucleotides in a chain termination reaction?
2. Why FDNB or Dansyl chloride cannot be used for sequencing a polypeptide chain beyond N-terminal residue?
3. How does sequencing by mass spectroscopy differ from that done by Edman's degradation?
4. What are the advantages of chemical sequencing of DNA over enzymatic sequencing?

Polymerase chain reactions

1. At which cycle of PCR two molecules of double stranded DNA of the exact expected size of the amplified DNA obtained and why?
2. Why should one use oligonucleotide primers of 17-30 nucleotides long?
3. What are the criteria that should be used for designing PCR primers?
4. What is a hot start method and why should it be used?
5. Why should one sequence the amplified DNA using Taq polymerase?
6. One μg of human DNA would produce what amount of DNA after 20 cycles of PCR?
7. Why is RT-PCR semiquantitative and what is the correct procedure for accurate estimation of gene expression in a cell?

Methods for determining DNA-protein interactions

1. What are the major differences among the EMSA, DNase footprint analysis and ChIP assay?

2. Why is it necessary to label only one end of the DNA during DNase footpriniting?
3. What is the difference between the information obtained regarding DNA- protein interaction in gel mobility shift assay and DNase footprint?

Recombinant DNA techniques

1. When should one decide to do cDNA cloning?
2. What are the basic requirements of a vector?
3. What are different ways of transferring recombinant DNA to host cells?
4. Differentiate between, Southern, Northern and Western hybridization with respect to the target and the probe used.
5. Describe any one method of labeling the probe along the length.

FURTHER READING

Microscopy

- Allen, T.D. and Goldberg, M.W. (1993). High resolution SEM in cell biology. *Trends Cell Biol.* **3**, 203-208.
- Egner, A. and Hell, S. (2005). Fluorescence microscopy with super-resolved optical sections. *Trends Cell Biol.* **15**, 207-215.
- Hyatt, M.A. (2000). *Principles and Techniques of Electron Microscopy* (4th edn.). Cambridge University Press.
- Lacey, A. (ed.) (1999). *Light Microscopy in Biology: A Practical Approach.* Oxford University Press, New York.
- Matsumoto, B. (ed.) (2002). Methods in Cell Biology. *Cell Biological Applications of Confocal Microscopy.* Vol. **70**. Academic Press.
- Parton, R.M. and Read N.D. (1999). Calcium and pH imaging in living cells. In *Light Microscopy in Biology. A Practical Approach,* 2nd edn. (Lacey, A. ed.). Oxford University Press.
- Porter, K.R., Claude, A. Fullam, E.F. (1945). A study of tissue culture cells by electron microscopy. *J. Exp. Med.* **81**, 233–246.
- Salmon, E.D. (1995). VE-DIC light microscopy and the discovery of kinesin. *Trends Cell Biol.* **5**, 154–158.
- Zernike, F. (1955). How I discovered phase contrast. *Science* **121**, 345–349.

Centrifugation

- De Duve, C. (1975). Exploring cells with a centrifuge. *Science* **189,** 186–194.
- Graham, J. and Rickwood, D. (ed.) (1997). *Sub-cellular Fractionation: A Practical Approach.* Oxford University Press, New York.
- Rickwood, D. (1992). *Preparative Centrifugation: A Practical Approach.* IRL Press.

Spectroscopy

- Gordon D. B. (2005). Spectroscopic techniques. In Principles and Techniques of Biochemistry and Molecular Biology (eds. Wilson, K. and Walker, J.). Cambridge University Press, UK.
- Sheehan, D. (2000). Physical Biochemistry: Principles and Applications. John Wiley and Sons, Ltd., New York

NMR, CD, and X-ray Crystallography

- Sands, D. E. (1969). Introduction to crystallography. W. A. Benjamin, New York.
- Tickle, I., Sharff, A., Vinkovic, M., Yon, J. and Jhoti, H. (2004). High-throughput protein crystallography and drug discovery. *Chem. Soc. Rev.* **33**, 558–565.
- Wiencek, J. M. (1999). New strategies for protein crystal growth. *Annu. Rev. Biomed. Eng.* **1**, 505–534.

Biochemical Techniques

Electrophoresis

- Andrews, A.T. (1986). Electrophoresis (2nd edn.) Oxford University Press.
- Domon, B. and Aebersold, R. (2006). Mass spectrometry and protein analysis. *Science* **312**, 212–217.
- Janson, J-C. and Ryden, L. (eds.) (1989). *Protein Purification: Principles, High Resolution Methods and Applications*. VCH Publishers, New York.
- Laemmli, U.K. (1970). Cleavage of structural proteins during the assembly of the head of bacteriophage T4. *Nature* **227**, 680–685.
- O'Farrell, P.H. (1975). High-resolution two-dimensional electrophoresis of proteins. *J. Biol. Chem.* **250**, 4007–4021.
- O'Farrell, P.Z., Goodman, H.M. and O'Farrell, P.H. (1977). High resolution two-dimensional electrophoresis of basic as well as acidic proteins. *Cell* **12**, 1133–1142.
- Pal, J.K. and Modak, S.P. (1984). Immunochemical characterization and quantitative distribution of crystallins in the epithelium and differentiating fibre cell populations of chick embryonic lens. *Exp. Eye Res.* **39**, 415–434.
- Reece, R.J. (2004). Analysis of genes and genomes. John Wiley and Sons Ltd., U.K.
- Sambrook, J. and Russell, D. (2001). Molecular Cloning: A Laboratory Manual (3rd edn.). Cold Spring Harbor Laboratory Press, New York.
- Schwartz, D.C. and Cantor, C.R. (1984). Separation of yeast chromosome-sized DNAs by pulsed field gradient gel electrophoresis. *Cell* **37**, 67–75.
- Slater, G.W., Mayer, P. and Drouin, G. (1996). Migration of DNA through gels. *Methods Enzymol.* **270**, 272–295.
- Southern, E.M. (1975). Detection of specific sequences among DNA fragments separated by gel electrophoresis. *J. Mol. Biol.* **98**, 503–517.

- Thomas, P.S. (1980). Hybridization of denatured RNA and small DNA fragments transferred to nitrocellulose. *Proc. Natl. Acad. Sci. USA* **77**, 5201–5205.
- Towbin, H., Staehelin, T. and Gordon, J. (1979). Electrophoretic transfer of proteins from polyacrylamide gels to nitrocellulose: procedure and some applications. *Proc. Natl. Acad. Sci. USA* **83**, 4849–4853.
- Wrestler, J.C., Lipes, B.D., Birren, B.W. and Lai, E. (1996). Pulsed-field gel electrophoresis. *Methods Enzymol.* **270**, 255–272.
- Yates, J.R. (2000). Mass spectrometry: from genomics to proteomics. *Trends Genet.* **16**, 5–8.

Chromatography

- Doonan, S. (ed.) (1996). An introduction to protein chromatography and purification. In *Protein Purification Protocols, Methods in Molecular Biology* **59**, Humana Press, Totowa.
- Janson, J-C. and Ryden, L. (eds.) (1989). *Protein Purification: Principles, High Resolution Methods and Applications*. VCH Publishers, New York.
- Sheehan, D. (2000). *Physical Biochemistry: Principles and Applications*. John Wiley and Sons, Ltd., New York.

Radiolabelling and detection

- Slater, R.J. (2002). Radioisotopes in Biology- A practical approach, Oxford University, Oxford.
- Wilson, K. and Walker, J. (2005). Principles and Techniques of Biochemistry and Molecular Biology, Cambridge University Press.

Sequencing techniques

- Griffin, H.G. and Griffin, A.M. (1993). Dideoxy sequencing reactions using Sequenase version 2.0. *Methods Mol. Biol.* **23**, 103–108.
- Maxam, A.M. and Gilbert, W. (1977). A new method for sequencing DNA. *Proc. Natl. Acad. Sci. USA* **74**, 560–564.
- Ronaghi, M., Ehleen, M. and Nyrn, P. (1998). A sequencing method based on real time pyrophosphate. *Science* **281**, 363–365.
- Sanger, F. (1988). Sequences, sequences, sequences. *Ann. Rev. Biochem.* **57**, 1–28.
- Sanger, F., Nicklen, S. and Coulson, A.R. (1977). DNA sequencing with chain terminating inhibitors. *Proc. Natl. Acad. Sci. USA* **74**, 5463–5467.

Polymerase chain reaction

- Cline, J., Braman, J.C. and Hogrefe, H.H. (1996). PCR fidelity of Pfu DNA polymerase and other thermostable DNA polymerases. *Nucleic Acids Res.* **24**, 3546–3551.

- Erlich, H. (ed.) (1992). *PCR Technology: Principles and Applications for DNA Amplification*. W.H. freeman and Company.
- Foord, O.S. and Rose, E.A. (1994). Long-distance PCR. *PCR Methods and Appl.* **3**, S149–S161.
- Innis, M.A., Gelfand, D.H. and Sninsky, J.J. (eds.). (1999). *PCR Applications: Protocols for functional Genomics*. Academic Press, San Diego, CA, USA.
- Kellogg, D.E., Rybalkin, I., Chen, S., Mukhamedova, N., Vlasik, T., Siebert, P.D. and Chenchik, A. (1994). TaqStart antibody: 'hot start' PCR facilitated by a neutralizing monoclonal antibody directed against Taq DNA polymerase. *Biotechniques* **16**, 1134–1137.
- Marchuk, D., Drumm, M., Saulino, A. and Collins, F.S. (1991). Construction of T-vectors, a rapid and general system for direct cloning of unmodified PCR products. *Nucleic Acids Res.* **19**, 1154.
- McPherson, M.J. and Moller, S.G. (eds.) (2000). *PCR Basics: from background to Bench*. Bios, Oxford.
- Mullis, K.B. (1990). The unusual origin of the polymerase chain reaction. *Sci. Am.* **262**, 56–61.
- Mullis, K.B. and Faloona, F.A. (1987). Specific synthesis of DNA *in vitro* via a polymerase-catalyzed chain reaction. *Methods Enzymol.* **155**, 335–350.

Methods for determining DNA-protein interactions

- Choo, Y. and Klug, A. (1997). Physical basis of a protein-DNA recognition code. *Curr. Opin. Struct. Biol.* **7**, 117–125.
- Galas, D.J. and Schmitz, A. (1978). DNAse footprinting, a simple method for the detection of protein – DNA binding specificity. *Nucleic Acids Res.* **5**, 3157–3170.
- Sambrook, J. and Russell, D. (2001). Molecular Cloning: A Laboratory Manual (3rd edn.). Cold Spring Harbor Laboratory Press, New York.
- Orlando, V. (2000). Mapping chromosomal proteins *in vivo* by formaldehyde-crosslinked chromatin. *Trends Biochem. Sci.* **25**, 99–104.

Recombinant DNA Technology and its Applications

- Brown, T.A. (2001). Gene cloning and DNA analysis. An introduction. Blackwell Science, Oxford.
- Brown, T.A. (2007). *Genomes* 3. Garland Science, New York and London.
- Grunstein, M. and Wallis, J. (1979). Colony hybridization. *Methods Enzymol.* **68**, 379–389.
- Gubler, U. and Hoffman, B.J. (1983). A simple and very efficient method for generating cDNA libraries. *Gene* **25**, 263–269.
- Nathans, D. and Smith, H.O. (1975). Restriction endonucleases in the analysis of restructuring of DNA molecules. *Ann. Rev. Biochem.* **44**, 273–293.
- Primrose, S.B. and Twyman, R.M. (2007). Princples of gene manipulation and Genomics. Blackwell Publishing, Oxford, U.K.

Concept of Genome
and Its Organization

OVERVIEW

- Most of the organisms contain DNA as the genetic material, except some viruses which contain RNA instead.

- Genome size, types of DNA sequences, and organization of genome vary between prokaryotes and eukaryotes.

- Eukaryotic genomes are in the form of chromatin, and there has been an extensive compaction of the genetic material through extensive higher order structure formation leading to the chromosomes seen during mitosis.

- Organelle genomes are evolved from those of bacteria during their endosymbiotic origin and evolution.

5.1 NUCLEIC ACIDS AS THE GENETIC MATERIAL

Genome of every living organism consists of a long sequence of nucleic acids that provides the information required to construct that organism. This sequence is used by a complex series of interactions to produce all the proteins of the organism at an appropriate time and place. Genome, thus, contains the complete set of hereditary information for any organism and is functionally divided into small parts referred to as *genes*. Each gene is a sequence of nucleotides representing a single protein or RNA. Genome of living organisms may contain as few as 500 genes as in case of Mycoplasma, or as many as 30,000 genes as in case of human beings.

The idea that the *genetic material is a nucleic acid* was based on the experiments of transformation performed in 1928. The pathogenic bacterium *Pneumococcus* when injected into a mouse kills the mouse by causing pneumonia. The virulence of the bacterium is determined by its capsular polysaccharide due to which they have a smooth appearance when grown on nutrient agar plates. The non-pathogenic strain of *Pneumococcus* lacks this polysaccharide and hence, has a rough appearance (R) while grown on nutrient agar plates. These non-pathogenic bacteria do not kill a mouse. When smooth bacteria (S) are killed by heating, they lose their pathogenicity and therefore the ability to kill a mouse.

In an experiment, when a mouse was injected with smooth bacteria, the mouse died. When it was injected with heat-killed smooth bacteria, it survived. On the other hand, when mice were injected with non-pathogenic rough strain of pneumococci, mice survived. However, when heat-killed smooth bacteria were injected with live non-pathogenic rough bacteria, they developed pneumonia and died (Fig. 5.1).

More importantly, pathogenic pneumococci could be recovered from the dead mice. This meant that some property of the dead smooth bacteria transformed live non-pathogenic

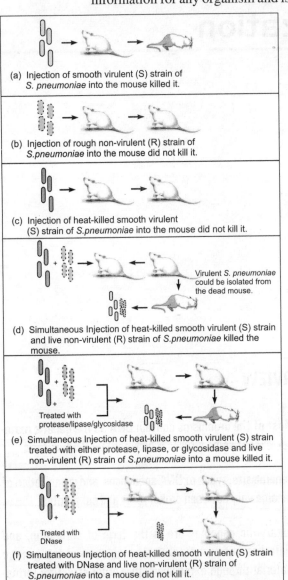

(a) Injection of smooth virulent (S) strain of *S. pneumoniae* into the mouse killed it.

(b) Injection of rough non-virulent (R) strain of *S.pneumoniae* into the mouse did not kill it.

(c) Injection of heat-killed smooth virulent (S) strain of *S.pneumoniae* into the mouse did not kill it.

Virulent *S. pneumoniae* could be isolated from the dead mouse.

(d) Simultaneous Injection of heat-killed smooth virulent (S) strain and live non-virulent (R) strain of *S.pneumoniae* killed the mouse.

Treated with protease/lipase/glycosidase

(e) Simultaneous Injection of heat-killed smooth virulent (S) strain treated with either protease, lipase, or glycosidase and live non-virulent (R) strain of *S.pneumoniae* into a mouse killed it.

Treated with DNase

(f) Simultaneous Injection of heat-killed smooth virulent (S) strain treated with DNase and live non-virulent (R) strain of *S.pneumoniae* into a mouse did not kill it.

Figure 5.1 Experimental strategy followed to demonstrate that the DNA is a genetic material. **(a)** to **(f)**, various experiments with results as indicated in the figure.

rough bacteria so that they became pathogenic and killed the mice. It was then necessary to identify the component from the heat-killed bacteria responsible for this transformation. This was referred to as the *transforming principle*. To identify the nature of this transforming principle, a mixture of heat-killed bacteria was treated separately with protease, lipase, and DNase before injecting it into the mice along with non-pathogenic rough bacteria. Lipase would degrade lipids, protease would degrade proteins, and DNase would destroy deoxyribonucleic acid (DNA). It was found that when the mixture of heat killed pathogenic bacteria was treated with DNase, non-pathogenic bacteria failed to transform indicating that the transforming principle was deoxyribonucleic acid (DNA).

Having shown DNA as the genetic material, the next step was to demonstrate that it provides the *information to build an entire organism*. Experimental evidence for this was provided by Hershey and Chase in 1952 (Fig. 5.2). The system used for this experiment was bacteriophage T2 and the bacterium *E. coli*. When phage particles are added to a bacterium, they adsorb to the surface. Some material enters into the bacterium and after sometime, the bacterium bursts open to release a large number of progeny phages.

In their experiment, Hershey and Chase infected bacteria with phage T2 that was radioactively labelled in their DNA component with ^{32}P or in their protein with

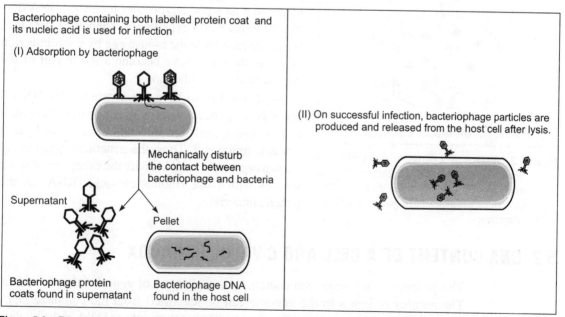

Figure 5.2 Experimental demonstration that DNA can direct the synthesis of bacteriophage particles. Bacteria were infected with T2 bacteriophages that had been radioactively labelled either in their DNA component (with ^{32}P) or in their protein component (with ^{35}S). The infected bacteria were agitated in a blender and the two fractions were separated. Radioactivity due to bacteriophage DNA was found in the infected bacteria and these could give rise to new progeny

^{35}S. After a few minutes of infecting the bacterial cells, they were agitated in a blender. Bacteria infected with phage settled down as a pellet, whereas the supernatant contained empty phage coats, which must have been released from the surface of bacteria. When both these fractions were checked for their radioactivity, ^{32}P label was found in infected bacteria, whereas the entire ^{35}S label of protein was present in the supernatant. On further incubation, the progeny phage particles were released from these bacteria from lysis. This experiment therefore directly demonstrated that only when the DNA from the parent phage which entered the bacteria is capable of giving rise to progeny phages, which is exactly the pattern of inheritance expected of the genetic material. This therefore re-enforces the conclusion that DNA is the genetic material.

Similar to the transformation experiments carried out in bacterial cells, experiments were also done with eukaryotes. The DNA corresponding to thymidine kinase (TK) gene was introduced into the recipient cells which were TK$^-$, i.e. they did not produce enzyme thymidine kinase. These cells when transfected with TK gene were successfully transformed to TK$^+$ phenotype (Fig. 5.3). Thus, even the eukaryotic cells acquired a new phenotype after receiving DNA during the transfection experiment. Thereafter, the DNA was introduced into the mouse egg by microinjection and was shown to have become a stable part of the genetic material of the mouse.

Such experiments have directly shown that DNA is not only the genetic material in eukaryotes, but also it can be transferred into different species, and yet it remains functional. Thus, the genetic material of all known organisms is DNA, with the exception of some viruses which use ribonucleic acid (RNA) as the genetic material.

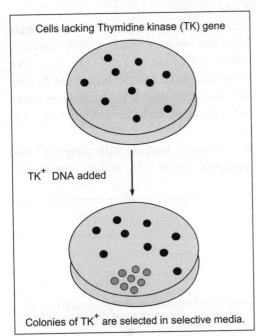

Figure 5.3 Addition of a particular purified DNA to eukaryotic cells leads to acquisition of a new phenotype. Thymidine kinase deficient (TK$^-$) cells on transfection with TK gene acquired TK$^+$ phenotype.

5.2 DNA CONTENT OF A CELL AND C-VALUE PARADOX

The genome of any organism contains a complete set of genes of that organism. The number of genes in the genome can be identified from the complete DNA sequence by defining open reading frames. The total amount of DNA in the haploid genome of any organism is a characteristic of that species and is known as *C-value*. There is an enormous variation in the range of C-values from less than 10^6 base pairs for Mycoplasma to more than 10^{11} base pairs for some plants and amphibians. It is observed that there is an increase in the genome size in each group of organisms as the complexity increases (Table 5.1).

Table 5.1 Lack of proportion between the genome size and the gene numbers, and percentage of protein-coding sequence in different organisms

Organism	Genome Size (Mb)	Number of Genes	Protein-Coding Sequence
Bacteria			
Mycoplasma genitalium	0.58	470	88%
H. inlfluenzae	1.8	1743	89%
E. coli	4.6	4288	88%
Yeast			
S. cerevisiae	12	6000	70%
S. pombe	12	4800	60%
Invertebrates			
C. elegans	97	19,000	25%
Drosophila	180	13,600	13%
Plants			
Arabidopsis thaliana	125	26,000	25%
Rice	440	30,000–50,000	~10%
Mammals			
Human	3200	30,000–40,000	1–1.5%

For example, Mycoplasma, the smallest prokaryote, has a genome of 10^6 base pairs. Bacteria which are more complex than Mycoplasma have a genome of approximately 4×10^6 base pairs. Yeast, the lowest unicellular eukaryote has a genome of 1.3×10^7 base pairs, and a further 2-fold increase in genome size is seen in case of the slime mould *Dictyostelium discoideum*, which can live in either unicellular or multicellular forms. A nematode worm *Caenorhabditis elegans* which is a multicellular organism is more complex than a slime mould and has a DNA content of 8×10^7 base pairs.

Further, in insects, birds, amphibians, and mammals, where the complexity goes on increasing, the DNA content also increases. Thus, it is clear that as the complexity increases, the amount of DNA (C-value) also increases. However, some peculiar observations have been made, e.g. the toad *Xenopus* (3.1×10^9 base pairs) and man (3.2×10^9 base pairs) have genomes of almost the same size. However, we know that man is much more complex in development than *Xenopus*. Additionally in some phyla, e.g. insects, amphibians, and plants, there are large variations in DNA content between organisms that do not vary much in complexity. A housefly has a genome of double the size of a fruit fly. In amphibians, the smallest genome is less than 10^9 base pairs, whereas the largest is about 10^{11} base pairs. Thus, the *C-value paradox* refers to the lack of co-relation between genome size and genetic complexity. There are two points of C-value paradox: 1. There is an excess of DNA than what is required to exhibit the complexity. 2. Within a phylum, there is a large variation in the DNA content between organisms that do not vary much in complexity.

It is therefore necessary to determine the organization of eukaryotic genomes to be able to answer these questions. Today, it is possible to study any genome by sequencing. However, in early 70s when the techniques of DNA sequencing were not known, the nature of eukaryotic genome was assessed by studying the kinetics of re-association of genomic DNA.

When a DNA solution is heated enough, the non-covalent forces (H-bonds) that hold the two strands together, weaken and finally break. This is referred to as *DNA denaturation* or *DNA melting*. To follow the melting curve, double stranded DNA is gradually exposed to high temperature, and the amount of melting is measured by the absorbance of DNA at 260nm. DNA absorbs at 260nm because of the aromatic bases, but when the two strands of DNA come together because of the close proximity of the bases of the two strands, some of the absorbance is quenched. When the two strands separate on melting, this quenching disappears and then absorbance rises by 30–40%. This is referred to as *hyperchromic shift*. The temperature at which the DNA strands are half denatured is called the *melting temperature* T_m (Fig. 5.4).

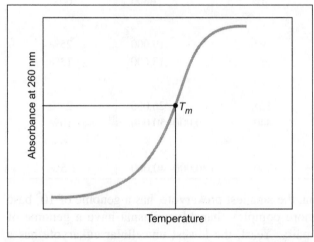

Figure 5.4 A typical melting curve of DNA. X-axis is temperature and Y-axis is absorbance at 260 nm. Temperature at which 50% of DNA denatures is referred to as T_m.

The GC content of DNA has a significant effect on T_m. Higher the GC, higher is its T_m. This is because GC pairs form three hydrogen bonds, and it requires higher temperature to melt them compared to AT pairs, which have only two hydrogen bonds. Therefore, the T_m of GC-rich DNA is higher compared to that of the AT-rich DNA. Among the prokaryotes, a large variation is seen between the GC content in the DNA, unlike in case of eukaryotes.

Once the two strands of DNA are separated, they can re-associate under proper conditions to form the double-stranded structure. This is referred to as *annealing* or *re-naturation*, and it follows a second order reassociation kinetics. Several factors contribute to re-naturation efficiency. Some of them are temperature, DNA concentration, re-naturation time, salt concentration, etc. The best temperature for re-naturation of DNA is about 25°C below its T_m. This temperature is low enough to promote re-naturation, but high enough to allow rapid diffusion of DNA molecules and to weaken bonding between mismatched sequences. The concentration of DNA in solutions is also important. Since renaturation of DNA depends on random collision; higher the concentration of DNA, higher is the probability that two complementary strands will encounter each other within a given time. Therefore, higher the concentration of DNA, faster is the annealing. As far as the renaturation time is considered, longer the time allowed for annealing, more will it occur. Thus, the rate of reaction is governed by the concentration of DNA (C) that is single stranded and time (t) and follows a second-order kinetics.

Britten and Kohne coined the term C_0t (pronounced as *cot*). C_0t is the product of the initial DNA concentration (C_0) in moles of nucleotides per liter and time (t) in seconds. If all other factors are equal, the extent of re-naturation of complementary strands in a DNA solution depends on C_0t. When a C_0t curve for any given DNA is to be studied, the DNA needs to be first denatured completely and re-association of this DNA is followed over a range of C_0t values. At each C_0t value, the amount of DNA re-associated is estimated. Then, a graph is plotted of C_0t values against the fraction of DNA re-associated, and it is referred to as C_0t curve. When the C_0t curves of two DNAs are compared, it is also essential to remember that the size of the DNA fragments should be similar, as the size will affect the re-association kinetics.

Figure 5.5 shows C_0t curves for two different DNA sequences, namely poly U: poly A and *E. coli* genomic DNA. The pattern of each curve is similar, however the C_0t ranges at which renaturation occurs are different.

DNA which is homopolymeric (poly dU:poly dA) completes 0–100% re-association between the C_0t range of $0–10^{-4}$, whereas *E.coli* DNA requires the C_0t range for 100% re-association from $5 \times 10^{-1}–10^3$ moles of nucleotides per litre. Similarly, the $C_0t_{1/2}$ of the simple polymeric DNA (poly U: poly A) is 2×10^{-6}, whereas for *E. coli*, it is 9. This means that as the genome becomes more complex, the $C_0t_{1/2}$ of the DNA increases. Since the rate of re-association depends on the concentration of the complementary sequences, the simple sequence DNA will re-associate faster, whereas the complex DNA with different sequences will require higher C_0t values. Thus, the $C_0t_{1/2}$ calculated based on renaturation of any genomic DNA is proportional to its complexity. Therefore, the complexity of any DNA can be determined by comparing its $C_0t_{1/2}$ with that of a standard DNA of known complexity. Usually *E. coli* DNA is used as a standard. Its complexity is taken to be identical with the length of the genome. Therefore,

$$\frac{C_0t_{1/2}(\text{DNA of any genome})}{C_0t_{1/2}(\textit{E. coli} \text{ DNA})} = \frac{\text{Complexity of any genome}}{\text{Complexity of } \textit{E. coli} \text{ genome } (4.2 \times 10^6 \text{ bp})}$$

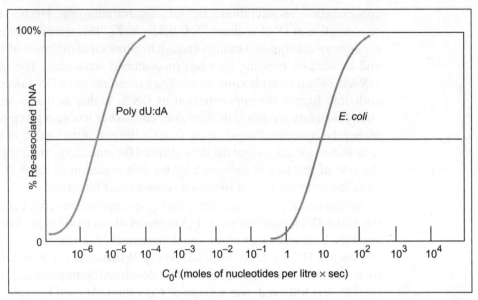

Figure 5.5 Re-association kinetics followed for simple homopolymeric DNA (polydU:dA) and *E. coli* DNA. X-axis is different C_0t values and Y-axis is percentage of DNA re-associated. Note that $C_0t_{1/2}$ of polydU:dA is 2×10^{-6} and for *E. coli*, it is 9.

When the re-association kinetics of the eukaryotic genome is performed (Fig. 5.6), the re-association occurs over a C_0t range of $10^{-4}–10^4$. This is a very broad range. Three distinct components are observed in this re-association kinetics. The fraction renaturing at a lower C_0t range is referred to as *fast component* and represents 25% of the DNA. It renatures between the C_0t values of 10^{-4} to 2×10^{-2} and has a $C_0t_{1/2}$ value of 0.0013. The next fraction is the *intermediate component* which represents 30% of DNA and renatures between C_0t values of $0.2 – 10^2$ and has a $C_0t_{1/2}$ value of 1.9. The last fraction to be renatured is the *slow component*, which is 45% of the total DNA, and renatures between the values of $10^2–10^4$ and has a $C_0t_{1/2}$ of 630.

To calculate the complexities of these fractions, each must be treated as an independent kinetic component, and its re-association is compared with the standard DNA. For example, the slow component represents 45% of the total DNA, so its concentration in the re-association reaction is 0.45 of the measured C_0. Therefore, the $C_0t_{1/2}$ applying to the slow fraction alone is $0.45 \times 630 = 283$. This means that if this slow component DNA is isolated as a pure component free of other fractions, it will renature with a $C_0t_{1/2}$ of 283. Suppose, under the conditions of re-association kinetics used, *E. coli* DNA used as a standard re-associates with a $C_0t_{1/2}$ of 4, using the formula stated above, the kinetic complexity of the slow fraction is 3×10^8 bp. Similarly for other components, the kinetic complexity is 6×10^5 bp for intermediate component, and it is 340 bp for the fast component. Suppose the chemical

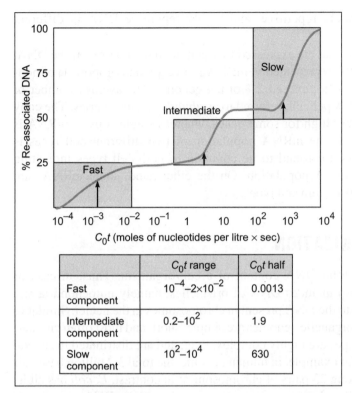

	C_0t range	C_0t half
Fast component	10^{-4}–2×10^{-2}	0.0013
Intermediate component	0.2–10^2	1.9
Slow component	10^2–10^4	630

Figure 5.6 Re-association kinetics of a eukaryotic DNA. X-axis has different C_0t values and Y-axis is percentage of DNA re-associated. Note that this C_0t curve reveals three distinct fractions, namely fast, intermediate, and slow. C_0t range for each factions and its $C_0t_{1/2}$ is mentioned.

complexity (haploid DNA content determined chemically) of the DNA used for this re-association kinetics is 7×10^8 bp, then the slow fraction which is 45% of the total genome will have a chemical complexity of 3.15×10^8 bp. Thus, the chemical complexity and kinetic complexity of this fraction is almost similar. This means that the slow component comprises of sequences that are unique in the genome.

On denaturation, each single stranded sequence is able to renature only with the corresponding complementary sequence. This component is referred to as *non-repetitive DNA* or *unique DNA*. The intermediate component occupies 30% of the genome, therefore, its chemical complexity is $0.3 \times 7 \times 10^8 = 2.1 \times 10^8$ bp. The kinetic complexity of this component is 6×10^5 bp. This means that the unique length of DNA corresponding to $C_0t_{1/2}$ of re-association is much shorter than the total length of the DNA, chemically occupied by this component of the genome. Thus, intermediate component consists of a sequence of 6×10^5 bp that is present in 350 copies in the genome so as to occupy chemically 2.1×10^8 bp. On renaturation, each single strand generated from this component has a compliment available from any one of the 350 copies. Sequence present in more than one copy in each genome is referred to as *repetitive DNA* and the number of copies present per genome are called *repetition frequency* (f).

Repetition frequency of each fraction is calculated by dividing chemical complexity by kinetic complexity. The fast component has a very low $C_0t_{1/2}$ of 0.0013. The kinetic complexity calculated for this fraction is 340 bp and therefore the repetition frequency is 5×10^5. This means that the sequences in the fast component are repeated several times in the genome and is referred to as *highly repetitive DNA*, whereas intermediate fraction is referred to as *moderately repetitive DNA*. Together, the complexity and the repetition frequency describes the properties of the sequence components of the genome. The prokaryotic genome solely consists of non-repetitive DNA, whereas different eukaryotic genomes have non-repetitive, moderately repetitive, and highly repetitive DNA in different proportions.

Most structural genes that are expressed by a cell lie in the non-repetitive DNA. The total amount of DNA represented in mRNA is a very small proportion of non-repetitive DNA and does not exceed 2% of the genome. The average number of molecules of each mRNA per cell, referred to as *abundance*, also varies. The copy number of mRNA is very high for some genes, whereas single copy of mRNA is made for some genes. If the mRNA population of two different cell types is compared, few mRNAs are found to be common to both cell types indicating overlap between the mRNA population. On the other hand, some mRNAs are produced exclusively by certain cell types.

5.3 GENOME AND ITS ORGANIZATION

Genome refers to the total DNA content of a cell/organism. Thus, in case of eukaryotes, genome also includes DNA of organelles, namely mitochondria and chloroplast, in addition to the DNA present in chromosomes in the nuclei. Similar to bacterial genomes, the organelle genomes are simple, short, and most often circular; while the nuclear genomes are highly complex, long, and are distributed in several linear chromosomes. For example, in human genome, the total 3.2 billion base pair DNA is distributed among 23 pairs of chromosomes. In contrast, *E. coli* has all its total genome of 4.6 million base pair present as a single circular DNA.

Although the genome is much larger in eukaryotes than in prokaryotes, the number of genes is not proportionately larger. Thus, the increase in the DNA content or genome in eukaryotes is largely due to extra-genic non-coding sequences. To understand this clear, one needs to know the definition of a gene. A gene is a stretch of DNA sequence that can give rise to a functional protein or an RNA (Fig. 5.7). In addition to the coding sequence, a gene also includes sequences that regulate its expression into a protein/RNA phenotype. The regulatory sequences that are located in the 5′- and 3′ flanking regions and also in the introns in eukaryotic genes, are called the promoters, enhancers, silencers and repressors.

With the advent of the whole genome sequencing technologies, a large number of genomes representing a wide variety of organisms, from simple to highly complex, have been sequenced. While comparing the genomes of these organisms, it turns out that most of the genes are conserved across species. The number of genes in organisms did not increase in proportion to their complexity (Table 5.1). For example, the most complex human genome contains approximately about 8 times (about 30,000) the number of genes present in a simple organism like *E. coli* (about 4,000), although the genome size has increased enormously (3.2 billion vs. 4.6 million). Thus, there has been an enormous addition and variation in the extra-genic non-coding DNA sequences as a function of evolution.

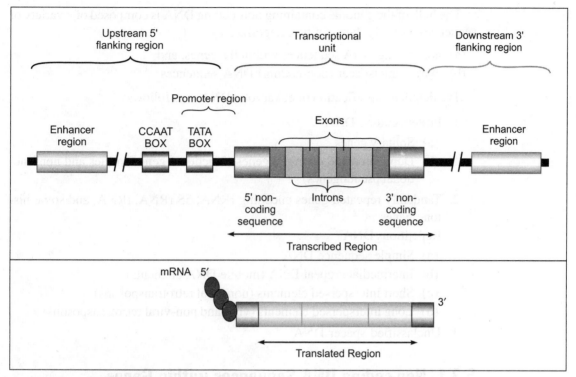

Figure 5.7 Structure of a eukaryotic gene. Gene is divided into an upstream 5′ flanking region, transcriptional unit and downstream 3′ flanking region. Regulatory elements such as promoters, enhancers reside in the flanking regions as well as within introns. After transcription, mRNA is processed and is translated. **(See Colour Plate 9)**

Although, in general, genome size reflects the complexity of the organism, this does not hold true for a few groups of organisms, namely amphibians and plants. Some of the organisms belonging to these two groups contain even higher quantity of DNA than that present in human genome. As referred earlier, this phenomenon is called the *C-value paradox*. These extra-genic sequences which were referred initially as *Junk DNA* by Francis H. Crick now appear to be very important from the point of view of genome organization and its function and evolution. In this manner, whole genome sequencing efforts have been yielding enormous information from the point of view of our discovery and understanding on

(a) many new genes, their regulation of expression under normal physiological conditions as well as during various pathological conditions leading to diseases, and

(b) non-coding sequences, their role in organization of genome and its function and evolution.

The bulk of the genome containing non-coding DNA is composed of a variety of sequences. Grossly they can be categorized as,

    (a) non-coding DNA sequences within the genes, and

    (b) extra-genic/spacer (non-coding) DNA sequences.

The detailed classification of eukaryotic DNA is as follows:

1. Protein coding DNA
   - (a) Solitary genes
   - (b) Duplicated and diverged genes (functional gene families and nonfunctional pseudogenes)
2. Tandemly repeated genes encoding rRNA, 5S rRNA, tRNA, and some histone genes.
3. Repetitious DNA
   - (a) Simple sequence DNA
   - (b) Intermediate repeat DNA (mobile DNA elements)
   - (c) Short interspersed elements (non-viral retrotransposons)
   - (d) Long interspersed elements (viral and non-viral retrotransposons)
4. Unclassified spacer DNA

5.3.1 Non-coding DNA Sequences within Genes

As indicated above, in most eukaryotic genes, the bulk is contributed by the non-coding sequences. These include introns and the flanking regulatory sequences. A few examples with the proportions of coding and non-coding sequences are presented in Table 5.1. Most of the eukaryotic genes are split, i.e. they have coding sequences (exons) interrupted by non-coding, intervening sequences called the *introns* (Fig. 5.7). Introns were initially discovered in adenoviruses by electron microscopy of mRNA-DNA hybrids, also called as heteroduplex analysis (Fig. 5.8) by Philip Sharp and Richard Roberts, who were subsequently awarded Nobel Prize. Now, it is known that most of the eukaryotic genes contain introns. Introns present in the primary mRNA transcripts are removed during post-transcriptional processing by splicing reactions leading to the generation of the mature mRNA containing only the coding sequences, the *exons*. The details are given in Chapter 8.

Many a times, it turns out that the introns contribute a major portion of a gene, and in higher organisms like human, introns account for about 25% of the total genome. The regulatory sequences in the genes mostly located at the 5′ flanking region also contribute a great deal to the total gene sequences. For more details on this topic, see Chapter 8.

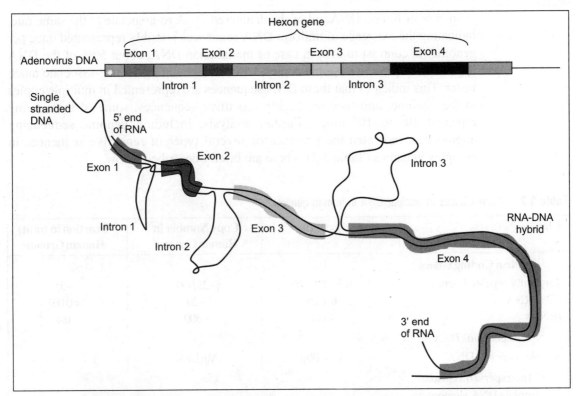

Figure 5.8   Structure of hexon gene. RNA-DNA hybrid of the same gene showing the introns in the gene.

5.3.2 Extra-genic Sequences

Major classes of extragenic DNA sequences (spacer sequences) belong to the repetitive DNA sequences. The existence of such sequences were initially demonstrated by Roy Britten and David Kohne using simple denaturation of double stranded genomic DNA fragments and their renaturation/re-association. By measuring the re-association kinetics, one can visualize different types of DNA sequences in a genome. Since DNA re-association is a bimolecular reaction between two strands of DNA, the rate of re-association is dependent on the concentration of DNA strands. Using DNA of two very different organisms, a simple prokaryote (e.g., *E. coli*) and a complex mammal for re-association kinetics, two very different profiles are obtained (Fig. 5.9).

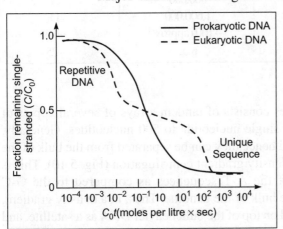

Figure 5.9   DNA re-association kinetics. Graph shows the re-association of the prokaryotic DNA and eukaryotic DNA indicating different types of sequences in the genome.

In case of *E. coli* DNA, all of the denatured DNA re-associate at the same rate, indicating the existence of uniform DNA sequences possibly represented once per genome. In contrast to this, in case of mammalian DNA, about 50% of the DNA appears to be present in single copies, while the remaining 50% re-associate much faster. This indicates that these DNA sequences are represented in multiple copies in the genome, and they are highly repetitive sequences, some of them being repeated 10^5 to 10^6 times. Further analysis, including genome sequencing approaches, indicated the presence of several types of repetitive sequences in complex genomes (Table 5.2). These are briefly described below.

Table 5.2 Major classes of nuclear DNA in human genome

Class	Length	Copy Number in Human Genome	Fraction in (unit) Human Genome
Protein Coding Genes			
Tandemly repeated genes	0.5–2200 kb	~25,000	~55
U2 snRNA	6.1 kb	~20	<0.001
rRNAs	43 kb	~300	0.4
Repetitious DNA			
Simple sequence DNA	1–500bp	Variable	~6
Interspersed repeats (mobile DNA elements)			
DNA transposons	2–3kb	300,000	3
LTR retrotransposons	6–11kb	440,000	8
Non-LTR retrotransposons			
LINEs	6–8kb	860,000	21
SINEs	100–400bp	1,600,000	13
Processed pseudogenes	Variable	1–100 (approx.)	~0.4
Unclassified spacer DNA	Variable	n.a.	~25

Simple-sequence repeats

This class of repetitive sequences consists of tandem arrays of several copies of short sequences that range from single nucleotide to 500 nucleotides. Generally, these sequences are A-T rich, and hence they can be separated from the bulk of the genomic DNA by simple CsCl-density gradient centrifugation (Fig. 5.10). This is due to the less dense nature of the A-T sequences as compared to the G-C sequences which is present in the bulk of the genome. Thus, in a density gradient, they separate out as a lighter band on top of the bulk heavier DNA as a satellite, and therefore called the *satellite DNA*. These sequences are highly repeated in the genomes (up to million times), and contribute almost up to 10% of the genome in eukaryotes. These sequences are not transcribed, and therefore do not carry any functional genetic information. However, their role in the structural organization of the genome as well as in evolution is very significant.

Retrotransposons and Transposons

These elements belong to the major class of repetitive sequences called the *interspersed repetitive sequences*. These are distributed throughout the genome, and in humans, they contribute about 45% of the genome. The bulk of the interspersed elements belong to the class of transposable elements, called the *retrotransposons*. As the name suggests, they transpose from one site to another in the genome via reverse transcription, i.e. through an RNA intermediate contrary to the other class which is called the *DNA transposons* (Fig. 5.11).

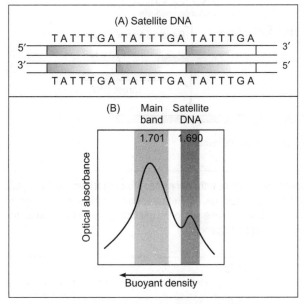

The three types of elements under retrotransposons are called SINEs (short interspersed elements), LINEs (long interspersed elements), and retrovirus-like elements, and they contribute to 21%, 13% and 8% of the human genome, respectively. SINEs are 100–300 bp long sequences, and approximately 1.5 million of them are interspersed over the entire genome. These sequences are transcribed but are not translated. LINEs on the other hand are longer (up to 8 kb) sequences that are repeated up to 8,50,000. These are transcribed and some of them also code for

Figure 5.10 Sequence of the satellite DNA. Satellite DNA is A-T rich (A) and thus in cesium chloride density gradient centrifugation, satellite DNA comes as a separate band of less density (B).

proteins (Fig. 5.12A). However, like SINEs, their functions are also unknown. The third type of retrotranspons are called *retrovirus-like elements*, and they resemble retro-viruses (Fig. 5.12B). They also transpose by reverse transcription, and their length ranges between 2 and 10 kb. There are about 4,50,000 copies of retrovirus-like elements in human genome. DNA transposons, which move in the genome through DNA copy (Fig. 5.11), represent about 3% of the human genome. Their length varies between 80 and 3,000 bp, and there are about 3,00,000 copies in human genome.

5.3.3 Gene Families

The total number of genes present in the human genome is approximately 30,000, and the coding sequences contribute to about 1–1.5% of the genome. A major part

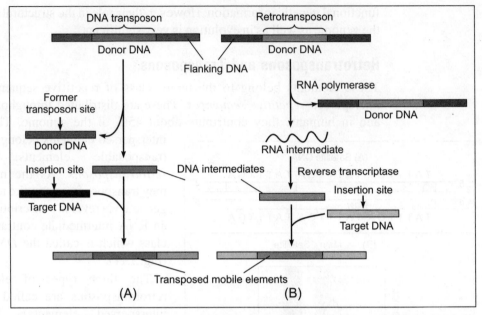

Figure 5.11 Two major classes of mobile DNA elements with their modes of transposition. **(A)** Eukaryotic DNA transposons move through a DNA intermediate. **(B)** Retrotransposons transpose through an RNA intermediate, i.e. they are transcribed from one site in the genome, reverse transcribed into DNA, and then get inserted into other sites in the genome.

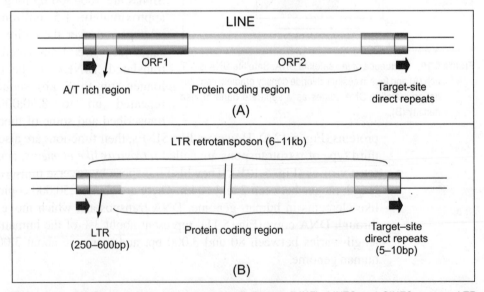

Figure 5.12 **(A)** Structure of a long interspersed element (LINE). LINES and SINES are non-LTR retrotransposons. A LINE contains two ORFs: ORF1 encodes a RNA-binding protein, and ORF2 encodes a bifunctional protein with reverse transcriptase and endonuclease activity. **(B)** Structure of LTR retrotransposons in eukaryotes. The protein coding region is flanked by LTRs.

of the gene sequences are contributed by *gene families* and *gene clusters* rather than single copy or *solitary genes*. The gene families which are originated through gene duplication during evolution are of tremendous significance for the functioning of higher organisms. For example, multiple copies of certain genes, such as histones, ribosomal RNAs, tRNAs, etc. are required for generating large quantities of their products by the cell. In another scenario, specific members of a group of similar genes called gene families are required during a particular stage of development in an organism. These have originated from a single gene through gene duplication and diversification through mutation. Thus, a set of duplicated genes which encode proteins with similar but non-identical amino acid sequences is called a gene family. The important examples are *globin genes* and *actin genes*.

In some cases, the members of the gene families are clustered in a single chromosome, while in other cases, they are distributed in different chromosomes. Some of these gene families and gene clusters are described in details below. Interestingly, in the evolutionary process, largely due to mutations, some of the genes in the gene families have become non-functional, and they are termed as *pseudogenes*. Further details about pseudogenes will be discussed along with the specific gene families.

In some cases, when the genes in a gene family code for products that are functionally related in general, but have a very weak sequence homology over a large segment, and without any significant conserved amino acid motif, they constitute a *gene superfamily*. For example, HLA genes, T-cell receptor genes, T4 and T8 genes which code for products of the immune system function and have domain structure similar to that of immunoglobulins (Ig), belong to Ig superfamily (Fig. 5.13).

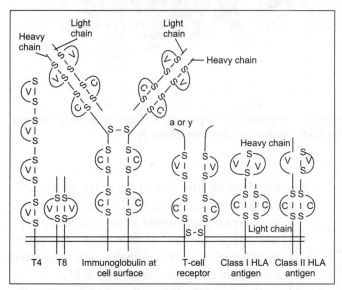

Figure 5.13 Various proteins belonging to the Immunoglobulin superfamily. All of these have similar types of domain structure.

Globin gene family

In an adult individual, hemoglobin is made up of a globin tetramer composed of 2α and 2β globin subunits and a heme molecule as the prosthetic group (Fig. 5.14). However, globins at various stages during development are not the same, although similar. Such varied members are encoded by individual genes and they together constitute globin gene families of two kinds, namely α- and β-globin gene families. The β globin gene family in human contains 5 functional genes called ε, Gγ, Aγ, δ,

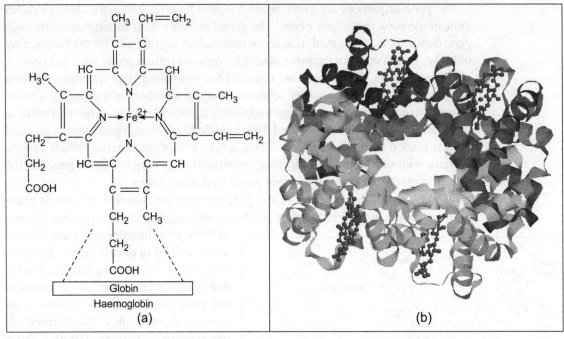

Figure 5.14 Structure of haemoglobin. Adult globin which is composed of two copies each of α and β subunits is linked to a molecule of heme as the prosthetic group giving rise to haemoglobin. **(See Colour Plate 9)**

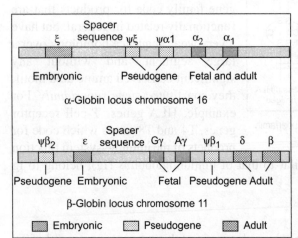

Figure 5.15 Structure of human α globin- and β globin clusters located on chromosome 11 and 16, respectively. They contain fetal, embryonic, and adult globin genes with a 5′ to 3′ polarity. There are pseudogenes in each cluster.

and β which give rise to the polypeptides with the same names, respectively. In addition, there are two non-functional gene sequences ($\psi\beta_1$ and $\psi\beta_2$) in the cluster that are called *pseudogenes*. These genes are clustered together in chromosome 11 (Fig. 5.15).

As seen in the figure, the globin genes of the β globin gene family are arranged in the cluster ($\psi\beta_2$, ε, $G\gamma$, $A\gamma$, $\psi\beta_1$, δ and β) in a 5′ to 3′ orientation according to the sequence of their expression during the life of the organism from development to adulthood. The synthesis of different globins at different times is of physiological importance. It is primarily because of their difference in property, i.e. varied affinity for oxygen. For example, $G\gamma$ and $A\gamma$, which are foetal hemoglobin, have higher affinity for oxygen than the adult β globin which has a weak affinity for oxygen. During foetal life, globins with high oxygen affinity are essential for drawing maximum oxygen from the maternal circulation in the placenta. Contrary to this, the adult globins with weak affinity for oxygen are ideal for maximum release of oxygen to the tissues, the muscles in particular, which require high level of oxygen during exercise.

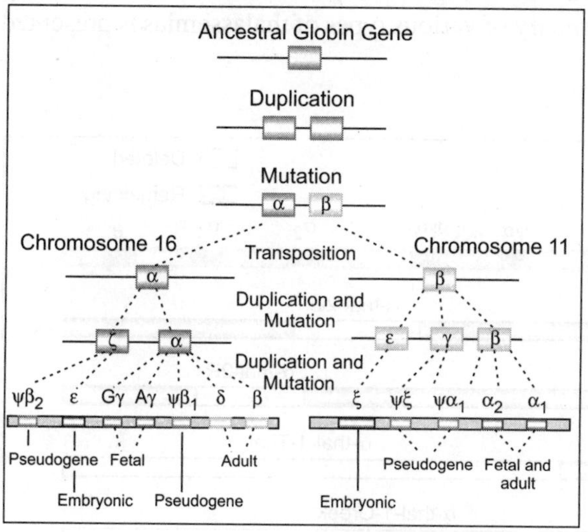

The α globin gene family in humans contains three functional members, namely ξ, α_2 and α_1, and two pseudogenes, $\psi\xi$ and $\psi\alpha_1$. These genes are organized as a cluster (ξ, $\psi\xi$, $\psi\alpha_1$, α_2, and α_1) in a 5′–3′ orientation on chromosome 16 in humans (Fig. 5.15). Like the members of the β globin gene cluster, the functional members of the α globin gene cluster are also sequentially expressed in embryonic, fetal, and adult tissues as per their polarity from the 5′ to the 3′ direction.

Various evidences from evolutionary studies indicate that these two clusters of globin genes have originated from a single ancestral gene through the processes of gene duplication and mutations (Fig. 5. 16). The possible mechanism for gene duplication is unequal crossing over that results due to similarities between the members in the globin gene clusters. Interestingly, some of the members called the pseudogenes, have become non-functional and they represent relics of evolution.

Pseudogenes can be grouped under two categories: *processed-* and *unprocessed pseudogenes*. Processed pseudogenes are intronless and lack regulatory sequences required for transcription. They are presumed to have originated by reverse transcription of the mature cytoplasmic mRNAs followed by their integration in the genome (Fig. 5.17). On the other hand, the unprocessed pseudogenes contain introns and are transcribed. However, they too do not give rise to functional protein products. It is presumed that the pseudogenes are important from the structural point of view, thereby they are continued to be parts of gene clusters.

Figure 5.16 Evolution of globin gene family from a single ancestral gene through gene duplication and mutations.

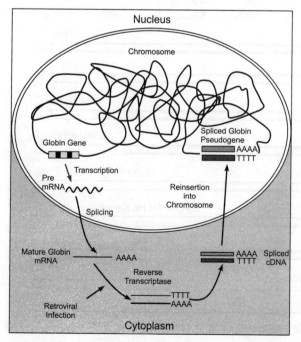

Figure 5.17 Possible mechanism of formation of pseudogenes; here globin gene cluster has been taken as an example.

The structural organization and the various members of the globin gene clusters are very important for function of the organism. Any perturbation in the form of partial deletion or mutation of the globin gene family members results in diseases called *thalassemias*. Thalassemias involving α and β clusters are called α and β thalassemias, respectively. A summary of various types of thalassemias is presented in Fig. 5.18.

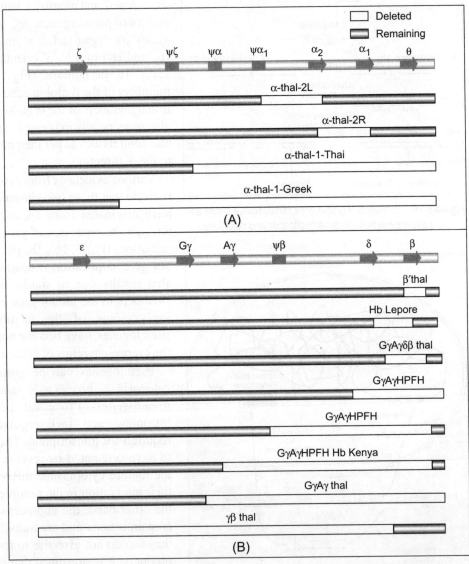

Figure 5.18 Deletions in globin gene clusters resulting in various types of α-thalassemia (A) and β-thalassemia (B).

Tandemly Repeated Gene Clusters/Arrays

As compared to the classical gene families, these group of genes, such as ribosomal RNA genes, tRNA genes, and histone genes are identical and give rise to identical gene products. These are arranged in tandem arrays. Ribosomal RNAs (rRNAs) are coded by a large number of identical rRNA genes which are arranged as clusters of tandem repeats. Such clusters occupy about 2 Mb of DNA (referred as ribosomal DNA or rDNA) and are located on the short arms of each of the 5 acrocentric chromosomes in humans, which together constitute the *nucleolar organizer region*. Due to the clustered organization of rRNA genes in eukaryotes, the rDNA can be separated from the rest of the genomic DNA by shearing followed by density gradient centrifugation. Such tandem clusters generate a circular restriction map.

The number of rRNA genes varies from seven in *E. coli* to several hundred in higher eukaryotes. Further, the rRNA gene products, i.e. the rRNAs contribute to the bulk of the total transcription products (80–90%) in a cell in both prokaryotes and eukaryotes, to fulfill the extensive requirement of rRNAs in the cells. The rRNA genes for small and large rRNAs, namely 28S, 5.8S, and 18S rRNA are arranged together and are transcribed in the form of a polycistronic RNA which is subsequently cleaved to give rise to the individual rRNAs (Fig. 5.19). The rRNA

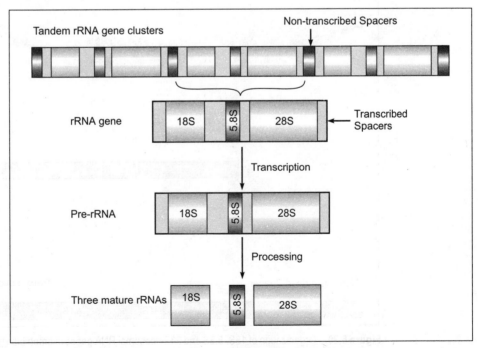

Figure 5.19   rRNA gene clusters. The figure shows processing of rRNA to form three mature rRNAs.

gene repeats are separated from each other by *non-transcribed spacer sequences*. Interestingly, while the coding sequences of rRNA genes are identical, the non-transcribed spacer sequences and lengths vary within the same species as well as different species.

The non-transcribed spacer sequences can be visualized by electron microscopy of the ribosomal RNA gene clusters during transcription, which look like a Christmas tree (Fig. 5.20) due to interruption of the transcription units by the non-transcribed spacers. The non-transcribed spacer sequence acts as the promoter for rRNA gene transcription. It contains two types of variable repeat sequences, *97 bp repeat* and *60–81 bp repeat*, alternating with a constant sequence called the *Bam island* (Fig. 5.21). This is so called because of its isolation by using the restriction enzyme BamH1. The variation in the number of repeat sequences contributes to the large variation of length of the non-transcribed spacers. The other major ribosomal RNA, 5S rRNA, is transcribed separately from genes that are part of a large gene family that is clustered on the long arm of chromosome 1 in humans.

Histones are produced from a large number of genes arranged in tandem clusters containing H1 to H4 in the genome. The number of histone genes and their orientation vary in different organisms (Fig. 5.22).

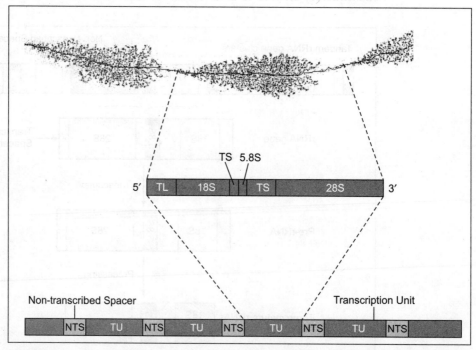

Figure 5.20 Organization of 18S, 5.8S and 28S ribosomal RNA genes in amphibian genomes. Christmas tree like appearance indicating transcription activity of the gene cluster interrupted by non-transcribed spacer sequences.

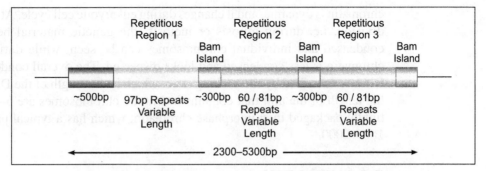

Figure 5.21   Structure of a non-transcribed spacer in the rRNA gene cluster.

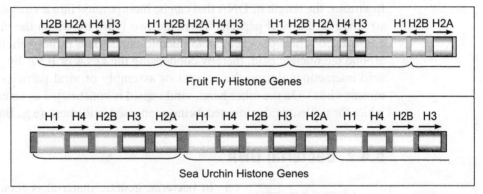

Figure 5.22   Arrangement of histone genes in the genomes of Drosophila and Sea urchin. (Adapted from: www.cbs.dtu.dk/staff/dave/roanoke/genetics980218.html)

5.4 PACKAGING OF DNA

In every cell, the genetic material exists as a compact mass confined to a limited volume. All its activities such as replication and transcription are accomplished within this phase. The organization of genetic material should be such that it can accommodate transitions between active and inactive state. From viruses to eukaryotic cells, the nucleic acid needs to be condensed so that it can be accommodated in a small compartment. For example, for phage T4, the length of the DNA is 5 μm and is accommodated in icosahedrons shaped phage of dimensions 0.065 × 0.10 μm; for *E.coli*, the DNA of size 1.3 mm is accommodated in a cylindrical shaped Coccobacillus of size 1.7 × 0.65 μm; whereas in a human cell, the nucleus has a diameter of 6 mm and it accommodates DNA spanning a length of 1.8 meters.

The condensed state of nucleic acid results from its binding to basic proteins. The positive charges of these proteins neutralize the negative charges of the nucleic acids. The structure of nucleo-protein complex is determined by the interaction of the proteins with the DNA. The genetic material in bacteria is seen in the form of a nucleoid, whereas in eukaryotic cells, it is seen as chromatin. The packaging of chromatin is dynamic, and it changes during eukaryotic cell cycle. At the time of cell division, i.e. during mitosis or meiosis, the genetic material becomes highly condensed and individual chromosomes can be seen, while during interphase, chromatin is less condensed and looks dispersed. The overall condensation of the DNA is described by the packing ratio, which is the length of the DNA divided by the length of the unit that contains it. Mitotic chromosomes are 5–10 times more tightly packaged than interphase chromatin, which has a typical packing ratio of 1000–2000.

5.4.1 Viral DNA

In viruses, the length of DNA that can be incorporated into a virus is limited by the structure of the viral particle. There are two types by which the viral capsids are constructed while packaging the nucleic acids. The protein shell is assembled around the nucleic acid, thereby condensing the DNA or RNA by protein-nucleic acid interactions during the process of assembly of viral particles (e.g., tobacco mosaic virus). On the other hand, viral capsid is constructed as an empty shell into which the nucleic acid is inserted, being condensed as it enters. (e.g., Bacteriophage λ).

5.4.2 Bacterial DNA

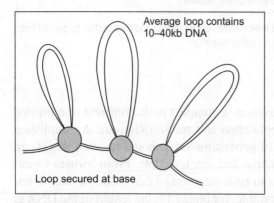

Average loop contains 10–40kb DNA

Loop secured at base

Figure 5.23   Bacterial genome is referred to as nucleoid and consists of loops of DNA, each of which is secured at the base.

In bacteria, genetic material is seen as a compact mass occupying one-third of the volume of the cell. Several DNA binding proteins are found to be associated with the DNA. The bacterial genome consists of a large number of loops of double-stranded DNA each of which is secured at the base to form an independent structural domain (Fig. 5.23). Each loop contains on an average around 40 Kb of DNA. The existence of separate domains permits different degrees of super-coiling to be maintained at different regions of the genome. In simple words, DNA in one structural domain can be used by transcriptional machinery independent of the transcriptional status of the DNA in another domain.

5.4.3 Eukaryotic DNA

In eukaryotic cells, as explained earlier, interphase chromatin is more dispersed compared to highly condensed mitotic chromosomes. Metaphase chromosomes have a protein scaffold to which the loops of super-coiled DNA are attached. The super-coils can be removed by nicking with an enzyme DNase, although the DNA remains in the form of 10 nm fibres, suggesting that the super-coiling is caused by the arrangement of fibre in a higher order structure. DNA is attached to the nuclear matrix at specific sequences called MARs (Matrix Attachment Regions) or SARs (Scaffold Attachment Regions). It is believed that attachment to the matrix is necessary for transcription and replication. MAR regions are AT rich but lack any consensus sequences.

The structure of interphase chromatin does not change visibly between divisions. This chromatin can grossly be divided into two types: *euchromatin* and *heterochromatin*. Euchromatin is less densely packed, has a relatively dispersed appearance in the nucleus, and occupies most of the region of the nucleus. Heterochromatin, on the other hand, is densely packed, mostly found at centromere and teleomere but occurs at other locations also. It passes through the cell cycle with little change in degree of condensation. The same fibre of DNA runs continuously between euchromatin and heterochromatin implying that these states simply represent different degree of condensation of the genetic material. The state of condensation of the genetic material is co-related with its activity. Heterochromatin which is highly condensed and replicates late in S-phase, has a reduced frequency of genetic recombination and also has a very low density of genes in it, whereas active genes are contained within the euchromatin. Thus, the location of genes in euchromatin is necessary for expression but is not sufficient.

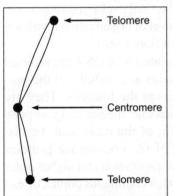

Figure 5.24 Chromosome has a centromere which holds two arms of the chromosome, and telomeres present at the ends are required for maintaining stability of DNA.

The two important regions in the chromosome (Fig. 5.24) structure are *centromeric region* and *telomeric region*. The centromeric region is responsible for segregation of chromosomes during mitosis. The regions flanking the centromere are often rich in satellite DNA sequences and display considerable amount of heterochromatin. The centromeric DNA binds specific proteins that are responsible for establishing the structure called *kinetochore* that attaches the chromosome to the microtubules. Centromeric sequences have been clearly identified only in yeast *S. cerevisiae* where they consist of short, conserved elements. Telomeres make the ends of chromosomes stable and are essential features in all chromosomes, which seal the ends. Almost all known telomeres consist of multiple repeats in which one strand has the general sequence $C_n(A/T)_m$, where $n > 1$ and $M = 1$ to 4. The other strand has $G_n(T/A)_m$ and it has a single protruding end that provides a template for addition of individual bases in a defined order. Enzyme telomerase is specifically required for replicating telomeres. This is described in Chapter 6. The telomeres stabilize the chromosome end by forming loops and leaving no free-ends.

5.4.4 Chromatin Organization

The fundamental subunit of chromatin has the same type of design in all eukaryotes. The nucleosome (Fig. 5.25) is the basic unit of chromatin, and it contains 200 bp of DNA associated with an octamer of small basic proteins called *histones*. They form the interior core and the DNA is wound around this core particle. Thus, nucleosome provides the first level of organization giving a packing ratio of six. The DNA in the nucleosome is visualized as 10 nm fibre and referred to as beads-on-string (Fig. 5.26) structure. The second level of organization is the coiling of the series of nucleosomes into a helical structure with a diameter of 30 nm and a

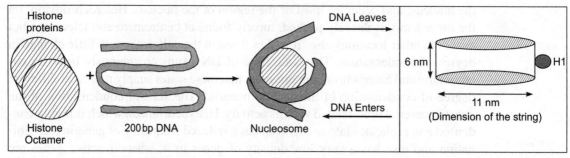

Figure 5.25 Nucleosome is the first stage of condensation of chromatin wherein 200 bp DNA associates with histone octamer.

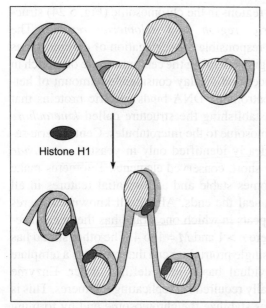

Figure 5.26 In presence of histone H1, more compact structure is formed. One histone H1 is associated with each nucleosome.

packing ratio of 40. The final packaging ratio is determined by the third level of organization involving packaging of the 30 nm fibre itself.

Other proteins associated with DNA are referred to as non-histone proteins and include all the proteins of chromatin except the histones. They are variable between tissues and species and comprise of a smaller proportion of the mass than the histones. The functions of the non-histone proteins include control of gene expression and higher order structure. In 10 nm fiber, a continuous duplex thread of DNA runs through the series of nucleosome particles. Individual nucleosomes can be obtained by treating chromatin with an enzyme micrococcal nuclease, which cuts the DNA between nucleosomes.

A nucleosome contains 200 bp of DNA associated with a histone octamer consisting of two copies each of H2A, H2B, H3, and H4. These are known as core histones. Histone H3 and H4 are among the most conserved proteins known.

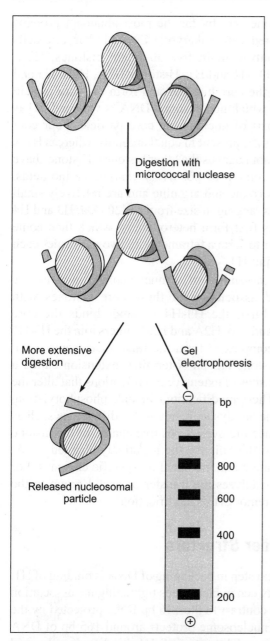

Histones H2A and H2B show species-specific variation in sequence. Histone H1 is different from the core histones, as it is present in half the amount and can be extracted readily from chromatin. Histone H1 can be removed without affecting the structure of nucleosome suggesting that its location is external to the particle. The shape of the nucleosome corresponds to a cylinder of diameter 11 nm and height of 6 nm (Fig. 5.25).

The DNA follows a symmetrical path around the octamer making two turns. The DNA enters and leaves the nucleosomes at points close to one another. Histone H1 is located in this region. When chromatin is digested with micrococcal nuclease, the DNA is cleaved into integral multiples of unit length. Fractionation by gel electrophoresis reveals the nucleosomal ladder (Fig. 5.27). When nucleosomes are fractionated on a sucrose gradient, they give a series of discrete peaks corresponding to monomers, dimers, trimers, and so on.

If the DNA is extracted from individual fraction and electrophoresed, each fraction yields a band of DNA whose size corresponds with the DNA in micrococcal nuclease ladder. The monomeric nucleosome contains the DNA of unit length, the nucleosome dimer contains the DNA of twice the unit length. Nucleosomal DNA is divided into the core DNA and the linker DNA. By assembling into the nucleosomes, the DNA is compacted approximately 6 folds and this DNA is wound approximately 1.65 times around the outside of the histone octamer like thread around the spool.

The length of DNA associated with each nucleosome can be determined using nuclease treatment. Core DNA has an invariant length of 146 bp and is resistant to digestion by nucleases. Linker DNA comprises the rest of the repeating unit, and the length can vary from 8–114 bp per nucleosome. The difference in linker DNA length reflects the differences in the nature of larger structures formed by nucleosomal DNA in each organism rather than differences in the nucleosomes itself.

Figure 5.27 Micrococcal nuclease digests chromatin and releases nucleosomal particles by cleaving into linker DNA. Early in digestion, oligonucleosomal particles are found and as the digestion progresses, more and more mononucleosomes accumulate. Electrophoresis of these particles reveals that 200bp DNA is associated with a mononucleosome.

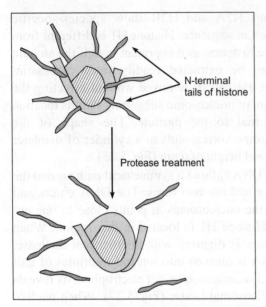

N-terminal tails of histone

Protease treatment

Figure 5.28 Histone proteins have free N-terminal ends, which can be removed by digestion by proteases. These tails are subjected to various modifications in relation to chromatin functions.

Histones are by far the most abundant proteins associated with eukaryotic DNA. Eukaryotic cells commonly contain five abundant histones: H2A, H2B, H3, H4 and H1. Histones H2A, H2B, H3 and H4 are the core histones and form the core protein around which nucleosomal DNA is wrapped. H1 is not a part of nucleosome core particle. Four core histones are present in equal amounts, whereas H1 is half as abundant as the other histones. Histones have a high content of positively charged amino acids, such as lysine and arginine and are relatively small proteins ranging in size from 11–20 kDa. H3 and H4 histones first form heterodimers, which then come together to form a tetramer with two molecules each of H3 and H4.

The assembly of a nucleosome involves the ordered association of these core histones with DNA. First, the H3-H4 tetramer binds the core DNA, and two, H2A and H2B dimers join the H3-H4 DNA complex to form the final nucleosome. Each core histone has an N-terminal extension called a *tail*, and these tails (Fig. 5.28) are the sites of extensive modifications that alter the function of individual nucleosome. These modifications include phosphorylation, acetylation, and methylation on serine and lysine residues. Modifications such as acetylation and phosphorylation reduce the overall positive charge of the histone tails, thereby reducing the affinity of the tails for the negatively charged DNA. Thus, histone modifications are dynamic and mediated by specific enzymes. For example, histone acetyl transferase catalyzes the transfer of acetyl group to the lysines, whereas histone deacetylase removes this modification.

5.4.5 Chromatin Higher Order Structure

Once nucleosomes are formed, the next step in packaging of DNA is binding of H1. H1 interacts with the linker DNA between nucleosomes tightening the association of DNA with nucleosomes. Thus, in contrast to the 146 bp DNA protected by the core histones, addition of H1 to the nucleosome protects around 165 bp of DNA from micrococcal nuclease digestion. The sites of H1 binding are located asymmetrically relative to the nucleosome, and H1 binds to the DNA where it enters and leaves the nucleosome. Binding of H1 stabilizes higher order chromatin structures. On addition of histone H1, 30 nm fibre structure is formed which is the next level of DNA compaction. It is proposed that in this 30 nm fibre, the nucleosomal DNA forms a super helix containing approximately six nucleosomes per turn and referred to as *solenoid structure* (Fig. 5.29a). An alternative model for

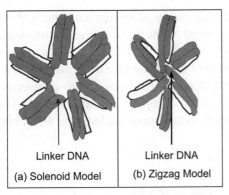

Linker DNA

(a) Solenoid Model

Linker DNA

(b) Zigzag Model

Figure 5.29 When 10 nm fiber undergoes condensation, a 30 nm fiber is formed which is either
of solenoid type (A) or zigzag type (B).

30 nm fiber is the *zigzag model* (Fig. 5.29b). In this, a zigzag pattern of nucleosomes is formed upon H1 addition. The N-terminal tails of core histones are essential for forming 30 nm fibers. Modification of the tails modulates this function and affects the ability of the nucleosome to form repressive higher order chromatin structure. Together, the packaging of DNA into nucleosomes and the 30 nm fibre results in approximately 40 fold compaction of DNA. Additional folding of 30 nm fibre is required to compact the DNA further. A 30 nm fibre forms loops of 40–90 kilobase that are held together at their bases by a proteinaceous structure referred to as *nuclear scaffold*.

The core histones are being the most conserved eukaryotic proteins, the nucleosomes formed by these are similar in eukaryotic cells. Several histone variants are found in eukaryotic cells. These can replace one of the four standard histones to form alternate histones and confer specialized functions to the nucleosomes into which they are incorporated. For example, a variant of histone H2A called *histone 2a.z* inhibits nucleosomes from forming repressive chromatin structures, thereby making the chromatin more compatible for transcription. Another example of histone variant is CENP-A which replaces histone H3 sub-unit associated with nucleosomes that include centromeric DNA.

Association of histone octamer with the DNA in the nucleosome is dynamic. This allows changes in nucleosome position and DNA association required for making DNA accessible for functions, such as replication, repair, etc. Thus, the association of any particular region of DNA with the histone octamer is not permanent and any individual region of the DNA can transiently be released from tight interaction with the octamer. This is referred to as *nucleosomal remodelling* and is brought about by a complex of various proteins. Nucleosome remodelling complexes facilitate changes in nucleosome location or their interaction with DNA using the energy of ATP hydrolysis. These remodelling complexes can either allow sliding of the histone octamer along the DNA, complete transfer of histone octamer from one DNA molecule to another or remodelling the nucleosome to allow increase access to the DNA.

There are multiple nucleosome remodelling complexes in any given cell. For example, SWI/SNF complex contains 8–11 subunits and can perform all three types of remodelling of nucleosome as stated above. Another nucleosome remodelling complex is Mi2/NuRD, which contains 8–10 subunits and can only bring about sliding of nucleosomes. The combination of histone N-terminal tail modification and nucleosome remodeling dramatically change the accessibility to DNA. The protein complexes involved in these modifications and those required to be recruited at the sites of active transcription will be discussed in Chapter 8.

The replication of DNA also requires dis-assembly of the nucleosome, and on DNA replication, the DNA is rapidly re-packaged on the nucleosome. Histones are extensively synthesized during the S-phase, and when the DNA is re-packaged in the nucleosomes, there is a random assortment of the old and the new histones. This nucleosome assembly also requires specific factors referred to as *histone chaperons*. These are negatively charged proteins that form complexes with either H3-H4 tetramers or H2A-H2B dimers and escort them to the site of nucleosome assembly.

5.5 ORGANELLE GENOME

Mitochondria and chloroplasts (in plants) in a cell contain their own genomes in addition to the nuclear genome. As we know from the evolutionary history, these two organelles which were once living as independent bacteria came into existence as cellular organelles in eukaryotic cells as endosymbionts (Fig. 5.30). It is proposed that a purple bacterium is the ancestor of mitochondria, while a relative of a cyanobacterium gave rise to the chloroplast, later during evolution. Thus, the

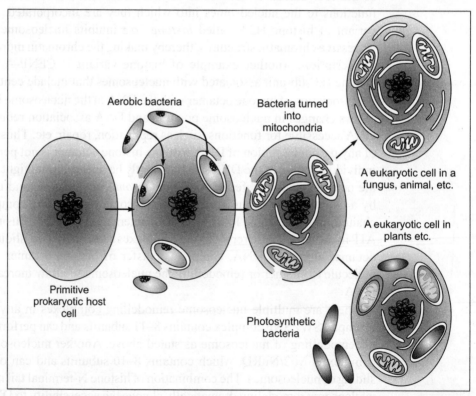

Figure 5.30 Endosymbiont theory of evolution of mitochondria and chloroplast. These two organelles which were once living as independent bacteria came into existence as cellular organelles in eukaryotic cells as endosymbionts. **(See Colour Plate 10)**

genomes of these organelles resemble more to the bacterial genome, although deviated a great deal for this adaptation.

During evolution, following the introduction of these organelles as endosymbionts in eukaryotic cells, there has been a reshuffling of genes between these organelles and the nucleus. For example, most of the original bacterial genes including those for biosynthesis of nucleotide, amino acid, and lipid were lost from the organelle DNA. These products are supplied by the nuclear genes. Further, some other genes present originally in the organelle DNA were transferred to the nuclear genome. Eventually, the genomes of the present-day organelles harbour only some of the most essential genes for their function, namely genes for some proteins, ribosomal RNAs, and transfer RNAs. However, for global function of the organelles, in addition to their own gene products made within them, many nuclear gene products synthesized in the cytoplasm are essential, and they are imported from the cytoplasm (Fig. 5.31). Therefore, the functions of the organelles are directed both

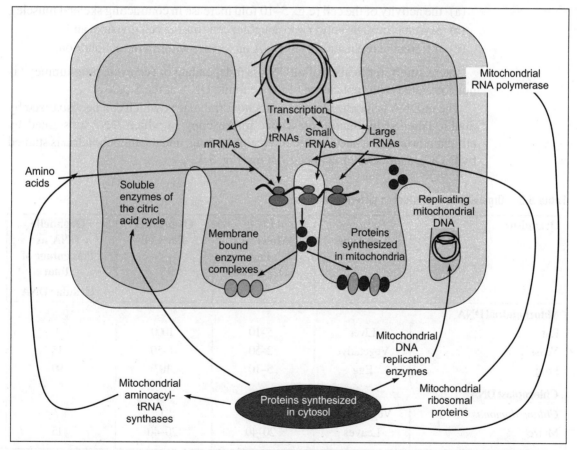

Figure 5.31 For global function of the organelles, in addition to their own gene products made within them, many nuclear gene products synthesized in the cytoplasm are essential, and they are imported from the cytoplasm.

by the nuclear genome and their own genomes. Another interesting feature of organelle genome is that they are maternally inherited.

5.5.1 Mitochondrial Genome

It is in the form of circular DNA. Each mitochondrion may have multiple copies of mitochondrial DNA (mtDNA). Thus, the total amount of mitochondrial DNA/genome in a cell depends on the following:

(a) the number of mitochondria per cell,

(b) the number of mtDNA per mitochondrion, and

(b) the size of the circular mtDNA.

Each of these eventually varies between different cell types (Table 5.3). The number of mitochondria per cell depends on the following:

(a) the activity of the cell (e.g., 5–10 fold increase in contracting skeletal muscle),

(b) asymmetric distribution into daughter cells during cell division, and

(c) differential replication of mtDNA molecules within a mitochondrion.

Interestingly, replication of mtDNA is independent of cell cycle programme; it is not correlated with replication of the genomic DNA at the S phase.

The mtDNA is localized in the mitochondrial matrix which can be visualized by double-labelled immunofluorescence microscopy in which DNA is stained by ethidium bromide, a nucleic acid intercalating drug, and the mitochondria is stained by $DiOC_6$, a mitochondria-specific stain (Fig. 5.32).

Table 5.3 Organelle DNA content in different organisms

Organism	Tissue or Cell Type	DNA Molecules Per Organelle	Organelles Per Cell	Organelle DNA as Percentage of Total Cellular DNA
Mitochondrial DNA				
Rat	Liver	5–10	1000	1
Yeast	Vegetative	2–50	1–50	15
Frog	Egg	5–10	107	99
Chloroplast DNA				
Chlamydomonas	Vegetatives	80	1	7
Maize	Leaves	20–40	20–40	15

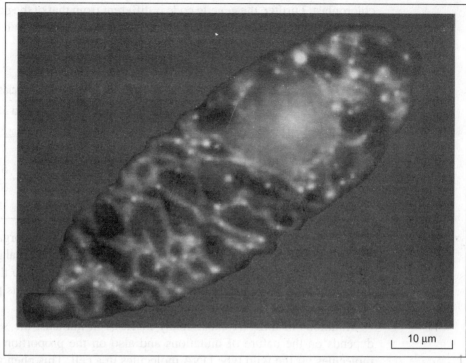

Figure 5.32 Dual staining reveals the multiple mitochondrial DNA molecules in a *Euglena gracilis* cell. Cells were treated with a mixture of two dyes: ethidium bromide, which binds to DNA and emits a red fluorescence, and DiOC$_6$, which is incorporated into mitochondria and emits a green fluorescence. Thus, the nucleus emits a red fluorescence, and areas rich in mitochondrial DNA fluoresce yellow— a combination of red DNA and green mitochondrial fluorescence. [From Y. Hayashi and K. Ueda, 1989, *J. Cell Sci.* **93**:565.] **(See Colour Plate 9)**

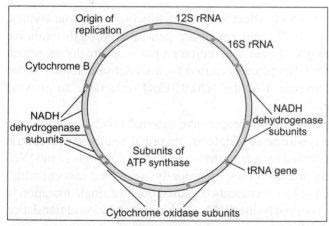

Figure 5.33 The organization of the human mitochondrial genome. It contains tRNA genes, different enzyme encoding genes and rRNA genes.

The human mtDNA was sequenced in 1981, and it was 16,569 nucleotides long. Based on information on tRNA sequences and partial amino acid sequences of mt proteins, the mt genes were mapped in human mtDNA (Fig. 5.33). Mitochondrial DNA encodes intron-less genes arranged on both strands of DNA. The human mt genome is unique in the following characteristics. It is compact, and the genes are densely packed. The codon usage is relaxed in the sense that there are only 22 tRNAs as compared to the 30 or more tRNAs in the cytosol or the chloroplast. Further, the genetic code is different from that of the nuclear genome: 4 of the 64 codons present in the nuclear genome have entirely different meaning (Table 5.4).

Table 5.4 In mitochondria, universal codon codes for different amino acids than in nucleus

Codon	Universal Code	Human Mitochondrial Code
UGA	STOP	Trp
AGA	Arg	STOP
AGG	Arg	STOP
AUA	Ile	Met

Interestingly, the genetic codes in mitochondria of different organisms also vary. It appears that the rate of nucleotide substitutions in mitochondrial genomes is much higher than that in the nuclear genomes. This is mostly due to the reduced stringency of the DNA replication and inefficient DNA repair. This property is often taken into account for studying evolution.

Mutations in mtDNA cause many diseases in humans, the severity of which depends on the nature of mutations and also on the proportion of mutant DNA molecules vs. the wild type DNA molecules in a cell. This phenomenon of mixed DNA molecules is known as *heteroplasmy*. The presence of mutant mtDNA however does not affect replication because most of the enzymes for DNA replication are nuclear genome-coded and the products are imported in the mitochondria. Further, deletion mutations give a replication advantage and cause faster replication of mtDNA. Accumulated mtDNA mutations however are linked with aging in mammals.

Further, some mutations which affect important functions in some tissues, particularly those with high ATP requirement are generated through oxidative phosphorylation. For example, Leber's hereditary optic neuropathy in which degeneration of optic nerve takes place is caused by a missense mutation in the mtDNA gene encoding sub-unit 4 of the NADH-CoQ reductase, an enzyme required for ATP production.

Two other diseases, namely chronic progressive external ophthalmologia (eye defects) and Kearns-Sayre syndrome (eye defects, abnormal heartbeat, and central nervous degeneration) are caused by any of the several large deletions in mtDNA. Another disease of 'ragged' muscle fibres (improperly assembled mitochondria) and associated uncontrolled jerky movements is caused due to a single mutation in the TψCG loop of the mitochondrial lysine tRNA, resulting in inhibition of translation of several mitochondrial proteins.

5.5.2 Chloroplast Genome

Unlike mtDNA, chloroplast DNA can be circular as well as linear depending on the organisms. For example, in Chlamydomonas, the chloroplast DNA is circular. On the other hand, chloroplast DNAs of higher plants which range in size between 120 to 160 kb are linear concatamers and recombination intermediates.

Several chloroplast genomes, particularly from higher plants, have been sequenced. In these higher plants, the chloroplast genome contains 120–135 genes. In a well-studied plant, *Arabidopsis thaliana,* chloroplast genome contains 130 genes. Among these, there are 76 protein coding genes and 54 genes coding for RNAs such as rRNAs and tRNAs. The genome map of liverwort chloroplast genome is shown in Fig. 5.34.

In general, there are striking similarities between the genes of chloroplast and bacteria. For instance, many genes are organized as polycistronic operons. Some of the chloroplast genes contain introns which are more similar to those of some specialized bacterial genes and mitochondrial genes from fungi and protozoans, rather than those of nuclear genes. From the evolutionary studies, it appears that many of the genes of the ancestral chloroplast endosymbionts have been transferred to the nuclear genomes.

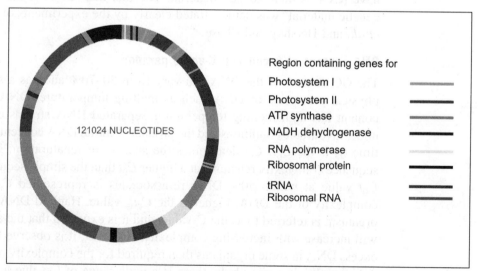

Figure 5.34 The organization of the liverwort chloroplast genome. **(See Colour Plate 10)**

In summary, from the comparative genomic studies, the following facts have emerged:

(a) chloroplasts in higher plants have arisen from photosynthetic bacteria,

(b) chloroplast genomes are fairly stable as compared to the mitochondrial genomes, in particular, and

(c) there has been substantial reshuffling of genes from the chloroplast genome to the nuclear genome, and they still indicate their bacterial origin.

In a cell, there are a large number of chloroplast DNAs. This phenomenon has been exploited in introducing thousands of copies of genetically engineered genes into each cell, thereby producing a high quantity of recombinant proteins. Further, chloroplast transformation has also been used for making transgenic plants with resistance to bacterial and fungal infections, herbicides, and drought. Production of human therapeutics and pharmaceuticals through this approach is likely to materialize in future.

SUMMARY

Nucleic acids as the genetic material

Genetic material of most of the organisms is made up of DNA, though some viruses have RNA as their genetic material. The fact that 'DNA is the most important genetic material' was demonstrated clearly by the experiments of Griffith, Avery *et al.*, and Hershey and Chase.

DNA content of a cell and C-value paradox

The GC content of the DNA can vary from 30–70% and has a strong effect on physical properties of DNA such as melting temperature. DNA with high GC content has high melting temperature. Separated DNA strands can re-associate under favourable conditions and the product of initial DNA concentrations (C_0) and time (t) is the C_0t. C_0t determines the success of renaturation; DNA with high sequence complexity renatures at a higher C_0t than the simple sequence DNA. The C_0t value at which 50% DNA re-associates is represented $C_0t_{1/2}$. Higher the complexity of the DNA, higher is the $C_0t_{1/2}$ value. Haploid DNA content of any organism is referred to as the C-value, and it is expected that the amount of DNA will increase with increasing complexity. However, it is observed that there is an excess DNA in some organisms than required for the complexity of that organism. Additionally, in some phyla, there is a wide range of C-value between different species whose apparent complexity does not vary much. The C-value paradox is probably explained by extra non-coding DNA in some organisms. This non-coding DNA mostly contains repetitive DNA as detected by C_0t analysis.

Genome and its organization

Genome refers to the total DNA content in a cell or an organism. Genome of eukaryotes is much more complex than that of prokaryotes, and it is due to the increased presence of non-coding sequences as a function of evolution. However, in some groups of organisms, namely amphibians and plants, the genome complexity and the total DNA content do not go hand-in-hand. This phenomenon is referred as C value paradox. A very small portion of the genome, e.g. about 2 to 3% in mammals actually belong to the coding sequence category. The remaining large chunk is contributed by various types of non-coding sequences, such as repetitive sequences including satellite DNA, LINES and SINES, transposons, retrotransposons, non-coding sequences within genes, spacer sequences, regulatory sequences of genes, etc.

The number of genes did not increase along with increased genome size of organisms during evolution. Genes are relatively conserved. On the other hand, the non-genic sequences are highly variable among species. Genes are organized either as solitary genes or gene families. Gene families consisting of many members of similar genes have evolved from a single ancestral gene through gene duplication and variation during evolution. The examples of various gene families are globin gene-, rRNA gene-, tRNA gene- and histone gene clusters. Some of them are identical and are arranged as tandem repeats, while others are non-identical and clustered at a single site. Unlike gene families, there are genes that are distantly related, yet they contain some conserved motifs. These belong to gene superfamilies. For example, immunoglobulins (Ig), T cell receptors, T4, T8 genes belong to Ig superfamily.

Packaging of DNA

Within the cell, DNA is organized into structures called chromosomes. In a prokaryotic cell, it is a circular chromosome, whereas in eukaryotes, it is a linear chromosome. In all cells, from viruses to man, DNA needs to undergo condensation, so that it can be accommodated in a small compartment. The eukaryotic DNA associated with protein is referred to as chromatin and the fundamental unit of chromatin is the nucleosome which is made up of 146 base pair of DNA and histone octamer containing 2 copies, each of core histones H2A, H2B, H3, and H4. The DNA packaged into a nucleosome then forms a more complex structure allowing further compaction of the DNA. This process requires histone H1. A more compact form of chromatin is called 30 nm fiber. The interaction of DNA with histones in the nucleosomes is dynamic, thereby allowing DNA binding proteins to interact with DNA. Modification of histone N-terminal tails alters the accessibility of chromatin. These modifications by reducing the charge on the histone help transient dissociation of histones from DNA. Nucleosome remodelling complexes additionally increase the accessibility of DNA in the nucleosomes. This is achieved either by sliding of histone octamer along the DNA, transfer of the histone octamer from one DNA molecule to another, or remodelling of the DNA-protein interactions within the nucleosomes. Nucleosomes are assembled immediately after replication. This requires histone chaperones that escort the core histones to the replication fork and allow the assembly of replicated DNA into nucleosomes.

Organelle genome

In eukaryotes, besides the nuclear genome, there exist organelle genomes for mitochondria and chloroplast. Based on the characteristics of the genomes of these two organelles, it is hypothesized that they have originated from the respective ancestral bacteria during evolution and have been incorporated in cells as endosymbionts. During evolutions, they have lost some of the genes, some others have been incorporated into the nuclear genome. Thus, although they have their genome, many of the required gene products are of nuclear origin, and they are imported from the cytoplasm. Mitochondrial DNA is small and is circular like that of a bacterium, while chloroplast DNA can be linear as well. Chloroplast genome is much more stable than that of the mitochondria.

SHORT ANSWER QUESTIONS

1. Compare the experiments performed by Avery and colleagues and by Hershey and Chase to demonstrate that DNA is the genetic material.
2. What is $C_0t_{1/2}$? What information does it give regarding the complexity of the DNA?
3. Define a eukaryotic gene. Provide a labelled sketch justifying the definition.
4. "The number of genes has not increased proportionately with the increased genome size of organisms." Justify the statement.
5. What are gene families? What is the implication of gene families in the genome with reference to the functions of organisms?
6. What are pseudogenes? Why are they still present in genomes?
7. Which is the type of sequences in the genome that satellite DNAs belong to? Why do they get separated from the genomic DNA while centrifugation?
8. Distinguish between transposons and retrotransposons. What is their evolutionary significance?
9. What is the composition of a nucleosome? Draw a nucleosome structure giving the dimensions of the molecule.
10. Differentiate between euchromatin and heterochromatin.
11. Why are N-terminal tails of histones important in regulating gene expression?
12. Why are nucleosomal remodelling complexes important?
13. Mitochondria and chloroplasts have evolved from their free-living ancestors as endosymbionts. Justify this hypothesis by analysing their genomes.

FURTHER READING

- Alberts, B., Johnson, A., Lewis, J., Raff, M., Roberts, K., and Walter, P. (2002). *Molecular Biology of the Cell*, 4th ed., Garland Science (Taylor & Francis Group), New York.
- Avery, O.T., MacLeod, C.M., and McCarty, M. (1944). Studies on the chemical nature of the substances inducing transformation of Pneumococcal types. *J. Exptl. Med.* **98**, 451–460.
- Belmont, A.S., Deitzels, Nye, A.C., Strukov, Y.G., and Tumbar, T. (1999). Large scale chromatin structure and function. *Curr. Opin. Cell. Biol.* **11**, 307–311.
- Bendich, A.J. (2004). Circular chloroplast chromosomes: the grand elusion. *Plant Cell* **16**, 1661–1666.
- Chan, D.C. (2006). Mitochondria: dynamic organelles in disease, aging and development. *Cell* **125**, 1241–1252.
- Charlesworth, B., Sniegowski, P., and Stephan, W. (1994). The evolutionary dynamics of repetitive DNA in eukaryotes. *Nature*, **371**, 215–220.
- Cooper, G. M. and Hausman, R. E. (2004). *The Cell–A Molecular Approach*, 3rd ed., ASH Press, Sinauev Assoc.Inc. Sunderland, MA.
- Daniell, H., Khan, M.S., and Allison, L. (2002). Milestones in chloroplast genetic engineering: an environmentally friendly era in biotechnology. *Trends Plant Sci.* **7**, 84-91.
- Gray, M.W., Burger, G., and Lang, B.F. (2001). The origin and early evolution of mitochondria. *Genome Biol.* **2**, 1018.1–1018.5.
- Hartwell, L., Hood, L. Goldberg, M.L. et al. (2000). *Genetics: from genes to genomes*, McGraw Hill, Boston.
- Henikoff, S., Greene, E.A., Pietrokovsky, S., Bork, P., Attwood, T.K., and Hood, L. (1997). Gene families: the taxonomy of protein paralogs and chimeras. *Science* **278**, 609–614.
- Hershey, A.D. and Chase, M. (1952). Independent functions of viral protein and nucleic acid in growth of bacteriophage. *J. Gen. physiol.* **36**, 39–56.
- International Human Genome Consortium (2004). Finishing the euchromatic sequence of the human genome. *Nature* **431**, 931–945.
- Kazazian, H.H., Jr. (2004). Mobile elements: drivers of genome evolution. *Science* **303**, 1626–1632.
- Kornberg, R.D. and Lorch, Y. (1992). Chromatin structure and transcription. *Ann. Rev. Cell Biol.* **8**, 563–587.
- Lewin, B. (2004). *Genes VIII*, International Edition, Prentice Hall, NJ, USA.
- Lodish, H., Berk, A., Kaiser, C.A., Krieger, M., Scott, M.P., Bretscher, A., Ploegh, H., and Matsudaira, P. (2008). *Molecular Cell Biology*, 6th ed., W. H. Freeman and Co., New York.

■ Strachan, T. and Read, A.P. (2005). Human Molecular Genetics, BIOS Scientific, Oxford.

■ Sudbery, P. (2002). *Human Molecular Genetics*, 2nd ed. Prentice Hall, England.

■ Woodcock, C.L. and Dimitrov, S. (2001). Higher order structure of chromatin and chromosomes. *Curr. Opin. Genet. Dev.* **11**, 130–135.

DNA Replication

6

OVERVIEW

- Every cell replicates its DNA before it divides. This chapter describes in detail the DNA replication as it happens in both prokaryotic and eukaryotic cells.

- It highlights the similarities as well as the differences in the process of replication in prokaryotes and eukaryotes.

- This chapter further describes the way in which the replication is regulated in prokaryotic and eukaryotic cells.

All cells need to duplicate the DNA before they divide. This job must be finished quickly with great accuracy. In both prokaryotes and eukaryotes, DNA replication is achieved by co-operation of many proteins, and it proceeds through three steps of *initiation*, *elongation,* and *termination*. Additionally, in eukaryotes, DNA is replicated as a linear molecule and as chromatin in which it is associated with proteins. The event of DNA replication is highly regulated to ensure that the DNA is replicated only once before the cell divides.

6.1 GENERAL FEATURES OF DNA REPLICATION

When a cell divides into two, each daughter cell receives the exact copy or replica of the parental DNA. Therefore, DNA is replicated prior to the cell division. The process of DNA replication is semi-conservative. This means, at the end of replication when two molecules of DNA are synthesized, each has one strand of parental origin and another newly synthesized. In a simple model of DNA replication, double stranded DNA is denatured and each strand is copied separately.

DNA replication always initiates from a specific region in the DNA referred to as the origin of replication. Once initiated, it can proceed in one direction (unidirectional) or in both directions (bidirectional). If a circular DNA replicates bidirectionally, the replication would finish in half the time than that required for unidirectional synthesis. The speed of replication in bacteria is around 1000 nucleotides per second. Therefore, once initiated bidirectionally from the origin,. it would take around 40 minutes to replicate 4×10^6 nucleotides long circular DNA.

When an eukaryotic cell divides resulting into two cells, the entire event is called a *cell cycle* and it is divided into four parts, namely G1, S, G2, and M. G1 and G2 are the preparatory phases for a cell to enter S and M phase, respectively. The act of DNA replication is restricted to a synthetic phase (S phase) that lasts for around 7–8 hours and then follows mitotic phase leading to cell division. The speed of replication in eukaryotes is much slower, around 50 nucleotides per second. With this slow speed, replication of entire DNA molecule would take several hours and would not finish in S phase.

If we consider that an average chromosome is composed of DNA molecule containing 150×10^6 base pairs, the time required to replicate this long molecule from one end to the other with single origin of replication would take around 800 hours. However, for most of the cells, S phase lasts for 7–8 hours during which all the chromosomes in the nucleus must replicate. To overcome this problem, cells have adopted the strategy of initiating replication at several positions along the DNA so that the replication of entire DNA would finish in a short time.

The DNA molecule in which a single act of replication can take place and is under the control of single origin is referred to as a *replicon*. Therefore, each replicon must contain an origin of replication, and a terminator for replication.

Bacterial chromosome is a single replicon bearing a single origin of replication and a single terminator, whereas eukaryotic DNA has multiple replicons.

DNA replication is basically similar in eukaryotes and prokaryotes. The major difference is that eukaryotic DNA is replicated as a linear molecule, and not as a bare DNA but as chromatin, in uhich the DNA is complexed with proteins called histones. We would first understand the DNA replication in prokaryotes.

6.2 REPLICATION IN PROKARYOTES

In semi-conservative mode of replication (Fig. 6.1), two parental strands unwind, get separated from each other, and are used as templates for synthesis of new strands following the rule of usual base pairing, i.e. A base pairs with T, and G with C. However, this model presents more questions than answers because two strands of DNA have opposite polarity. This means that the replication machinery is able to make DNA in both 3 ′-5′ and 5′-3′ direction.

In reality, all known replication machines can make DNA only in the 5′-3′ direction. That is, it inserts 5′ most nucleotide first and extends the chain towards the 3′ end by adding the nucleotides to the 3′ end of a growing chain. This means when replication begins at one point on DNA, both strands cannot be replicated continuously. To replicate both strands simultaneously, two strands with opposite polarity would have to separate totally from each other and then replicate in the 5′ to 3′ direction initiating the event at two far ends. However, cells cannot afford to allow two strands to completely separate from each other. To overcome this problem, one of the strands is made continuously in 5′-3′ direction, while another strand is made discontinuously in the same direction (Fig. 6.2).

The strand where DNA synthesis proceeds continuously is called as the *leading strand* and the other where DNA synthesis takes place discontinuously is called as the *lagging strand*. The discontinuity of synthesis comes about because its direction of synthesis is opposite to the direction in which replication fork moves. Therefore, as the fork opens up and exposes a region of DNA to replicate, the lagging strand grows away from the fork. Therefore, to replicate this newly opened region, the only way out is to restart DNA syn-

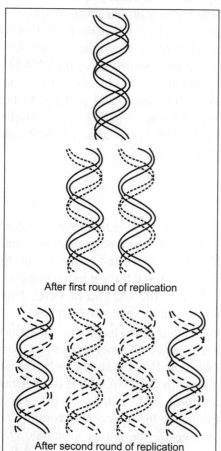

After first round of replication

After second round of replication

Figure 6.1 Semi-conservative replication. Each DNA strand serves as a template for the synthesis of a new strand, producing two new DNA molecules, each one with a new strand and an old (parental) strand.

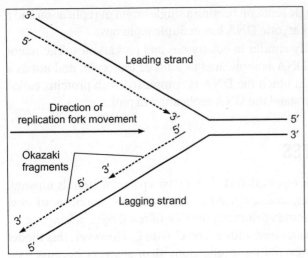

Figure 6.2 Leading and lagging strand synthesis. DNA strand is always synthesized in the direction 5′ to 3′. The leading strand is synthesized continuously and the lagging strand is synthesized discontinuously, in short pieces as Okazaki fragments in a direction opposite to that of the replication fork movement.

thesis at the fork behind the piece of DNA that has already been made. This starting and restarting of DNA synthesis occurs over and over again. The short pieces of DNA thus created would have to be joined together to produce the continuous strand. These short fragments made discontinuously are called as *Okazaki fragments*. The average length of Okazaki fragments is about 1200 nucleotides.

The DNA replication process requires co-operation of many proteins. These include initiator proteins that recognize the origin region from where the replication has to begin, proteins to synthesize new DNA molecule, proteins to open up the DNA helix to be copied, proteins to join short matured Okazaki fragments, and proteins to help relieve helical winding and tangling problems. DNA replication can be divided into three steps, namely initiation, elongation, and termination.

6.2.1 Initiation

Initiation of replication begins with identifying the origin region in the DNA and synthesis of primer necessary to begin polymerization of a polynucleotide chain. The origin region called OriC consists of 245 nucleotides and contains four nine nucleotides long consensus sequence TTATCCACA and three thirteen nucleotide long consensus sequence GATCTNTTNTTTT at the left end of OriC. DnaA protein recognizes and binds to the nine nucleotide long sequence (dna box).

A huge complex of DnaA binds to the origin along with HU protein and induces bending in the DNA as well as destabilization of adjacent 13 mer repeats and causes local DNA melting. This allows the next protein, DnaB, to bind to the melted region. DnaC binds to DnaB protein and helps deliver it to the DNA. DnaB is a helicase, which can unwind the DNA in the direction 5′-3′. At this stage, the pre-priming complex is ready. Finally DnaG, a primase binds to the pre-priming complex and converts it to the primosome, which can synthesize the primer to initiate DNA replication (Fig. 6.3). DnaG is a RNA polymerase that synthesizes small 8–10 nucleotides long RNA at the priming site. Even in eukaryotes, several proteins are required for identifying the origin region (referred to as ORC, origin recognition complex) and initiating the process.

6.2.2 Elongation

Once a primer is in place, DNA synthesis can begin. This includes two distinct operations, leading strand synthesis and lagging strand synthesis. The enzymes involved in DNA synthesis are referred to as DNA polymerases. All DNA polymerases require a template, a primer, deoxynucleotide triphosphates, magnesium, and ATP for the catalytic reaction. Polymerization reaction involves adding a nucleotide to the growing 3′ OH end in a template-dependent fashion (Fig. 6.4a).

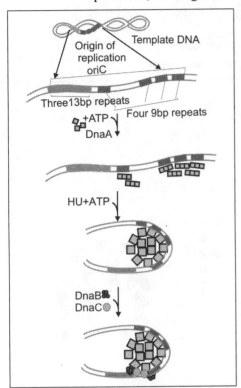

Figure 6.3 Initiation of DNA replication. A complex of protein DnaA binds to the four nine base pair repeats in the origin of replication and induces bending in the DNA along with the protein HU. DnaB is then delivered to the site by DnaC. DnaB is a helicase, which unwinds the duplex in the region of thirteen base pair repeat in the origin region. Prepriming complex is ready. DnaG then joins this complex to synthesize the primer. **(See Colour Plate 12)**

The precursor for DNA synthesis is a nucleotide triphosphate, which loses the two terminal phosphate groups as pyrophosphate in the reaction and the polynucleotide chain grows by one nucleotide in the direction 5′-3′. Polymerase then scrutinizes the base pair that it has added and in case it is wrong, excises it out with its proofreading activity.

There are three DNA polymerases in *E. coli*, namely DNA polymerase I, II, and III. Of these, DNA polymerase III is a major polymerase in DNA replication. DNA polymerase I also has a role to play in DNA synthesis. DNA Polymerase III is a complex enzyme (Fig. 6.4b) of several subunits and has two catalytic activities, namely 5′-3′ polymerase and 3′-5′ exonuclease, and it is assembled at the site of replication. It does not exist as a holoenzyme in the cell. The core enzyme consists of sub-unit $\alpha\epsilon\theta$, wherein 5′-3′ polymerase activity is in the α subunit, and 3′-5′ exonuclease is with the ϵ subunit, and these two are held together by the θ subunit. The core by itself is a poor polymerase. It puts together about 10 nucleotides and then falls off the template. Again, it has to re-associate with the template and start synthesis. However, in reality, the replication fork moves with the speed of 1000 nucleotides per second.

Another subunit called β ring associates with the core and makes it a highly processive enzyme, allowing it to remain engaged with the template while polymerizing at least 50,000 nucleotides before it stops. The β subunit forms a dimer that is ring shaped. This ring fits around the DNA template and interacts with the α subunit of the core to tether the polymerase and the template together. Thus, the

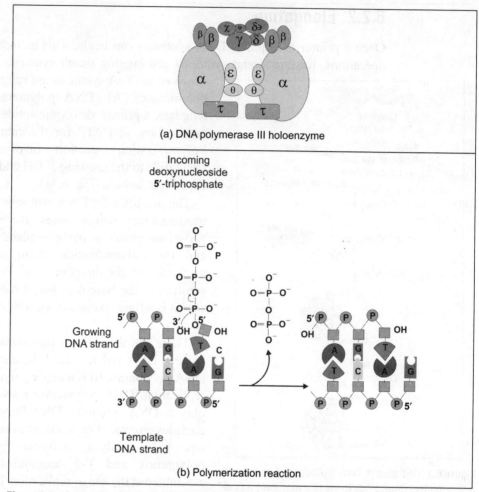

(a) DNA polymerase III holoenzyme

(b) Polymerization reaction

Figure 6.4 **(a)** Polymerization reaction. Polymerization reaction involves adding a base to the growing polynucleotide chain in a template-dependent manner. Incoming deoxynucleotide-5′-triphosphate (dNTP) is incorporated by enzyme DNA polymearse at the 3′OH end by forming a phosphodiester bond and a pyrophosphate is released. **(b)** DNA polymerase III involved in DNA replication at the replication fork is a multi-subunit enzyme. Holoenzyme has two core polymearses each having three subunits, α, θ, and ε. Subunit Tau holds two cores together and the gamma complex is associated with the core engaged in lagging strand synthesis.

enzyme can stay on template for a long time. β subunit is therefore called as the *clamp of polymerase*. The clamp needs help from another subunit called the γ complex to load on the DNA template. This complex is named as *clamp loader* and consists of many subunits such as γδδ′χψ. This clamp loader has both loading and unloading activities for β clamp.

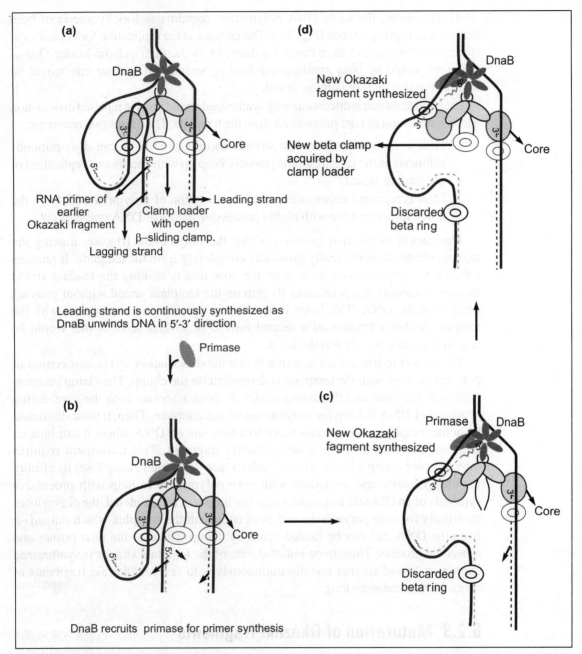

(a)

DnaB

3'-

5'-

3'-

Core

RNA primer of
earlier
Okazaki fragment

Clamp loader
with open
β–sliding clamp.

Leading strand

Lagging strand

Leading strand is continuously synthesized as
DnaB unwinds DNA in 5'-3' direction

Primase

(b)

DnaB

3'-

3'-

5'-

5'-

Core

DnaB recruits primase for primer synthesis

(d)

DnaB

New Okazaki
fragment synthesized

5'-

3'-

3'-

Core

New beta clamp
acquired by
clamp loader

Discarded
beta ring

(c)

Primase DnaB

New Okazaki
fagment synthesized

5'-

3'-

3'-

Core

Discarded
beta ring

Figure 6.5   At the replication fork, the synthesis of leading and lagging strand is coordinated. At the replication fork, the events
are coordinated by DNA polymearse III holoenzyme with two cores, one engaged in leading strand synthesis and the other
in lagging strand synthesis. Since the leading strand moves faster than the lagging stand, the lagging strand is looped so
that the same enzyme can synthesize both the strands steadily. Arrows indicate the 3' of newly synthesized strands.
(See Colour Plate 11)

At replication, the same DNA polymerase coordinates fork synthesis of both leading and lagging strands (Fig. 6.5). The enzyme at the replication fork consists of two core polymerases linked through a dimer of τ subunit to a clamp loader. One of the cores would be busy synthesizing leading strand and another one would be involved in synthesis of lagging strand.

Lagging strand that is discontinuously synthesized would involve repeated dissociation and re-association of core polymerase from the template. This raises two questions:

1. How is the synthesis of two strands coordinated? How can discontinuous synthesis of the lagging strand possibly keep up with continuous replication of the leading strand?
2. How is repeated dissociation and re-association of Polymerase III from the template compatible with highly processive nature of DNA replication?

The answer to the first question is that the Polymerase III core, making the lagging strand, does not really dissociate completely from the template. It remains tethered to it by its association with the core that is making the leading strand through τ subunit. So, it releases its grip on the template strand without straying away from the DNA. This helps it find the next primer and re-associate with the template within a fraction of a second instead of several seconds that would be required if it completely left the DNA.

The answer to the second question is that the dissociation and re-association of polymerase core with the template is dependent on the clamp. The clamp interacts both with the core and the clamp loader. It must associate with the core during synthesis of DNA to keep the polymerase on the template. Then, it must dissociate from the template so that it can move to a new site on DNA where it can interact with another core to make a new Okazaki fragment. This movement requires interaction of clamp with the clamp loader. Once the loaded clamp loses its affinity for clamp loader and associates with core polymerase to help with processive synthesis of an Okazaki fragment. Once the fragment is completed, the clamp loses its affinity for core polymerase and associates with the complex which unloads it from the DNA and can be loaded again by clamp loader to the next primer and repeat the process. Thus, once initiated, one of the template strands is synthesized continuously and another one discontinuously with several Okazaki fragments of about 1200 nucleotides long.

6.2.3 Maturation of Okazaki fragments

The Okazaki fragments need to be matured in the continuous DNA strand (Fig. 6.6). This involves two steps, viz.

(i) each Okazaki fragment has a small RNA primer at the 5′ end, which needs to be removed and replaced by DNA, and
(ii) all these Okazaki fragments need to be joined to each other.

Colour Plate 9

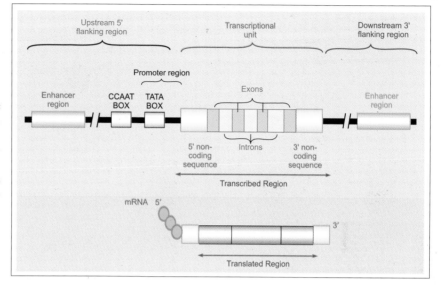

Structure of a eukaryotic gene. Gene is divided into an upstream 5′ flanking region, transcriptional unit and downstream 3′ flanking region. Regulatory elements such as promoters, enhancers reside in the flanking regions as well as within introns. After transcription, mRNA is processed and is translated. (Chapter 5, p. 141)

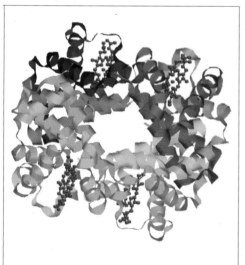

Structure of haemoglobin. Adult globin which is composed of two copies each of α and β subunits is linked to a molecule of heme as the prosthetic group giving rise to haemoglobin. (Chapter 5, p. 148).

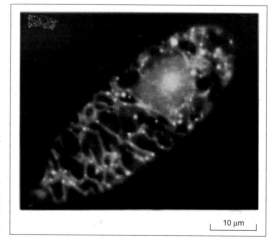

Dual staining reveals the multiple mitochondrial DNA molecules in a *Euglena gracilis* cell. Cells were treated with a mixture of two dyes: ethidium bromide, which binds to DNA and emits a red fluorescence, and $DiOC_6$, which is incorporated into mitochondria and emits a green fluorescence. Thus, the nucleus emits a red fluorescence, and areas rich in mitochondrial DNA fluoresce yellow—a combination of red DNA and green mitochondrial fluorescence. (Chapter 5, p. 163)

Colour Plate 10

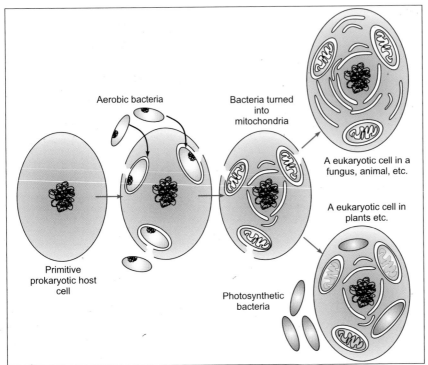

Aerobic bacteria

Bacteria turned into mitochondria

Endosymbiont theory of evolution of mitochondria and chloroplast. These two organelles which were once living as independent bacteria came into existence as cellular organelles in eukaryotic cells as endosymbionts. (Chapter 5, p. 160)

A eukaryotic cell in a fungus, animal, etc.

A eukaryotic cell in plants etc.

Primitive prokaryotic host cell

Photosynthetic bacteria

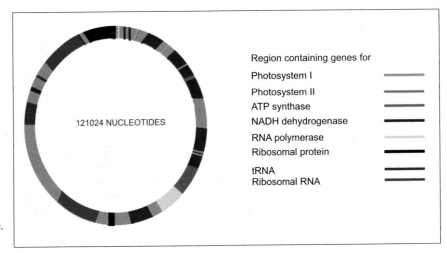

121024 NUCLEOTIDES

Region containing genes for

Photosystem I

Photosystem II

ATP synthase

NADH dehydrogenase

RNA polymerase

Ribosomal protein

tRNA
Ribosomal RNA

The organization of the liverwort chloroplast genome. (Chapter 5, p. 165).

Colour Plate II

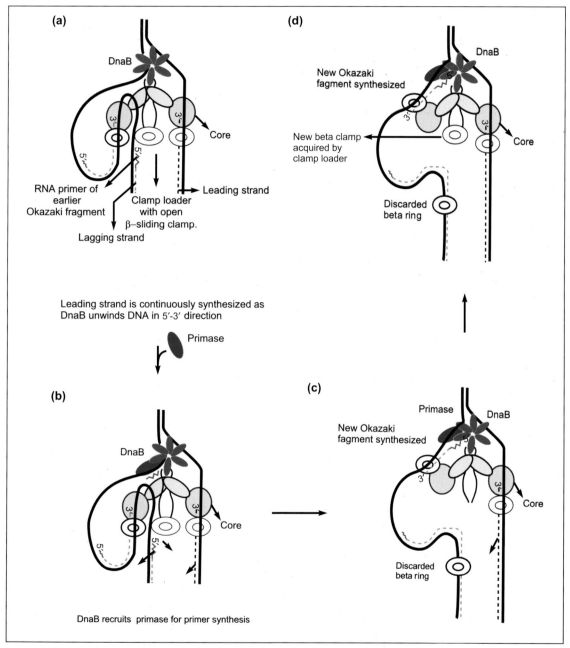

(a)

DnaB

Core

RNA primer of earlier Okazaki fragment

Clamp loader with open β–sliding clamp.

Leading strand

Lagging strand

Leading strand is continuously synthesized as DnaB unwinds DNA in 5'-3' direction

Primase

(b)

DnaB

Core

DnaB recruits primase for primer synthesis

(c)

Primase   DnaB

New Okazaki fagment synthesized

Core

Discarded beta ring

(d)

DnaB

New Okazaki fagment synthesized

New beta clamp acquired by clamp loader

Core

Discarded beta ring

At the replication fork, the synthesis of leading and lagging strand is coordinated. At the replication fork, the events are coordinated by DNA polymearse III holoenzyme with two cores, one engaged in leading strand synthesis and the other in lagging strand synthesis. Since the leading strand moves faster than the lagging stand, the lagging strand is looped so that the same enzyme can synthesize both the strands steadily. Arrows indicate the 3′ of newly synthesized strands. (Chapter 6, p. 177)

Colour Plate 12

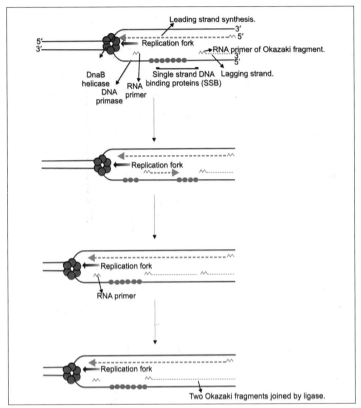

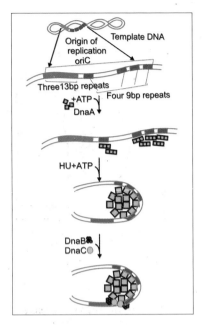

Maturation of Okazaki fragments. Every Okazaki fragment has a RNA primer at the 5′. Maturation of these fragments involves removal of the RNA primer, replacing it with DNA, and joining of the two Okazaki fragments as they are formed. (Chapter 6, p. 179)

Initiation of DNA replication. A complex of protein DnaA binds to the four nine base pair repeats in the origin of replication and induces bending in the DNA along with the protein HU. DnaB is then delivered to the site by DnaC. DnaB is a helicase, which unwinds the duplex in the region of thirteen base pair repeat in the origin region. Prepriming complex is ready. DnaG then joins this complex to synthesize the primer. (Chapter 6, p. 175)

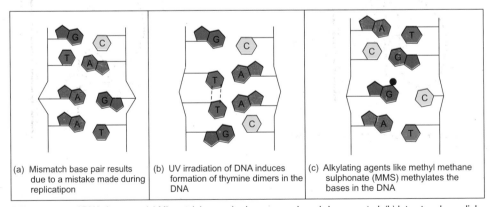

(a) Mismatch base pair results due to a mistake made during replicatipon

(b) UV irradiation of DNA induces formation of thymine dimers in the DNA

(c) Alkylating agents like methyl methane sulphonate (MMS) methylates the bases in the DNA

Different types of DNA damages. (a) Mismatch base pair when a wrong base is incorporated, (b) Intrastrand crosslink– Cyclobutane pyrimidine ring formed on UV irradiation, and (c) Base modification, methylation of a base by alkylating agent methyl methane sulphonate. (Chapter 7, p. 189)

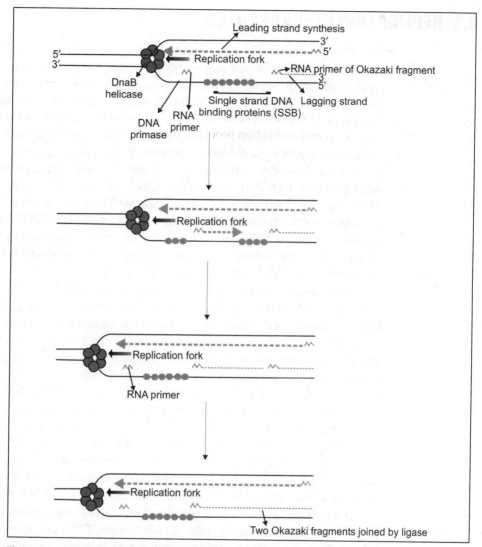

Figure 6.6 Maturation of Okazaki fragments. Every Okazaki fragment has a RNA primer at the 5′. Maturation of these fragments involves removal of the RNA primer, replacing it with DNA, and joining of the two Okazaki fragments as they are formed. **(See Colour Plate 12)**

Enzymes DNA polymerase I, and DNA ligase do these jobs, respectively. DNA polymerase I has three catalytic activities, namely 5′-3′ polymerase, 3′-5′ exonuclease and 5′-3′ exonuclease. The synthesis of each Okazaki fragment ends when the DNA polymerase runs into the RNA primer attached to the 5′ end of the previous fragment. DNA polymerase I with 5′-3′ exonuclease removes ribonucleotides, simultaneously extending the 3′ end of Okazaki fragment. Finally, the nick is sealed by ligase by making a phosphodiester bond between 3′OH and 5′PO$_4$ ends.

6.3 REPLICATION IN EUKARYOTES

The DNA molecules in eukaryotic cells are considerably larger than those in bacteria and are organized into nucleoprotein complex. The essential features of replication are the same in eukaryotes and prokaryotes and many of the protein complexes are structurally and functionally conserved. As explained earlier, eukaryotic DNA has several origins of replication spaced 30,000 to 3,00,000 base pairs apart and replication proceeds bidirectionally from these origins. Initiation of replication requires a multisubunit protein, the origin recognition complex (ORC) which binds to origin of replication and interacts with many other proteins. Two other proteins, CDC6 (Cell Division Cycle 6) and CDT1 (Cdc 10 Dependent Transcript 1), bind to ORC and load complex of another protein called minichromosome maintenance proteins (MCM). CDC6 and CDT1 are functionally similar to DnaC, which loads DnaB at the origin of replication in bacteria. MCM proteins are a ring-shaped replicative helicase similar to DnaB helicase of *E. coli* and unwinds the DNA at the origin of replication.

Like bacterial cells, eukaryotic cells have several polymerases: polymerase α, β, γ, δ, ε, η, ζ, etc. Of these, polymerase α, δ, and sometimes ε are involved in nuclear DNA replication. Polymerase γ is involved in mitochondrial DNA replication and the remaining (β, η, ζ) are primarily involved in different DNA repair processes. DNA polymerase α is a multisubunit enzyme. One of the subunits has primase activity and the other sub-unit contains polymerase activity. However, this polymerase has no proofreading activity, making it unsuitable for high fidelity replication. This polymerase is basically involved in synthesis of short primers containing RNA and DNA called i-DNA (initiator DNA), once on leading strand and many times on the lagging strand for Okazaki fragments. DNA polymerase δ then extends these primers. This enzyme is associated and stimulated by a protein called PCNA (Proliferating Cell Nuclear Antigen) whose three-dimensional structure is very similar to that of the β subunit of *E. coli* DNA polymerase III.

DNA polymerase δ has 3′-5′ proofreading activity and appears to carry out both leading and lagging strand synthesis. Sometimes, DNA polymerase ε can replace DNA polymerase δ. Thus, at the replication fork in eukaryotic cell, the replication is co-ordinated by DNA polymerase α and two molecules of DNA polymerase δ. During the same time, when DNA is being synthesized, lot of histone proteins are made by the cell, as both the newly synthesized DNA and the parental DNA, must form nucleoprotein complex on completion of replication. Separate class of proteins called *chromatin assembly proteins* carry out the assembly of DNA into nucleoprotein complex called *chromatin*.

6.4 TERMINATION OF REPLICATION

Like OriC which defines the origin of replication, there are sequences called *Ter sequences* present on the DNA which are responsible for terminating actively the

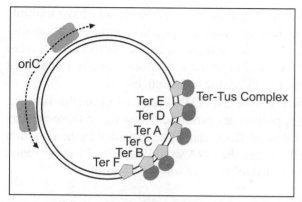

Figure 6.7 Termination of replication. Ter sequences are involved in termination of DNA replication. Ter proteins bind to these Ter sequences and knock off the helicase leading the replication fork. Once the helicase is lost, the fork's movement is stopped.

process of replication. These are 22 base pair long sequences that bind specific proteins called TUS (terminator utilization substances). There are six Ter sequences in all. Of these, three Ter sequences work as terminator for the fork moving in a clockwise direction from OriC, and the remaining three work as terminator for the fork moving in anti-clockwise direction (Fig. 6.7). Thus, these Ter sequences are orientation-specific with respect to replication fork. Tus protein is a contra-helicase. The replication fork is led by helicase, continuously unwinding the DNA in the 5′-3′ direction. If the helicase is removed, replication fork cannot proceed. Tus protein being contra-helicase knocks off the helicase, thus terminating the progression of replication fork.

On completion of replication of circular DNA, two daughter duplexes remain entangled, and these need to be separated so that they can be passed on to the two daughter cells. The two daughter duplexes formed during replication remain entwined as two interlocked rings, called *catenates*. These are unlinked or decatenated (Fig. 6.8) by the enzyme called topoisomerase IV.

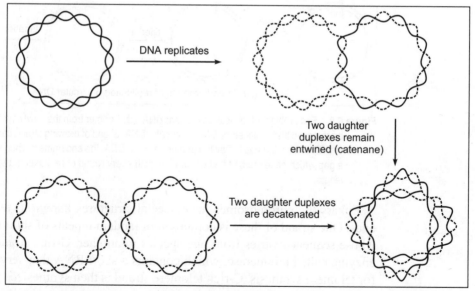

Figure 6.8 Decatenation of two daughter DNA molecules formed. On replication of circular DNA, two daughter molecules are formed which remain entangled with each other. They are separated (decatenated) by the enzyme topoisomerase.

Eukaryotic chromosomes are not circular. They are linear and have multiple replicons, so replication forks from neighbouring replicons approach one another just as the two replication forks of a bacterial chromosome approach each other near the termination point. Eukaryotic chromosome also forms catenates that must be dis-entangled, and this is done by eukaryotic topoisomerase II.

Eukaryotes face another difficulty at the end of DNA replication and that is filling the gaps left when the first RNA primers are removed. Removal of these primers leaves a gap which cannot be filled, as DNA cannot be extended in the direction 3'-5' (Fig. 6.9). If this is left unfilled, then the DNA would shorten with every round of replication. A special enzyme called *telomerase* solves this problem.

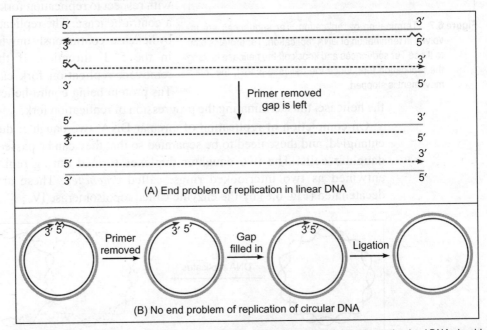

(A) End problem of replication in linear DNA

(B) No end problem of replication of circular DNA

Figure 6.9   End problem of replication in linear DNA. RNA primer from the newly synthesized DNA should be removed and replaced by DNA. In circular DNA, 3' end of growing strand can prime the synthesis of DNA to fill in the gap left by the primer. In linear DNA, the end primers when removed would leave a gap which needs to be filled up, or else, with every round of replication, the DNA will shorten in length.

Eukaryotic chromosomes have special structures known as telomeres at their ends. One strand of these is composed of tandem repeats of short GC-rich regions whose sequence varies from one species to another. G-rich strand is made by an enzyme called telomerase, which contains a short RNA that serves as a template for telomere synthesis. C-rich telomere strand is then synthesized by RNA primed DNA synthesis. This ensures that chromosome ends are re-built and do not suffer shortening with each round of replication (Fig. 6.10).

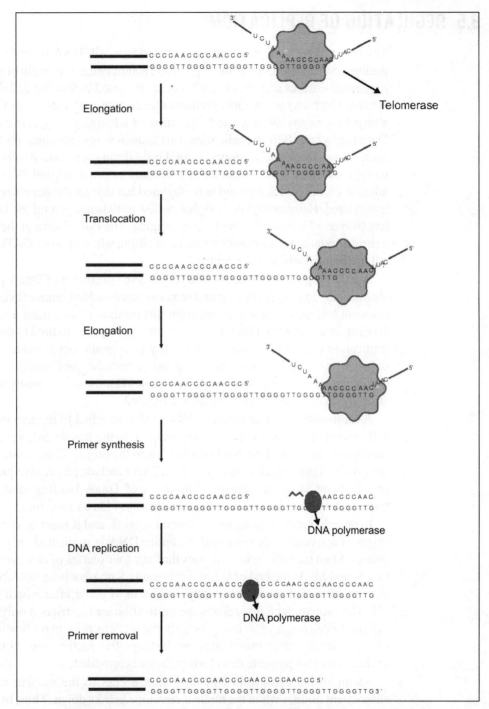

Figure 6.10 DNA ends are synthesized in eukaryotes by telomerase. Problem of end replication of linear DNA is solved by telomerase. Telomerase is a ribozyme containing a catalytic RNA and a protein.

6.5 REGULATION OF REPLICATION

The event of replication is tightly regulated in a cell. DNA is replicated only once during a cell division. Since the replication is initiated at the origin of replication, the regulation ensures that the origin is not re-activated before the cell divides. This is achieved by many ways. OriC contains eleven copies of a sequence GATC which is a target for methylation at the N^6 position of adenine by a specific enzyme called Dam methylase. Before replication, this sequence is methylated on the adenine on each strand. Replication inserts normal base during replication, which needs to be methylated later. Thus, replication generates hemimethylated DNA, wherein the adenine on the parental strand is methylated but that on the daughter strand is non-methylated. Hemimethylated origins cannot initiate again until the Dam methylase has converted it into fully methylated origins. The GATC sites at the origin remain hemimethylated for 13 minutes after replication, whereas other GATC sites begin to get methylated within 1.5 minutes.

Another region that behaves like origin is the promoter of DnaA gene. This also shows delay in re-methylation and remains repressed in hemimethylated state. This does not let transcription proceed from this promoter that causes a reduction in the level of DnaA protein. DnaA is the first protein required to bind to the origin before replication can initiate, and unavailability of it would not activate the origin. The delay in re-methylation of these regions is because they remain sequestered by binding of the protein called SeqA. This protein binds to hemimethylated DNA very strongly and prevents it from being methylated.

Additionally, hemimethylated DNA is also attached to the membrane, whereas fully methylated DNA is not. While attached to the membrane, a protein bound to membrane does not allow binding of DnaA to the origin. Thus, several mechanisms are involved in regulating the replication. These include physical sequestration of the origin, delay in remethylation, inhibition of DnaA binding, and repression of transcription from DnaA reducing the amount of DnaA protein.

In an eukaryotic cell, there are several origins, and it must be ensured that each origin is activated only once and the entire DNA is replicated only once during S phase. When the cell divides, it goes through four phases of cell cycle, namely G1, G2, S, and M. At the end of G1, the cell enters S phase where DNA replicates, then the cell enters G2 phase and finally divides in M phase after which it again enters G1. The protein, which regulates the replication and restricts it only to S phase, is called as *licensing factor*. It is present in the nucleus prior to replication. One round of replication either inactivates or destroys the factor, and another round of replication is not possible until further factor is provided.

Factor in the cytoplasm can only gain access to the nuclear material at the subsequent mitosis when the nuclear envelope breaks down. Thus, by removing the necessary components after replication, it prevents more than one cycle of replication from occurring. It also provides a feedback loop that makes the initiation of replication dependent on passing through the cell division.

The candidate proteins for being a licensing factor are Cdc6 and MCM proteins. Recall that in an eukaryotic cell, presence of Cdc6 at the origin allows MCM proteins to bind to origin, only after which the replication can initiate. When origin of replication enters S phase, it is in pre-replication complex containing ORC, Cdc6, and MCM proteins. When initiation occurs, Cdc6 and MCM are displaced, thereby leaving origin with only ORC complex. Cdc6 is rapidly degraded on displacement and is not available for loading of MCM proteins, necessary for activation of origins. This ensures that each origin is used only once during a single S phase.

SUMMARY

Replication of DNA, in both prokaryotes and eukaryotes, is semi-conservative and occurs with very high fidelity. It is carried out in three distinct phases, namely initiation, elongation, and termination.

Initiation involves recognition of the origin sequence by a complex of proteins. Helicase unwinds the DNA at the origin, primase synthesizes the primer, and the replication begins on both the strands. Replication fork moves bidirectionally from the origin of replication.

During elongation, DNA polymerase is the principle enzyme. In *E.coli*, DNA polymerase III is a major replicase, whereas in eukaryotes, it is DNA polymerase δ. Each polymerase needs a primer, which in *E coil* is provided by DnaG and in eukaryotes by DNA polymerase α. DNA polymerases work in the 5'-3' direction.

At the replication fork one strand is synthesized continuously called the leading strand and another strand is synthesized discontinuously as Okazaki fragments called the lagging strand. These Okazaki fragments are subsequently matured and then ligated by DNA ligase. Fidelity of replication is maintained by correct base selection by polymerase and by 3'-5' proofreading activity, which is a part of DNA polymerase. In case an error is still left in the replicated DNA, it is corrected by a specific repair system meant to remove the mismatch.

Termination of replication in *E. coli* is mediated by Ter sequences and Tus protein that bind to these sequences and stop the moving replication forks by removing DNA helicase, which guides the replication fork. Termination in case of eukaryotes is still not clearly understood. The replication is tightly regulated and coupled to the cell division. This is achieved in *E. coli* mainly by methylation status of origin of replication, and in eukaryotic cells by a protein called licensing factor.

SHORT ANSWER QUESTIONS

1. If *E. coli* DNA is of the size 4.8×10^6 base pair, how long would it take to replicate this DNA at the rate of 1000 bases per second?
2. Why are the mutations in the following genes lethal to *E. coli*: DnaB, DnaG and DnaA?

3. What is OriC? Why is it important for replication?
4. What are Ter sequences? How do they contribute to termination of replication?
5. Explain the major differences in the replication events of prokaryotes and eukaryotes.
6. Polymerase III is the main polymerase for DNA replication in *E. coli*. Can it be replaced by Polymerase I? Explain.
7. "Both DNA polymerase α and δ are necessary for replication in eukaryotes." Justify the statement.
8. Why is the rate of mutation in *E. coli* in the range of 10^{-8}–10^{-10}, when the rate of error in DNA polymerase is in the range of 10^{-6}–10^{-7}?
9. Why is it necessary to regulate the event of DNA replication?

FURTHER READING

- Baker, T.A. and Wickner, S.H. (1992). Genetics and enzymology of DNA replication in *E. coli. Annu. Rev. Genet.* **26**, 447–477.
- Bell, S.P. and Dutta, A. (2002). DNA replication in eukaryotic cells. *Annu. Rev.Biochem* **71**, 333–371.
- Benkowic, J.J., Valentine, A.M., and Salinas, F. (2001). Replisome mediated DNA replication. *Annu. Rev.Biochem*, **70**, 181–208.
- Boye, E., Lobner-Olsen, A., and Skarstad, K. (2000). Limiting DNA replication to once and only once. *EMBO rep.* **1**, 479–483.
- Devay, M.J. and O'donnell, M. (2000). Mechanisms of DNA replication. *Current. Opin. Chem. Biol.* **4**, 581–586.
- Hubscher, U., Maja, G., and Spadori, S. (2004). Eukaryotic DNA polymerases. *Annu. Rev.Biochem.* **71,** 133–163.
- Joyce, C.M. and Steitz, T.A. (1994). Function and structure relationships in DNA polymerases. *Annu. Rev.Biochem*, **63,** 777–822.
- Mott, M. and Berger, J. (2007). DNA replication intiation mechanism and regulation in bacteria. *Nature reviews Microbiology*, **5**, 343–354.
- Radman, N. and Wagner, R. (1988). The high fidelity of DNA replication. *Scientific American* 40–46.

DNA Damage, Repair, and Recombination

7

OVERVIEW

- This chapter will make you understand the importance of maintaining the integrity of DNA.

- Different types of damages induced in the DNA and their consequences, if not rectified from the DNA molecule, are described.

- Different DNA repair pathways employed by a cell to remove the damages from DNA are explained.

- You will further understand how the variation in the genetic information is generated by the events of recombination, which occurs in different modes.

Survival of every cell demands that the integrity of genetic material is accurately maintained. In a cell, DNA is continuously exposed to DNA damaging agents. Injury to DNA, either by endogenous or exogenous DNA damaging agents is reduced by DNA repair systems that recognize and rectify the damages from the DNA. Failure of these repair systems to remove damages from DNA leads to induction of mutations, i.e. changes in DNA sequence.

7.1 TYPES OF DNA DAMAGES

A prokaryotic cell has a single copy of DNA, while an eukaryotic cell has two copies of genomic DNA. These DNA molecules and their integrity is very important for the survival and propagation of the cell. Proteins and RNA can be quickly replaced on damage by using information encoded in the DNA, but DNA molecules are irreplaceable and hence damages in them need to be repaired. DNA in a cell can be damaged by a variety of processes: some spontaneous, and others mediated by environmental agents. Occasionally, damage can occur due to error in DNA replication.

Different types of DNA damages include: base deletion or insertion, base substitution, single-strand and double-strand breaks, and inter-strand and intra-strand crosslinks. Base deletion, insertion, modification, and substitution would change the open reading frames, and if left uncorrected and replicated and transmitted to future cell generations, would introduce permanent changes in the nucleotide sequence of DNA, which are called *mutations*. If such a change occurs in non-coding DNA, it may be a silent mutation. If coding DNA is changed, these mutations can be deleterious. In mammals, there is a strong correlation between accumulation of mutations and cancer. However, a cell is armed with several DNA repair systems, which normally rectify these lesions and restitute the genome.

DNA damage can be induced by either chemical agents such as methyl methane sulphonate (MMS), ethyl methane sulphonate (EMS), Benz-o-pyrene, etc. or physical agents such as ionizing radiations (X-rays, γ-rays), ultraviolet radiations (UV rays), etc. Let us first understand a few different types of DNA damages induced by these agents (Fig. 7.1).

Alkylating agents such as methyl methane sulphonate (MMS) attack bases in the DNA and transfer methyl group onto them. Frequently, the sites attacked by MMS are N7 (nitrogen in 7th position) of guanine and N3 (nitrogen in 3rd position) of adenine. Different alkylating agents have different preferences for these targets. What is the consequence of methylation by MMS? N7 alkylation of guanine does not change the base pairing and is harmless, but N3 alkylation of adenine cannot base pair with any other base. As a result, DNA polymerase stalls at N3 adenine, unable to incorporate any base.

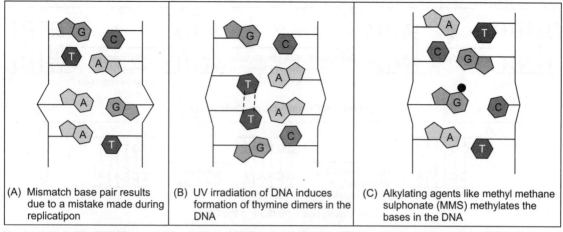

(A) Mismatch base pair results due to a mistake made during replicatipon

(B) UV irradiation of DNA induces formation of thymine dimers in the DNA

(C) Alkylating agents like methyl methane sulphonate (MMS) methylates the bases in the DNA

Figure 7.1   Different types of DNA damages. (A) Mismatch base pair when a wrong base is incorporated, (B) Intrastrand crosslink–Cyclobutane pyrimidine ring formed on UV irradiation, and (C) Base modification, methylation of a base by alkylating agent methyl methane sulphonate. **(See Colour Plate 12)**

All of the nitrogen and oxygen atoms involved in base pairing are also subjected to alkylation, which can disrupt the base pairing and lead to mutation. UV radiations induce different types of DNA damages among which pyrimidine dimers are formed with highest frequency followed by pyrimidine 6-4 pyrimidone. Adjacent pyrimidines on DNA strand are cross-linked forming pyrimidine dimers, which interrupt base pairing between the two DNA strands. Formation of pyrimidine 6-4 pyrimidone induces a lot of structural distortion as it involves cross-linking between 6th carbon of one pyrimidine and 4th carbon of adjacent pyrimidine in the DNA double helix. Both these lesions also block DNA replication.

The considerably more energetic gamma rays and X-rays can interact directly with the DNA molecule and modify the bases. They also ionize the molecules generating free radicals, which can interact with DNA and induce single- or double-strand breaks. Single-strand breaks are not serious, as they are easily repaired; but double-strand breaks pose a major threat, as the continuity of DNA molecule is lost. These therefore need to be repaired quickly. All these different types of damages, if left unrepaired, lead to induction of mutations in the DNA.

7.2 MULTIPLE REPAIR PATHWAYS

Every cell has several repair systems to cope up with the variety of damages that are induced in the DNA. The cell uses two basic strategies to do it. One is directly to revert the damage, the second is to remove the damaged section of DNA and fill it in with new undamaged DNA. Repair systems belonging to the first category are

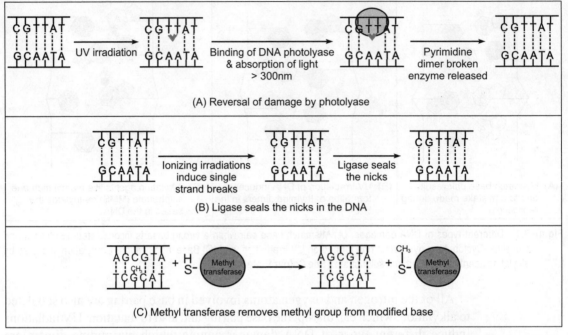

Figure 7.2 Single step repair systems. (A) Reversal of pyrimidine dimer by photolyase, (B) Single-strand breaks are sealed by ligase by forming a phosphodiester bond, and (C) Removal of methyl group from modified base by methyl transferase.

also referred to as single step repair systems. These include repair mediated by enzymes ligase, photolyase, and methyl transferase, etc. (Fig. 7.2). Single-strand breaks induced in the DNA are simply joined by the enzyme ligase. A phosphodiester bond is formed between 3'OH and 5'PO$_4$ termini separated by a nick. The enzyme photolyase recognizes the pyrimidine dimer and reverses the dimer using light as the source of energy. The enzyme binds to the dimer, absorbs light, which activates the enzyme to break the bond between two pyrimidines. This restores pyrimidines to their original independent state and the enzyme then dissociates from the DNA. Methyl transferase is a peculiar enzyme, which removes methyl group from modified bases. This enzyme accepts methyl group onto itself and gets inactivated. It is therefore referred to as the *suicidal enzyme*.

Methyl transferases are highly base-specific enzymes and are induced on exposure of the cells to alkylating agents. The proportion of DNA damages that can be repaired by these direct reversal systems is small. Therefore, other types of repair systems are necessary to rectify the damage in DNA. These are referred as multistep repair systems and involve removing the damaged part of DNA and replacing it with new undamaged DNA. Most important among these repair systems is *excision repair system*, either a base excision repair (BER) which repairs damage to the bases, or a nucleotide excision repair system (NER) which deals with a variety of DNA adducts, often distorting DNA double helix.

7.2.1 Nucleotide Excision Repair

Nucleotide excision repair (Fig. 7.3) system removes most of the bulky adducts from the DNA. The enzyme system is called as excinuclease complex or UvrABC complex, which has three proteins, UvrA, B, and C, participating in the repair

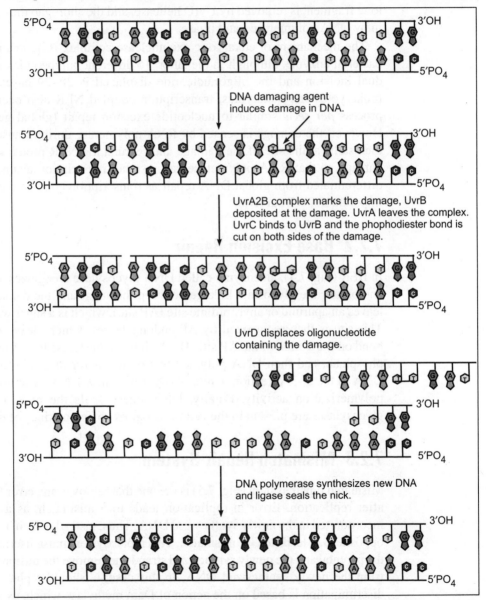

Figure 7.3 Nucleotide excision repair system in *E. coli*. Major DNA repair system in both prokaryotes and eukaryotes. Damage is recognized and removed by UVR ABC excinuclease complex. DNA polymerase and ligase replace the missing DNA together.

process. To begin with, $UvrA_2B$ scans the DNA and binds to the site of lesion. The UvrA dimer then dissociates leaving a tight UvrB-DNA complex, which is recognized by UvrC. On binding of UvrC to UvrB, UvrB makes incision at the 5th phosphodiester bond on the 3′ side followed by a UvrC mediated incision at the 8th phosphodiester bond on the 5′ side of the damage. The resulting 12–13 nucleotide long fragment is removed by UvrD helicase, and the short gap thus created is filled in by DNA polymerase I and ligase.

The mechanism of eukaryotic excinuclease complex is quite similar to that of bacterial enzyme, although in the latter, 16 different polypeptides are required for dual incision and the oligonucleotide displaced is 28–29 bases long. In both prokaryotes and eukaryotes, transcription coupled NER also occurs. The repair process *per se* is similar to nucleotide excision repair (global genomic repair). However, the damage is sensed by RNA polymerase during transcription, which then stalls at the site of the damage and recruits the NER proteins to remove the damage so that RNA polymerase can resume transcription again. This repair is differentiated from global DNA repair as transcription coupled repair and is very important for cell survival.

7.2.2 Base Excision Repair

In BER (Fig. 7.4), an enzyme called DNA glycosylase recognizes a damaged base and breaks the glycosidic linkage between the sugar and the damaged base. This leaves an apurinic or apyrimidinic site (AP site), which is a sugar without the base. This site is then recognized by AP endonucleases, which cleave phosphodiester bond on the 5′ side of the AP site. DNA phosphodiesterase then removes AP sugar phosphate, and then DNA polymerase I performs repair synthesis by degrading DNA in the 5′-3′ direction, while filling it with new DNA by template-dependent polymerization activity. Finally, DNA ligase seals the nick. Different DNA glycosylases are present in the cell to recognize different kinds of damaged bases.

7.2.3 Mismatch Repair System

Mismatch repair system (Fig. 7.5) is the one that removes any error left in the DNA after replication. Error in replication leads to a mismatch, as a wrong base is incorporated in the newly synthesized DNA. Thus, instead of C, if T is incorporated opposite G, it leads to G:T mismatch. This repair system must discriminate between the parental and the newly synthesized strand and remove the mismatch specifically from the newly synthesized DNA. In this case, C should replace T. The strand discrimination is based on the action of Dam methylase which, as you will recall, methylates N^6 position of all adenines in the sequence GATC in the newly synthesized DNA. The transient unmethylated state of the newly synthesized strand permits the new strand to be distinguished from the parental strand.

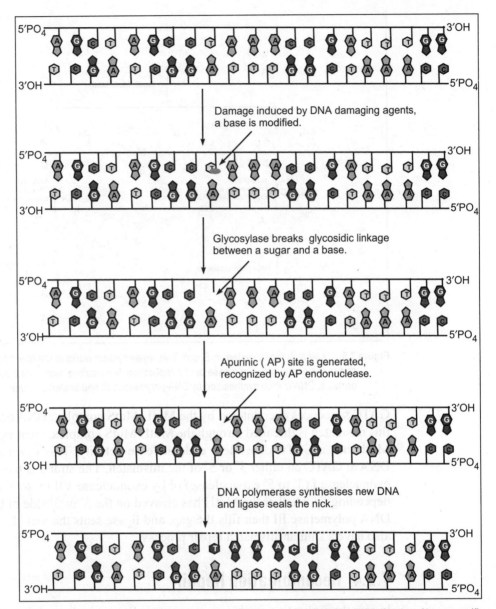

Figure 7.4 Base excision repair system in *E. coli*. Repair is catalyzed by concerted action of base-specific glycosylase, AP endonuclease, exonuclease, DNA polymerase, and ligase.

Replication mismatches in the vicinity of unmethylated GATC sequences are repaired using the information in the methylated parent strand.

Mismatch repair system includes around 12 proteins that function either in strand discrimination or in the repair process itself. MutL protein forms a complex with MutS, and this complex binds to mismatches. MutH protein binds to MutL and to

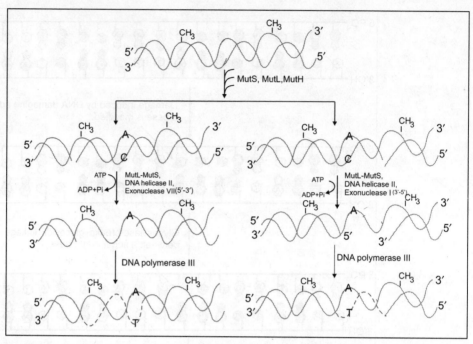

Figure 7.5 Mismatch repair system in *E. coli*. This repair system works to remove the damage resulting due to misincorporation of a base during replication. Mismatch is recognized and removed by Mut proteins. DNA is then synthesized by DNA polymerase III and sealed by ligase.

GATC sequences encountered by the MutL-MutS complex. DNA on both sides of this mismatch is threaded through the MutL-MutS complex, creating a DNA loop. MutH has a GATC-dependent endonuclease activity. This enzyme cleaves the DNA at GATC on either 3′ or 5′ of the mismatch. The strand is then degraded by exonuclease I (3′ to 5′ exonuclease) or by exonuclease VII (5′ to 3′ exonuclease), depending on whether the MutH has cleaved on the 3′ or 5′ side of the mismatch. DNA polymerase III then fills the gap, and ligase seals the nick. Eukaryotic cells also have a similar mismatch repair process.

7.2.4 Recombination Repair

In case the replicating machinery encounters damage in the template, the replication fork stalls and then either of the two following events can happen.

DNA polymerase leaves a gap in the daughter strand as it fails to incorporate correct base opposite the damage. This gap needs to be filled; it is done by *daughter strand gap repair* or *recombination repair*. This recombination repair will be explained later in recombination section.

A second option is to follow error prone repair pathway called *translesion synthesis*. This is carried out by DNA polymearse V and involves improper base pairing and therefore is error prone. Though error prone and therefore mutagenic, the cell chooses this pathway under a desperate situation. When the damage in DNA is extensive and the replication forks are stalled, translesion synthesis allows a few mutant cells to survive. Normally, this repair pathway would not be functional and is used only under critical situation by the cells. Eukaryotic cells also have a polymerase for translesion synthesis.

7.2.5 Double-Strand Break Repair

A pathway called *double-strand break repair* or *non-homologous end joining* repairs double-strand breaks in the DNA (Fig. 7.6). The key components of this system are DNA–PK (DNA protein kinase) and ligase. DNA-PK is a heterodimer of DNA targeting subunit Ku and catalytic subunit DNA-PKcs. DNA-PK binds to DNA ends and protects them from exonucleolytic attack, aligns them to facilitate

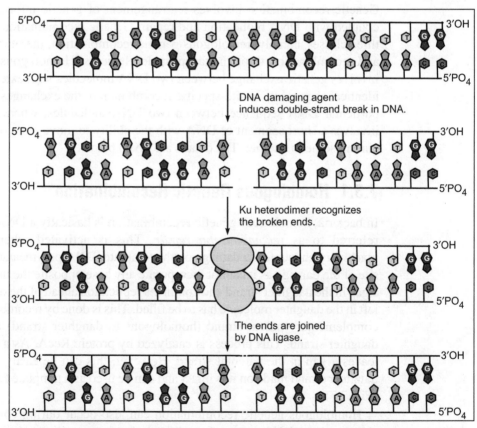

Figure 7.6 Double-strand break repair system in *E. coli*. Double-strand breaks are recognized by Ku heterodimer and broken ends are ligated together by ligase.

their ligation, and also helps recruit ligase. Ligase then forms the phosphodiester bond between 3′OH and 5′PO$_4$ termini and seals the nicks in both the strands.

Defects in DNA repair system makes most of the cells highly prone to mutagenesis and carcinogenesis on exposure to DNA damaging agents. In humans, defects in different DNA repair systems lead to development of different syndromes. Defective nucleotide excision repair system results in a recessive disorder called *Xeroderma pigmentosum*. Individuals with this disorder are hypersensitive to sunlight, especially UV, and are highly prone to development of skin tumors. On the other hand, people with defect in transcription coupled repair suffer from *Cockayne syndrome*, whereas those with defect in double-strand break repair suffer from *Nijmejan breakage syndrome*.

7.3 DNA RECOMBINATION

Genetic recombination involves rearrangement of genetic information, and it is coupled with DNA replication and repair. Genetic recombination events fall into three classes, namely homologous genetic recombination, meiotic recombination, and recombination through transposable elements. Homologous recombination involves genetic exchange between two DNA molecules with extended regions of identical sequences. In site-specific recombination, the exchanges occur only at a particular DNA sequence between two DNA molecules, whereas transposition involves a short segment of DNA with the ability to move from one location in a chromosome to another. The details are given below.

7.3.1 Homologous Genetic Recombination

In bacteria, homologous genetic recombination is basically a DNA repair process referred to as *recombination repair*. This is activated when a replicating machinery encounters a damage on the template. DNA polymearse III stalls at the site of the damage as it cannot incorporate any base opposite the damage. It leaves a gap in the daughter strand and continues replication ahead of the damage. The gap left in the daughter molecule has to be filled. This is done by recombination between complementary parent strand (homologous to daughter strand) and the gapped daughter strand. This process is catalyzed by protein RecA. As a consequence, it leaves a gap in the parental strand. However, this can be easily repaired, as the daughter strand made on it is intact and can be used as a template for filling the gap (Fig. 7.7).

Homologous genetic recombination can also occur during conjugation, when chromosomal DNA is transferred from donor to recipient during conjugation, although it happens very rarely in natural populations. Recombination during conjugation contributes to genetic diversity.

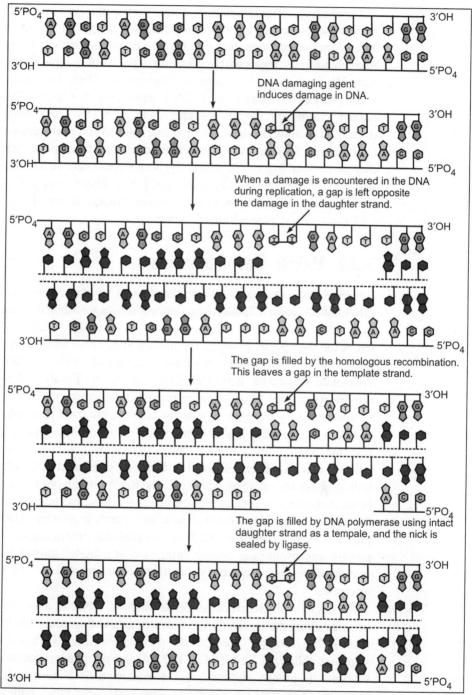

Figure 7.7 Recombination repair in *E. coli*. This system is activated when the damage is encountered during replication. Gap in the daughter strand is repaired by homologous recombination mediated by RecA.

In eukaryotes, homologous recombination is involved in DNA repair like in case of bacteria, however, recombination occurs with highest frequency during meiosis. Meiosis involves production of haploid gametes, sperm or ova, from diploid germ line cell. Meiosis begins with replication of the DNA in diploid germ line cell, so that each DNA molecule is present in four copies. Then, the cell goes through two rounds of cell division without DNA replication in between. This reduces DNA content to a single molecule in each gamete. After DNA replication during prophase of the first meiotic division, the resulting sister chromatids remain associated at the centromeres. At this stage, the genetic exchange takes place between the closely associated homologous chromatids by homologous genetic recombination, a process involving the breakage and rejoining of DNA. This exchange is also referred to as *crossing over*. This is not entirely a random process, though it can occur with equal probability at any point along the chromosome.

7.3.2 Meiotic Recombination

Recombination during meiosis initiates when there is a double-strand break in the DNA molecule. First, homologous chromosomes are aligned. A double-strand break in a DNA molecule is enlarged by an exonuclease, leaving a single-stranded extension with 3′OH group at the broken end. This end invades the intact duplex DNA, and it is followed by branch migration and replication that creates a pair of crossover structures, called *Holliday junctions*. Finally, cleavage of the two crossovers creates two complete recombinant products (Fig. 7.8). The detailed process can vary from one species to another, but most steps are common.

When the Holliday junctions are resolved, recombinant products are generated. Enzymes that promote various steps of homologous recombination have been isolated and characterized from both prokaryotes and eukaryotes. RecBCD has both helicase and nuclease activities. The RecA protein promotes all the major steps in the homologous recombination process. These include pairing of two DNAs, formation of Holliday intermediates, and branch migration. The RuvA and RuvB proteins form a complex that binds to Holliday intermediates, displacing RecA protein and promoting branch migration at a higher rate than RecA. Holliday intermediates are resolved by nucleases called resolvases; RuvC is one of these nucleases.

To begin with, RecBCD enzyme binds to DNA at a broken end and moves inwards along the double helix, unwinding and degrading the DNA using energy from ATP hydrolysis. The activity of enzyme is altered when it interacts with the sequence called *chi* 5′GCTGGTGG3′. At this point, degradation of strand with a 3′ terminus is greatly reduced, but degradation of 5′ terminal strand is greatly increased. This process generates a single-stranded DNA with a 3′ end, which is used during subsequent steps in recombination. RecA protein assembles cooperatively on DNA in a large number.

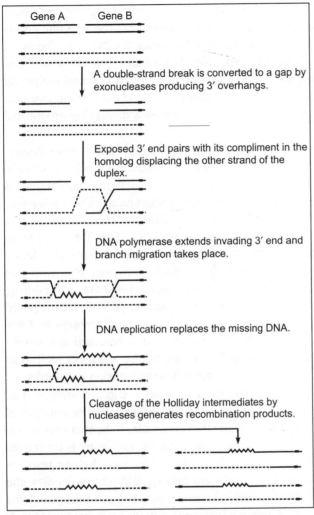

Gene A   Gene B

A double-strand break is converted to a gap by exonucleases producing 3′ overhangs.

Exposed 3′ end pairs with its compliment in the homolog displacing the other strand of the duplex.

DNA polymerase extends invading 3′ end and branch migration takes place.

DNA replication replaces the missing DNA.

Cleavage of the Holliday intermediates by nucleases generates recombination products.

Figure 7.8 Recombination during meiosis. Homologous recombination in meosis proceeds by double-strand break repair pathway.

Thousands of monomers bind to single-stranded DNA, formed by the action of RecBCD enzyme. RecF, RecO, and RecR proteins regulate the assembly and disassembly of RecA filaments on DNA. RecA binds to single-stranded DNA to form a nucleoprotein complex. This complex takes up a homologous duplex DNA and aligns it with the bound single-strand. Strands are then exchanged between two DNAs to create hybrid DNA. The exchange occurs at the rate of 6 bp/second and progresses in the direction 5′-3′ relative to the single-stranded DNA within the RecA filament. At this stage, Holliday intermediates are formed.

When the duplex DNA is incorporated within the RecA filament and aligned with bound single-stranded DNA over regions of several base pairs, one strand of the duplex switches the pairing partners. Because DNA is a helical structure, continued strand exchange requires an ordered rotation of the two aligned DNAs. Once a Holliday junction is formed, a host of enzymes, topoisomerases, the RuvAB branch migration protein, a resolvase, other nucleases, DNA polymerase I and III, and DNA ligase are required to complete the recombination. The RuvC protein cleaves Holliday intermediates to generate full length unbranched chromosome products.

Although the Holliday model for homologous recombination explains crossing over during meiotic recombination, another model called *double-strand break model* for homologous recombination has been proposed, especially to explain the phenomenon of gene conversion. According to this model, homologous recombination initiates with a double-strand cut that breaks one of the partners in the recombination into two pieces, one strand in each half of the molecule is shortened so that each end has a 3′ overhang (Fig. 7.8). One of these overhangs invades the homologous DNA molecule, setting up a Holliday junction which can migrate along the heteroduplex if the invading strand is extended by a DNA polymerase. To complete the heteroduplex, the other broken strand, not involved in Holliday junction, is also extended. Both DNA syntheses involve extension of strands from the partner that had a first double-strand cut, using the equivalent regions of the uncut partner as templates. This means that the polynucleotide segment removed from the cut partner is replaced by copies of DNA from the uncut partner. After ligation, the resulting heteroduplex has a pair of Holliday structure which can be resolved, leading to gene conversion or giving a standard reciprocal strand exchange. Thus, during meiotic recombination, gene conversion can take place which would result in 4 haploid products of meiosis displaying an unusual segregation pattern.

Site-specific recombination results in precise DNA rearrangements: homologous genetic recombination can involve any two homologous sequences, but site-specific recombination is limited to specific sequences. Each site-specific recombination system consists of an enzyme called *recombinase* and a short unique DNA sequence. A specific recombinase recognizes and binds to each of the two recombination sites on two different DNA molecules or within the same DNA. One DNA strand in each site is cleaved at a specific point and the recombinase becomes covalently linked to DNA at the cleavage site through a phosphotyrosine. This transient protein-DNA linkage preserves high-energy phosphodiester bond that is lost during cleavage of the DNA so that high-energy molecules such as ATP are not required subsequently. The cleaved DNA strands are rejoined to new partners to form Holliday intermediates. To complete, the reaction process is repeated at a second point within each of two recombination sites. The exchange is always reciprocal and precise, regenerating the recombination sites when the reaction is complete.

Recombinase works as a site-specific nuclease and a ligase together. The outcome of site-specific recombination depends on the location and orientation of the recombination sites. In a double-stranded DNA molecule, if the two sites are on the same DNA molecule and are in opposite orientation, the end result is inversion. On the other hand, if the two sites are in the same orientation, the end result is deletion of the intervening DNA. If the sites are on two different molecules, the recombination is intermolecular and results in insertion. A very good example of this kind of insertion is bacteriophage λ, when it lysogenises in the host chromosome. A recombinase from bacteriophage λ called *integrase* acts at recombination sites on the phage and the bacterial DNAs called attP and attB, respectively, and brings about insertion of phage DNA into the bacterial chromosome.

7.3.3 Other Recombination Events

The third type of recombination is the one that allows the movement of transposable elements or transposons (Fig. 7.9). These segments of DNA found in virtually all cells move or jump from one place on a chromosome to another on the same or different chromosomes. Sequence homology is not required for this movement. The new location is determined more or less randomly. Two types of transposons are: *insertion sequences* and *complex transposons*. Insertion sequences have only

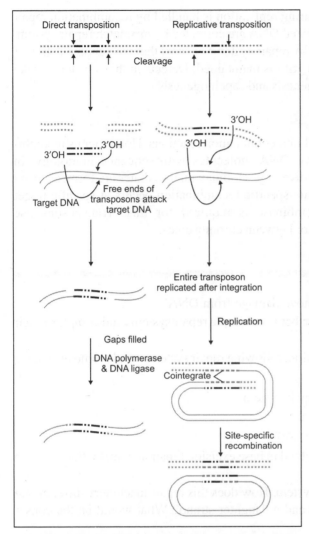

Figure 7.9 Recombination by transposition. It occurs either by direct insertion of transposon in target DNA or by replicative mode, wherein transposon is replicated and then inserted in the target DNA.

sequences required for transposition, and the gene for the enzyme transposes that promotes transposition. Complex transposons contain one or more genes in addition to those needed for transposition.

There are two general pathways for transposition in bacteria. In direct or simple transposition, cuts on each side of transposon excise it, and the transposon moves to a new location. This leaves a double-strand break in the donor DNA that must be repaired. At the target site, a staggered cut is made, the transposon is inserted into the break. Then, DNA replication fills in the gaps to duplicate the target site sequence. In replicative transposition, the entire transposon is replicated, leaving a copy behind at the donor location.

Additionally, two recombination events are known. In one, recombination events lead to rearrangement to create new genes which are required to be expressed in particular circumstances as in case of immunoglobulins. Even for switching expression of pre-exisiting gene to another, rearrangement is involved, e.g. Yeast mating type genes. Another type of recombination is used by RNA viruses called the *copy choice mechanism*. In this, RNA polymerase switches from one template to another, while synthesizing RNA. This results in synthesis of new RNA molecule containing sequence information from two different templates.

SUMMARY

DNA damage and repair

Several DNA repair processes exist, both in prokaryotes and eukaryotes, to protect their DNA from any kind of damage. The most important among them is excision repair system (BER and NER), which rectifies most of the damages from the

DNA. Damage encountered during replication is handled by recombination repair system, and errors left in replicated DNA are corrected by mismatch repair system. Thus, the vast repertoire of DNA repair systems ensures that the genome is free of any DNA alteration and its iltegrity is maintained. Defects in these systems make the cells highly prone to mutagenesis and capcinogenesis.

Recombination

DNA sequences are rearranged in recombination reactions. Homologous recombination can occur between two DNA molecules with sequence homology. In meiosis, this generates genetic diversity. Holliday intermediates are formed during homologous recombination. Site-specific recombination occurs at specific target sequences and specific recombinase is required for this. Transposons use recombination to move within or between chromosomes.

SHORT ANSWER QUESTIONS

1. Why is it necessary to remove damage from DNA?
2. What is the basic difference between simple repair systems and complex repair systems?
3. Under which specific situation, following repair systems would remove damage from DNA?
   (a) Nucleotide excision repair system
   (b) Mismatch repair system
   (c) Recombination repair system
4. Differentiate the steps involved in base excision repair and nucleotide excision repair in *E. coli*.
5. Explain mismatch repair system. How does this repair machinery differentiate between a parental strand and a daughter strand? What would be the consequence if it fails to do so?
6. Why is recombination necessary?
7. How does formation of Holliday intermediates in homologous recombination differ from site-specific recombination?
8. List the important steps in homologous recombination in which RecA participates.
9. How is the transposon-mediated recombination different from site-specific recombination?
10. When is the process of recombination repair activated? Explain the steps involved.

FURTHER READING

- Barnes, D.E. and Lindall, T. (2004). Repair and genetic consequences of endogenous DNA base damage in mammalian cells. *Annu. Rev.Genet.* **38**, 445–476.
- Cohen, S.Ṅ. and Shapiro, J.A. (1980). Transposable genetic elements. *Scientific American* **242**, 240–249.
- Gellert, M. (2002). V (D) J recombination: RAG proteins, repair factors and regulation. *Annu. Rev. Biochem.* **71**, 101–132.
- Haber, J.E. (2004). Partners and Pathways: repairing a double strand break. *Trends Genet.* **16**, 259–264.
- Haren, L., Ton-Hoang, B., and Chandler, M. (1999). Integrating DNA: Transposases and retroviral integrases. *Annu. Rev. Microbiol.* **53**, 245–281.
- Kunkal, T.A. and Erie, D.A. (2005). DNA mismatch repair. *Annu. Rev.Biochem* **74**, 681–710.
- Lindahl, T. and Wood, R.D. (1999). Quality control by DNA repair. *Science* **286**, 1897–1905.
- Lusetti, S.L. and Cox, M.M. (2002). The bacterial RecA protein and the recombinational DNA repair of stalled replication forks. *Annu. Rev.Biochem* **71**, 71–100.
- McCullough, A.K., Doddson, L.L., and Lloyd, R.S. (1999). Initiation of base excision repair. Glycosylase mechanism and structures. *Annu. Rev. Biochem.* **68**, 255–286.
- McGlann, P. and Lloyd, R.G. (2002). Recombinational repair and restart of damaged replication forks. *Nat. Rev. Mol.Cell. Biol.* **3**, 859–870.
- Sankar, A. (1996). DNA excision repair. *Annu. Rev.Biochem* **65**, 43–81.
- Sankar, A., Lindsey-Boltz, L.A., Unsal-Kacmaz, K., and Linn, S. (2004). Molecular mechanisms of mammalian DNA repair and DNA damage checkpoints. *Annu. Rev.Biochem* **73**, 39–85.
- Wood, R.D., Mitchell, M., Sgouros, J., and Lindall, T. (2001). Human DNA repair genes. *Science* **291**, 1284–1289.
- Yang, W. (2003). Damage repair DNA polymerases. *Curr. Opin.Struct. Biol.* **13**, 23–30.

8

OVERVIEW

- Transcription is the first step in gene expression and it is a vital control point in the expression of several genes. This chapter initially describes the transcriptional machinery in prokaryotes and eukaryotes focusing on RNA polymerase, the enzyme that catalyzes transcription, and moves on to explain the process of transcription in detail.

- The process of transcription is finely regulated, implying that some genes are turned on or off depending on the need of their products.

- In prokaryotes, as RNA polymerase transcribes the genes, ribosomes bind to mRNA and translate it to protein. However, in eukaryotes, the compartments in which transcription and translation occur are different. Transcription occurs in nucleus, whereas translation takes place in cytoplasm. Therefore, unlike in case of prokaryotes, transcription and translation are not coupled in eukaryotes.

- The transcripts are excessively processed in eukaryotes before they are translated. This chapter will make you understand the processing of all the three types of transcripts, namely rRNA, tRNA, and mRNA at both the 3′ and 5′ end and also along the length so as to generate mature transcripts.

The first step of gene expression is transcription, and RNA polymerase plays a major role in transcription. Three different types of RNA molecules are synthesized during transcription, namely messenger RNA (mRNA), ribosomal RNA (rRNA), and transfer RNA (tRNA). In prokaryotes, a single RNA polymerase transcribes all the three RNAs, whereas in eukaryotes, three different RNA polymerases, namely Polymerase I, II, and III are involved in the synthesis of three different RNA molecules.

Transcription process involves three different steps: *initiation*, *elongation*, and *termination*. Initiation begins at a specific sequence called the *promoter*. Every type of RNA polymerase recognizes a different promoter for initiation. In eukaryotes, additionally, initiation proteins called *transcription factors* are required to be assembled at the promoter sequence before RNA polymerase can initiate the event. During elongation, RNA polymerase synthesizes a polyribonucleotide using DNA as the template. In prokaryotes, transcription is terminated either by rho-dependent or rho-independent manner, whereas in eukaryotes, the three polymerases have different modes of termination of transcription. In prokaryotes, the mature rRNA and tRNA are produced from precursor RNA molecules, while mRNA is immediately used for protein synthesis. In eukaryotes, all the three RNAs are synthesized as precursor RNAs and undergo excessive processing at both the ends and along the length.

Transcription is highly regulated in both prokaryotes and eukaryotes. This means that all genes are not transcriptionally on all the time, but are switched on by the cell whenever it is required. This regulation is achieved by specific DNA–protein interaction. Eukaryotic cells have additional gene-specific transcription factors or activators, which boost up and control transcription. Additionally, eukaryotic genes also exist as chromatin, where the DNA is coiled around the protein complex. To make genes in chromatin accessible for transcription, chromatin remodelling is necessary, which is achieved by specific proteins.

Transcription is the first step in gene expression and it is a major control point in the expression of several genes. As discussed earlier, RNA polymerases play a major role in transcription, and these enzymes were discovered, as early as around 1960 from bacteria, animals, and plants. Subunit structure of these enzymes varies substantially in prokaryotes and eukaryotes, though the catalytic reaction catalyzed in both is the same, i.e. synthesis of polyriobonucleotide (RNA) chain in a template-dependent manner. RNA synthesis always occurs in a fixed direction, from 5' to 3' end of the RNA molecule. Only one of the strands of DNA is transcribed into RNA. The DNA strand which has the same sequence as the RNA is called *coding* or *sense strand*, while the other which is used as a template by the RNA polymerase is called *antisense* or *non-coding strand*.

Three main types of RNA molecules are synthesized during transcription, namely messenger RNA (mRNA), ribosomal RNA (rRNA), and transfer RNA (tRNA). mRNA is translated into protein by translation assembly, rRNA is a major constituent of ribosomes involved in protein synthesis, and tRNA participates in translation process. In addition, especially in eukaryotic cells, many other non-coding RNAs, such as snRNA, snoRNA, gRNA, miRNA are synthesized by transcription, and they participate in processing of the transcripts and regulating the stability of RNA. In prokaryotes, the transcripts when synthesized are immediately used for translation, whereas in eukaryotes, all the transcripts are processed extensively before they become functional.

8.1 TRANSCRIPTION MACHINERY IN PROKARYOTES

The *E. coli* RNA polymerase is one of the largest enzymes in the cell. It is composed of a *core*, which contains the basic transcription machinery, and a *sigma* (σ) *factor*, which directs the core to transcribe specific genes. The core enzyme has β and β′ subunits of MW 150 and 160 kDa, respectively, and two α subunits of 40 kDa. The sigma factor joins the core to make the *holoenzyme* (Fig. 8.1A), which efficiently transcribes RNA from DNA.

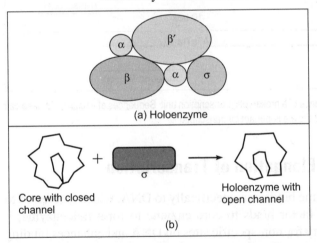

(a) Holoenzyme

Core with closed channel

σ

Holoenzyme with open channel

(b)

Figure 8.1 RNA polymerase: **(a)** Holoenzyme subunits; **(b)** Core polymerase on combining with sigma changes its conformation.

For the initiation of transcription, RNA polymerase binds to a specific site on DNA called the *promoter sequence*. The σ factor allows initiation of transcription by causing the holoenzyme RNA polymerase to bind tightly to a specific promoter. Core enzyme has an affinity for binding to DNA non-specifically, however σ factor confers specificity of binding tightly to promoter elements. Only σ factor can therefore select which genes will be transcribed. After sigma has participated in the initiation, it leaves the core to carry on with the elongation process. Differential binding to DNA by core and holoenzyme is due to the change in the shape induced by binding of σ to the core. The core polymerase resembles the right hand with the thumb and fingers just touching at the upper left corner to form a circle. This produces a channel of about 25 Å, just enough to accommodate a double helical DNA. In the holoenzyme, by contrast, the fingers and the thumb do not touch, leaving the channel open (Fig. 8.1B).

Because of the closed channel of core polymerase, it cannot get around the DNA to grasp it, instead the outside of the hand simply binds weakly and non-specifically to DNA. In contrast, the open channel holoenzyme can grasp the DNA and bind tightly. On initiating the transcription, sigma dissociates from holoenzyme leaving the core firmly clamped around the DNA so that it can transcribe the DNA processively. RNA polymerase binds to a specific site upstream of the start point of transcription, generally assigned as position +1. In accordance with this, promoter sequences are assigned a negative number reflecting the distance upstream from the start of transcription. By comparing several promoters from *E. coli* and phages, two regions were found to be common to all the promoters. One was a sequence of 6 to 7 base pairs, centered approximately 10 bp upstream of the start of transcription called *–10 box*. The second was also a short sequence, centered approximately 35 bp upstream of the transcription start site known as *–35 box*. Both –10 and –35 sequences are recognized by σ factor. (Fig. 8.2).

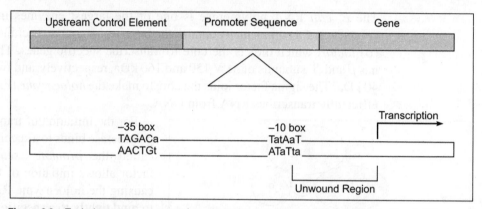

Figure 8.2 Typical promoter sequence of a prokaryotic transcription unit. Sequences at –10 and –35 base pair upstream of transcription start site are important for transcription.

8.1.1 Initiation and Elongation of Transcription

RNA polymerase core enzyme binds non-specifically to DNA, which is referred to as *loose binding*. When σ factor binds to core enzyme to form holoenzyme, it markedly reduces the affinity for non-specific sites on DNA and enhances binding to correct promoter binding sites. For the antisense strand to become accessible for base pairing, the DNA duplex must be unwound by the polymerase. The initial unwinding of DNA results in the formation of an *open complex* with the polymerase. Polymerase initially incorporates the first two nucleotides and forms a phosphodiester bond between them. First nine bases are added without the enzyme movement along the DNA. At this stage, there is a possibility that the chain would be aborted. This process of *abortive initiation* is very important for the overall rate of transcription, since it has a major role in determining how long the polymerase takes to leave the promoter (*promoter clearance*) and allows another RNA polymerase to initiate a further round of transcription (Fig. 8. 3).

When initiation succeeds, the enzyme releases the σ factor and forms a *ternary complex* of RNA polymerase-DNA and newly synthesized (nascent) RNA chain.

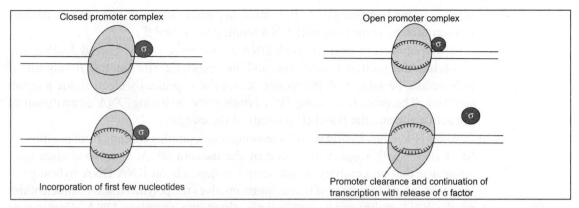

Figure 8.3 Initiation of transcription. It proceeds via formation of a closed complex and an open complex. Once the transcription begins, sigma factor dissociates and RNA polymerase moves ahead clearing the promoter.

Polymerase then steadily progresses along the DNA, allowing *reinitiation of transcription* from the promoter by another RNA polymerase holoenzyme. The enzyme takes around 1-2 seconds to clear the promoter. The region of unwound DNA called *transcription bubble* appears to move along the DNA with the polymerase. The size of the transcription bubble remains constant at around 17 base pairs. Growing RNA chain forms a hybrid of about 12 bases with template DNA strand. Enzyme moves at an average rate of 40 nucleotides per minute, however, this can vary depending on local DNA sequences. To maintain the transcription bubble, DNA is unwound ahead of the bubble and is wound again behind the bubble.

During elongation, enzyme would continuously go on adding ribonucleotides to the growing RNA chain in a template-dependent manner. α subunit coded by gene rpoA serves to assemble RNA polymerase holoenzyme. It has two distinct domains, N-terminal domain and C-terminal domain, which fold independently and are tethered together by a flexible linker (Fig. 8.4). N-terminal domain is involved in assembly of holoenzyme, whereas C-terminal domain recognizes and binds to 'up elements' in the promoter stimulating transcription. β subunit coded by gene rpoB binds nucleotides at the active site of the RNA polymerase where phosphodiester bonds are formed. This subunit is the *catalytic center* of RNA polymerase.

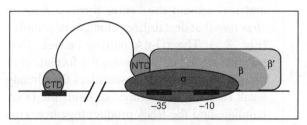

Figure 8.4 The α subunit of RNA polymerase has two distinct domains, namely N-terminal (NTD) and C-terminal domain (CTD) which fold independently and are tethered together by a flexible linker.

Antibiotics *rifampicin* and *streptolydigins*, which inhibit transcription initiation and elongation, respectively, bind to β subunit. Another subunit β' present in the core enzyme and encoded by gene rpoC binds two Zn^{+2} ions and participates in catalytic function of the enzyme. This subunit is responsible for binding to the template DNA. Polyanion heparin binds to β' subunit and inhibits transcription by competing with DNA binding site of the β'.

RNA polymerase interacts with DNA at two sites. One is a *weak binding site* involving the melted DNA zone and the catalytic site on the β subunit of polymerase. Protein DNA interaction at this site is primarily electrostatic and salt sensitive. The other is a *strong DNA binding* site involving DNA downstream of the active site and the β and β' subunits of the enzyme.

The RNA–DNA hybrid within the elongation complex extends from position –1 to –8 or –9 with respect to 3' end of the nascent RNA. *In vitro* studies have suggested that processivity of transcription depends on RNA-DNA hybrid of at least 9 bp long. Elongation of transcription involves the polymerization of nucleotides as the RNA polymerase core travels along the template DNA. Continuous movement on the template DNA requires that the DNA ahead of advancing polymerase is unwound and is wound again once the transcription is over. This introduces a strain into the template DNA that is relaxed by *topoisomerases*.

8.1.2 Termination of Transcription

When the polymerase reaches the terminator at the end of the template, it falls off the template terminating the transcription. Two kinds of terminators are identified in *E. coli*. One is called as *intrinsic terminator* which functions without any help from other proteins, whereas the other is called as *rho-dependent terminator*, requiring the factor 'rho' for termination of transcription.

Intrinsic terminator contains two elements, an inverted repeat followed immediately by a T-rich region in the non-template strand of the gene. Inverted repeat by virtue of being symmetrical around the center forms a *hairpin structure* that destabilizes elongation complex (Fig. 8.5). The rU-dA pairing causes RNA polymerase to stall, allowing the formation of hairpin which further destabilizes the already weak rU-dA base pairing that holds the DNA template and the RNA product together. This destabilization results in the dissociation of RNA from its template, terminating transcription.

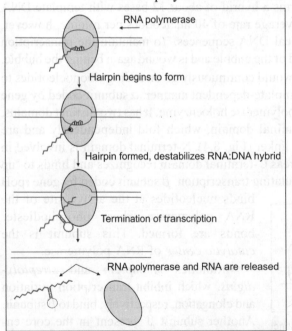

RNA polymerase

Hairpin begins to form

Hairpin formed, destabilizes RNA:DNA hybrid

Termination of transcription

RNA polymerase and RNA are released

Figure 8.5 Rho-independent termination of transcription. Termination of transcription is influenced by formation of a hairpin structure in the transcript.

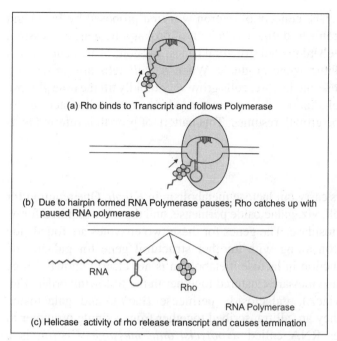

(a) Rho binds to Transcript and follows Polymerase

(b) Due to hairpin formed RNA Polymerase pauses; Rho catches up with paused RNA polymerase

RNA

Rho

RNA Polymerase

(c) Helicase activity of rho release transcript and causes termination

Figure 8.6 Rho-dependent termination of transcription. This mode of termination of transcription makes use of the factor rho which, with its helicase activity, unwinds RNA-DNA hybrid helping termination of transcription.

On the other hand, rho-dependent terminator consists of an inverted repeat, which causes hairpin formation in the transcript but no string of T's. Rho binds to a growing transcript (Fig. 8.6) upstream of termination site at a rho-loading site, a sequence of 60-100 nucleotides rich in cytosine residues. Rho consists of a hexamer of identical subunits, each of which has an ATPase activity. Binding of rho to the RNA activates the ATPase which supplies energy for rho to propel the hexamer in a 5'-3' direction following RNA polymerase. The chase continues till RNA polymerase stalls at the terminator. On encountering stalled RNA polymerase, rho unwinds the RNA-DNA hybrid within the transcription bubble, releasing the RNA and terminating the transcription.

8.1.3 Regulation of Transcription

A prokaryotic genome contains around 3000 genes. Some of them are active all the time, since their products are required by the cell constantly. Others are activated only when their products are in demand and are therefore turned off most of the time. For example, the enzymes involved in metabolism of lactose would be present only when lactose is available as the sole carbon source, and the most favourite energy source 'glucose' is absent. Therefore, the genes coding for lactose metabolizing enzymes are mostly turned off and are switched on only when necessary. Cell cannot afford to keep all its genes active all the time simply because synthesis of RNA, storage of RNA, and its translation are energy-demanding processes. In case a cell turns on all its genes all the time, it would be drained of its energy such that it would not be able to compete with more efficient organisms. Thus, control of gene expression, which involves controlling the time, space, and amount of transcription being performed, is essential for life. Different strategies are being used by the cell to achieve this differential regulation of gene expression.

Concept of Operon

One of the strategies is to group functionally related genes together so that they are regulated together. Such a group of contiguous coordinately controlled genes is referred to as an *operon*. The concept of operon was first proposed by Jacob and Monod in 1961. An operon was defined as an unit of prokaryotic gene expression, which includes coordinately regulated structural genes, and control elements, which are recognized by regulatory gene products. When *E. coli* cells are grown in a medium containing glucose and lactose, cells grow very rapidly till the time glucose gets exhausted. Then, cells induce the enzymes needed to metabolise lactose. As those enzymes appear, the growth resumes. This pattern of growth is referred to as *diauxic growth*.

Inducible Operon

Principally, two enzymes carry out lactose metabolism in *E. coli*. One required for its transport inside the cell, viz. galactoside permease, and the other required for its breakdown, viz. β-galactosidase. The genes for these two enzymes are found side by side in the lac operon along with another structural gene for galactoside transacetylase whose function in lactose metabolism is not clearly known. Three genes coding for three enzymes are clustered together in the following order (Fig. 8.7): β-galactosidase (lacZ), galactoside permease (lacY), and galactoside transacetylase (lacA). They are all transcribed together from a single promoter to produce one messenger RNA called a *polycistronic message*. Cistron is a synonym for a gene. Therefore, a polycistronic message is simply a message with information from more than one gene. Each cistron has its own ribosome binding site so that each can be translated independent of the other. Thus, three genes coding for three enzymes of lactose metabolism constitute the structural genes of the operon. Regulatory genes involve control elements, such as an *operator sequence* situated close to the promoter and a gene lacI whose product recognizes the control element.

The product of gene lacI is called as *repressor* molecule. Under normal conditions, the repressor is bound at the operator, thereby, repressing the operon. Only when the lactose present in the medium is to be used, the repression is lifted by removing the repressor bound at the operator. Thus, when repressor is bound at the operator, the operon is turned off; and when it is removed from the operator, it is turned on. This kind of regulation is called as *negative regulation*. It implies that the operon will be on when a regulator is physically not present on the DNA (Fig. 8.7).

The lacI gene encodes the lac repressor, which is active as a tetramer of identical subunits. It has a very strong affinity for the lac-operator binding site. It also has a generally high affinity for DNA. In the absence of lactose, repressor occupies the operator-binding site. Promoter from where the transcription begins is adjacent to the operator site. Both the lac repressor and the RNA polymerase can bind simultaneously to the lac operator and promoter, respectively. Two competing hypotheses seek to explain the mechanism of repression of lactose operon. One is that the RNA polymerase can bind to the promoter in presence of repressor bound

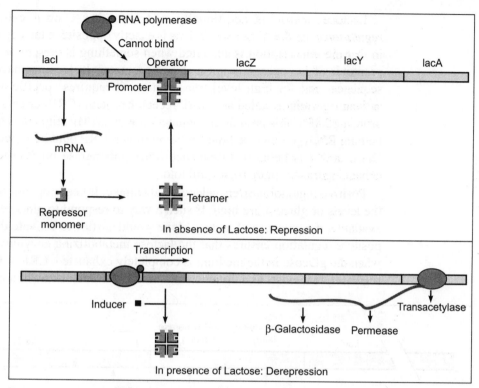

Figure 8.7 Negative regulation of *lac* operon. *Lac* operon is negatively regulated by a repressor molecule, which binds to the operator and keeps the operon turned off. In presence of lactose, repressor is inactivated and the operon is turned on.

at the operator, but the repressor inhibits transition from abortive transcription to processive transcription. The other hypothesis is that the repressor, by binding to operator, denies access to the RNA polymerase to the promoter. In absence of lactose, repressor blocks all but a very low level of transcription of lac operon. When lactose is added to the cells, the low basal level of permease allows the uptake of lactose, and β-galactosidase catalyzes the conversion of lactose to allolactose, which acts as an inducer.

Allolactose, by binding to the repressor, changes its conformation reducing its affinity for the lac operator. Removal of lac repressor from the operator allows RNA polymerase binding to promoter to rapidly begin the transcription of lac operon. Thus, the addition of lactose or a synthetic inducer such as isopropyl-β-D-thiogalactopyranoside (IPTG) very rapidly stimulates the transcription of lactose operon structural genes. Synthetic inducer only activates the transcription but is not metabolized itself. Such inducers are called as *gratuitous inducers*. Lactose, on the other hand, would be metabolized by β-galactosidase to glucose and galactose. Subsequent removal of the inducer leads to an immediate inhibition of induced transcription, since the free lac repressor quickly reoccupies the operator site, inhibiting transcription.

Lactose operon is additionally under another control called as *positive regulation* (Fig. 8.8 A, B, and C). This is exactly opposite to the negative regulation in that the transcription is activated when something is present on the DNA. The promoter of lac operon is not a strong promoter. It does not have a strong −35 sequence, and for high level transcription, it requires specific activators. This activator protein is called as cAMP receptor protein (CRP) or catabolite activator protein (CAP). This protein in conjunction with cAMP stimulates transcription by helping RNA polymerase bind to the promoter. CRP binding induces 90° bend in DNA, and it is believed to enhance RNA polymerase binding to the promoter enhancing transcription by several fold.

Positive regulation offers selective advantage. It keeps operon turned off when the levels of glucose are high. If such a way to respond to glucose levels was not available, the presence of lactose alone would suffice to activate the operon. This positive regulation ensures that the lactose metabolizing enzymes are made only when the glucose in the medium is completely exhausted. CRP or CAP, a positive regulator, that exists as a dimer cannot bind to DNA on its own, nor can it regulate

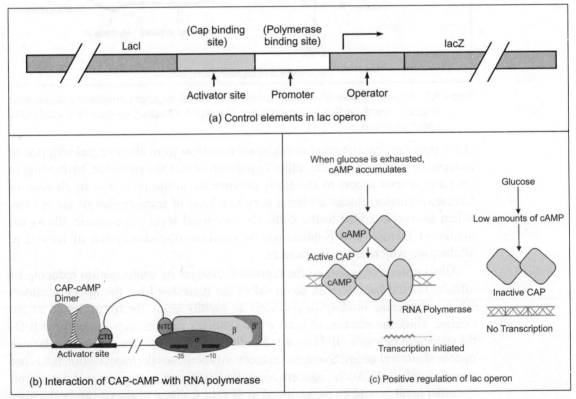

Figure 8.8 Positive regulation of *lac* operon: **(a)** Control region of lactose operon. **(b)** Interaction between positive regulator CAP-cAMP with RNA polymerase bound at promoter. **(c)** CAP by itself is inactive, in presence of cAMP it is activated. Active complex binds to DNA and activates transcription.

transcription. It needs to bind to cAMP to act as a transcription activator. Levels of cAMP are low when there is enough glucose. When glucose is used up, the levels of cAMP increase and then CRP can bind to cAMP to form an active complex that can bind to DNA to activate the transcription.

Thus, both negative and positive regulations ensure that the cell makes lactose metabolizing enzymes when lactose is to be metabolized and glucose is completely exhausted.

Repressible Operon

Two types of operons are functional in *E. coli: inducible operons* and *repressible operons*. Inducible operons are functional when the substance their gene products metabolize is present in the medium, e.g. lactose operon, whereas repressible operons are those that are functional only when the substance their gene products synthesize is absent in the medium, e.g. tryptophan operon.

Tryptophan operon contains the genes for the enzymes required to make the amino acid tryptophan. Like lac operon, it is under the negative control by repressor. However, there is a fundamental difference. Lac operon codes for catabolic enzymes that break down the substance. Such operons tend to be turned on by that substance, in this case, by lactose. On the other hand, trp operon codes for anabolic enzymes that build a substance and that substance turns off such operons (Fig. 8.9). When there is enough tryptophan available in the medium, the trp operon is repressed. Normally, an *aporepressor* by itself cannot bind to operator and block the transcription. However, in presence of tryptophan, binding of tryptophan to aporepressor changes its conformation such that it can bind to the operator and block the transcription, thereby repressing the operon.

The trp operon also has an additional control called *attenuation*, not seen in the lactose operon.

Attenuation imposes an extra level of control on an operon over and above the repressor operator system. It operates by causing premature termination (Fig. 8.10 A and B) of transcription of the operon when the operon's products are abundant. This additional control is necessary as the repression of the trp operon is weak, and in presence of repressor alone, the fully repressed level of transcription is only 70 fold lower than fully derepressed level. The attenuation permits another 10-fold control over operon's activity. Thus, a combination of repression and attenuation controls the operon over 700 fold. This is very important, as synthesis of tryptophan requires considerable energy.

Attenuation works as follows. Two loci, the trp leader and trp attenuator, lying in between the operator and the first gene trpE are involved in attenuation control. Attenuator contains a transcription stop signal, an inverted repeat followed by a string of A's. Because of the inverted repeat, the transcript in this region forms a hairpin structure. We already know that a hairpin followed by a string of U's in a transcript causes destabilization between the transcript and the DNA, and thus

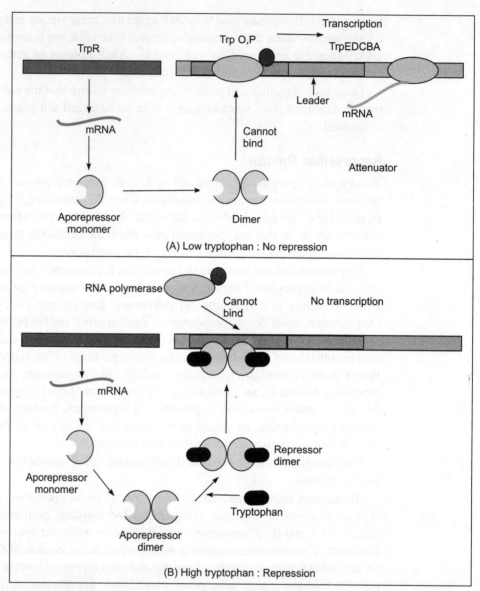

Figure 8.9 Negative control of tryptophan operon. (A) Tryptophan operon is on in absence of enough tryptophan in the medium. (B) In presence of tryptophan, aporepressor binds to the operator and turns off the operon.

causes premature termination. When tryptophan is scarce, cells must defeat attenuation so that the operon is on. This means that something should prevent the hairpin from forming, which would destroy the termination signal, breaking down attenuation, and transcription would continue. Near the end of a leader sequence, two hairpins are formed, of which the second one which is followed by a run of U's is involved in termination.

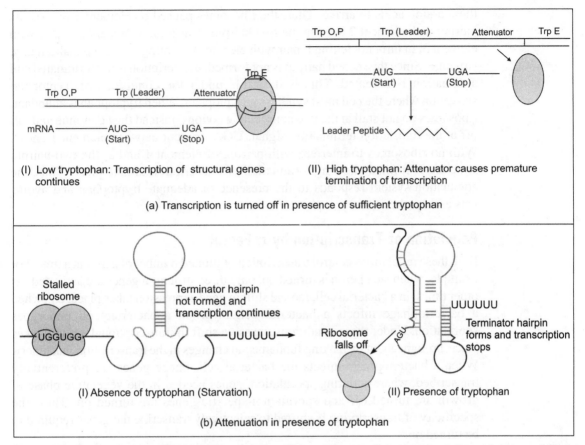

Figure 8.10 Regulation of tryptophan operon by attenuation. **(a)** In presence of sufficient tryptohan in the medium, transcription is terminated by attenuator. **(b)** Under tryptophan starvation, typical terminator hairpin is not formed and the transcription continues.

Two-hairpin arrangement is not the only available arrangement; another arrangement containing only one hairpin is also possible which contains elements from each of the two hairpins in the first structure. It means that the formation of this alternative hairpin precludes the formation of the other two hairpins, including the one adjacent to the run of U's which is a necessary part of terminator. The two-hairpin structures involve more base pairs than the alternative one-hairpin structure, therefore it is more stable. Why then is the less stable structure ever formed? Answer to this question can be given based on the sequence of the leader region (Fig. 8.10B).

The striking feature of this leader sequence is that there are two consecutive codons for tryptophan (UGG) that occur in a row in element 1 of the first potential hairpin. As soon as the trp leader region is transcribed, ribosome begins translating this emerging mRNA. Under the condition of tryptophan starvation where tryptophan is scarce, ribosome stalls at these codons waiting for charged tRNA with these amino acids to arrive. Thus, the ribosomes paused on element 1 prevent its pairing with element 2 to form the first hairpin. This allows element 2 to pair with element 3, in turn, not letting it pair with element 4 forming one-hairpin alternative structure. Since the second hairpin is not formed, transcription is not terminated and attenuation is defeated. This is desirable under the conditions of tryptophan starvation where the cell must synthesize tryptophan. When tryptophan is abundant, ribosomes do not stall at the two consecutive codons, instead they continue and fall off at the translation termination signal, UGA, present between element 1 and 2. With no ribosomes to interfere with pairing of element 1 and 2, the two-hairpin structure is formed allowing termination of transcription to take place. Thus, the attenuation system responds to the presence of adequate tryptophan and avoids wasteful synthesis of tryptophan.

Regulation of Transcription by σ Factor

Both these regulations control transcription of limited number of genes at a time. For example, when lac operon is turned on, only three structural genes are activated. At other times, in a bacterial cell, radical shift in gene expression takes place, e.g. when a bacteriophage infects a bacterial cell or when a bacterial cell undergoes sporulation. Under these conditions, massive shift in transcriptional activity takes place, brought about by making fundamental changes in the transcription machinery. When a bacteriophage infects the bacterial cell, phage genes are preferentially transcribed, whereas during sporulation, genes needed in the vegetative phase of growth are turned off and sporulation-specific genes are turned on. Thus, the specificity of transcription is shifted to selectively transcribe the genes required to be turned on.

The easiest way to change the specificity of RNA polymerase is by changing its sigma factor. When the bacterium *B. subtilis* sporulates, a new set of sporulation genes is turned on, while most vegetative genes are turned off. This switch is accomplished by several new sigma factors, e.g. σ^{29}, σ^{30}, and σ^{32} that displace vegetative sigma factor from the RNA polymerase and direct transcription of sporulation genes instead of vegetative genes. Each sigma factor has its preferred promoter sequence.

Another example is that of heat shock genes. When cells experience an increase in temperature, they mount a defense called *heat shock response*, to minimize the damage. Proteins called *molecular chaperons* are produced which bind to the partially unfolded proteins and help them fold properly again. They also produce proteases that degrade proteins, which are badly unfolded. Production of heat shock proteins is achieved by selectively transcribing heat shock genes by the use of a specific sigma factor σ^H that, on binding to RNA polymerase by displacing $\sigma^{\lambda 70}$ (σ^A), directs the enzyme to heat shock gene promoters.

In most of the bacterial infections, transcription of phage genes proceeds according to a temporal program in which *early genes* are transcribed first, then *middle genes*, and finally *late genes*. This switching is directed by a set of phage-encoded factors that associate with the host RNA polymerase and change its specificity for promoter recognition from early to middle to late. The host sigma factor transcribes early genes, and later, phage-encoded sigma factor changes it to middle and late genes. For example, transcription of phage SPO1 genes in infected *B. subtilis* cells begins using host RNA polymerase which transcribes early genes, the phage sigma factor produced then (gp 28) switches specificity to middle genes and yet another set of phage-encoded sigma factors gp 33 and gp 34 switch specificity to late gene transcription. Phage T7, on the other hand, instead of coding for a new sigma factor to change host polymerase's specificity from early to late genes, encodes a new RNA polymerase with absolute specificity for late phage genes.

Regulation of Transcription in Phage λ

Most of the phages are *virulent phages*, meaning when they replicate, they kill their host by lysing it. On the other hand, bacteriophage lambda (λ) is a *temperate phage*. When it infects the host cell, it can follow either of the two paths. First is the *lytic mode* where the infection progresses just as it would for virulent phage. It begins with phage DNA entering the host cell and serving as the template for transcription by host RNA polymerase. Phage mRNAs are then translated to yield phage proteins, the phage DNA replicates and progeny phages are assembled from this DNA and phage proteins. The infection ends when the host cell is lysed and phage progeny released.

In the other mode of infection, called *lysogenic mode*, initial events are similar to those occurring in lytic mode of life cycle. However, phage DNA does not replicate independently, instead, it is integrated into host DNA and progeny phage particles are not produced. Integrated phage DNA called the *prophage* replicates along with the host genome. On integration of lambda DNA in the host genome, only one phage protein called *repressor protein* is produced which prevents transcription of rest of the phage genes. A bacterium harbouring integrated phage genome is called as a *lysogen* and lysogenic state can exist indefinitely, however under certain conditions, lysogeny can be broken and phage can enter the lytic phase. The program of gene expression is controlled by transcriptional switches using *anti-termination*. Figure 8.11(A) outlines this scheme.

The host RNA polymerase holoenzyme first transcribes the immediate-early genes. Only two of the genes *cro* and *N* lie immediately downstream of promoter P_R and P_L, respectively. At this stage of lytic cycle, since no repressor is bound at operator (O_R and O_L) that govern these promoters, the transcription proceeds uninterrupted. When RNA polymerase reaches the end of these genes, it encounters rho-dependent terminators and stops short of delayed-early genes. Products of immediate early genes are crucial for further expression of the λ program. The cro gene product is a repressor that blocks transcription of repressor

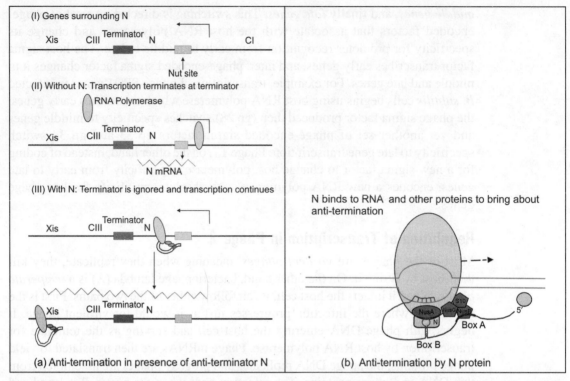

Figure 8.11 Anti-termination. **(a)** In bacteriophage lambda "N" one of the early proteins synthesized which allows RNA polymerase to read through terminator and continue transcription. **(b)** Many proteins are involved in forming a complex with the anti-terminator. This complex allows RNA polymerase to read through the terminator.

gene *cI* and therefore prevents synthesis of λ repressor protein. This is necessary for expression of the other phage genes, which would be blocked by λ repressor. The *N* gene product *N* is an anti-terminator that permits RNA polymerase to ignore the terminator at the ends of immediate-early genes and continue transcription into delayed-early genes.

Same promoters are used for both immediate-early and delayed-early transcripts. However, the switch does not involve a new sigma factor or a new RNA polymerase like in case of other phages, instead it involves extension of transcripts produced from the same promoters. The delayed-early genes are important in continuing the lytic cycle and in establishing lysogeny. Genes *O* and *P* are crucial for phage DNA replication, whereas *Q* gene product is another anti-terminator that permits transcription of late genes. Late genes are transcribed from promoter P_R, downstream to gene *Q*. The transcription from this promoter can continue only in the presence of *Q*. *N* gene product, though an anti-terminator, cannot act as a substitute for *Q*. It is specific for anti-termination after *cro* and *N*.

The late genes code for proteins that make head and tail and lyse the host cells so the progeny phage can escape. Both *N* and *Q* use different mechanism to achieve anti-termination. On transcription and translation of *N* protein, it binds to nut site (*N* utilization site) in the transcript and also interacts with a complex of host proteins bound to the RNA polymerase. This alters the polymerase which ignores the terminator and keeps on transcribing into delayed-early genes. The same mechanism applies to rightward transcription from promoter P_R. Five proteins (N, NusA, NusB, NusG, and S10) (Fig. 8.11B) collaborate in anti-termination at the Nut site. NusA and S10 and NusG bind to RNA polymerase and N and NusB bind to boxB and boxA regions of Nut site in the growing transcript. N and NusB bind to NusA and S10 respectively, tethering transcript to the polymerase. This alters the polymerase so that it reads through the terminator at the end of immediate-early genes.

Anti-termination by Q uses a different mechanism. Q utilization site (Fig. 8.12) (qut) overlaps the promoter P_R. This qut site also overlaps a pause site 16-17 bp downstream of the transcription initiation site. In contrast to N, Q binds directly to qut site, not to its transcript. In the absence of Q, RNA polymerase pauses for several minutes, then transcribes to the terminator, where it aborts late transcription. In presence of Q bound at qut site recognizes paused complex and alters it in such a way that RNA polymerase ignores terminator and continues transcription. Q-altered polymerase inhibits hairpin formation behind the polymerase, thereby inhibiting terminator activity.

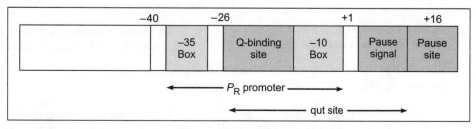

Figure 8.12 A map of P_R region of Lambda genome indicating the process of anti-termination by Q. Protein Q acts as an anti-terminator. It binds to qut site that overlaps promoter and the pause signal downstream of start site. Binding of Q alters the polymerase so that it ignores the terminator and transcribes late genes.

Establishing Lysogeny

The delayed-early genes help establish lysogeny in two ways. Some delayed-early gene products are needed for integration of phage DNA into host genome, a prerequisite for lysogeny. Second, the products of cII and cIII genes allow transcription of cI gene and therefore production of λ repressor, the central component in lysogeny. Two promoters control the cI gene, P_{RM} and P_{RE}. P_{RM} stands for promoter for repressor maintenance and P_{RE} stands for promoter for repressor establishment.

Transcription from P_{RE} gives an RNA product, which is antisense transcript of cro as well as sense transcript of cI, which on translation produces repressor. Antisense transcript helps establishing lysogeny, as it binds to cro RNA and blocks its translation. P_{RE} has an interesting requirement. It has -10 and -35 boxes which are different than the consensus sequences and require additional help from cII for RNA polymerase to bind and start transcription. cII also stimulates RNA polymerase to bind to another promoter P_{int}, promoter for int gene whose product is necessary for integration of phage DNA in bacterial genome during lysogeny. Product of gene cIII has an indirect role in establishing lysogeny. It protects cII from degradation by proteases. Thus, delayed-early products cII and cIII help establish lysogeny (Fig. 8.13).

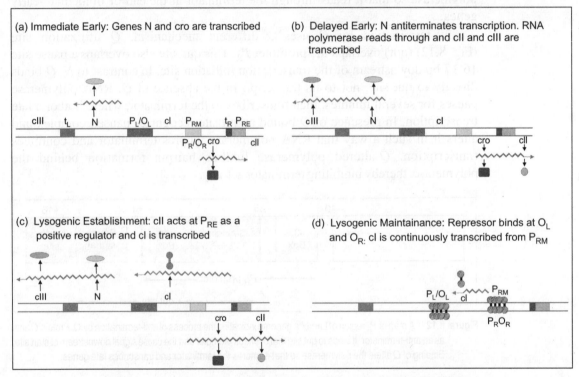

Figure 8.13 Lysogeny in λ: Many events are required to establish and maintain lysogeny.

Once λ repressor appears, it binds to operator O_R and O_L. This binding turns off transcription of *cro* and *N* from promoters P_R and P_L. Both O_R and O_L are subdivided into three parts. O_R region is more interesting, as it controls transcription from two promoters, namely P_R and P_{RM}. Three binding sites in O_R region are called $O_R 1$, $O_R 2$, and $O_R 3$, and they bind repressor with different affinities. Repressor binds more tightly to $O_R 1$ than to $O_R 2$ and $O_R 3$. Binding of $O_R 1$ and $O_R 2$ is cooperative which means that as soon as repressor binds to $O_R 1$, it facilitates binding to $O_R 2$. Cooperative binding to $O_R 3$ does not happen. Repressor molecule is dumb-bell shaped (Fig. 8.14A and B) with two domains, viz. amino and carboxyl domain joined by a linker.

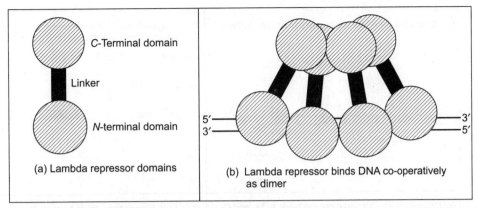

Figure 8.14 Repressor of bacteriophage λ: **(a)** λ repressor has a carboxyl terminal domain and N-terminal domain joined by a linker. **(b)** Repressor binds cooperatively to the target DNA as a dimer.

Amino domain is DNA binding domain, whereas carboxy domain is used for dimerisation and cooperative binding of repressor. Once repressor dimer is bound at both O_R1 and O_R2 repressor occupying O_R2 lies very close to promoter P_{RM} such that the RNA polymerase at P_{RM} and repressor at O_R2 touch each other. This protein-protein contact is necessary for RNA polymerase to initiate transcription from this promoter. The transcript from this promoter continuously makes repressor. At very high concentration of repressor, it binds to O_R3, thereby blocking its own transcription from P_{RM}. Thus, repressor controls promoter P_R negatively, and simultaneously activates transcription from P_{RM} by positive regulation. At high concentrations, it autoregulates.

Cro protein, on the other hand, binds to the same region as that bound by repressor but in opposite orientation. This means *cro* first binds to O_R3 and shuts off synthesis of repressor from P_{RM}. Then, it binds to O_R2 and O_R3 shutting all the transcription from P_R and P_L. This prevents transcription of all the early genes including *c*II and *c*III. Without product of these genes, repressor synthesis cannot be established from P_{RE}. Lytic infection is then ensured. Thus, the life cycle is decided depending upon who wins the fate of λ, whether *c*I or *cro*.

*c*I gene codes for repressor which blocks O_R1, O_R2, O_L1, and O_L2, turning off all early transcription including transcription of *cro* gene. This leads to lysogeny. On the other hand, *cro* gene codes for cro which blocks O_R3 and O_R2, turning off repressor transcription. This leads to lytic infection (Fig. 8.15).

Whichever gene product appears first in higher concentration to block its competitor's synthesis wins the race and determines the fate of the cell. The winner is actually determined by *c*II concentration, which in turn is determined by protease concentration in the cell. Under nutritionally rich conditions, there are many proteases present in the cell which degrade *c*II, thereby not allowing repressor synthesis to be established. On the other hand, under starved conditions, very few

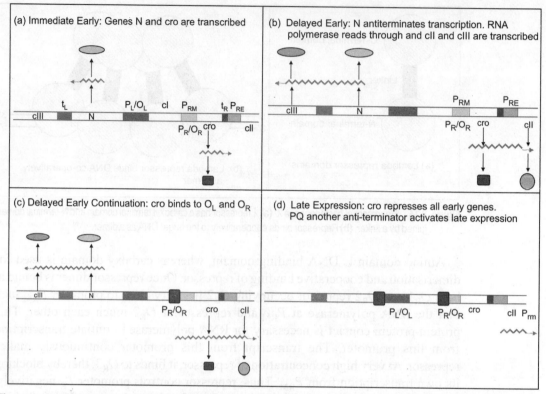

(a) Immediate Early: Genes N and cro are transcribed

(b) Delayed Early: N antiterminates transcription. RNA polymerase reads through and cII and cIII are transcribed

(c) Delayed Early Continuation: cro binds to O_L and O_R

(d) Late Expression: cro represses all early genes. PQ another anti-terminator activates late expression

Figure 8.15 The lytic cycle of bacteriophage λ begins with expression of protein *cro*, which also prevents establishment of lysogeny.

proteases are present in the cell, *c*II can then protect *c*III and initiate synthesis of repressor required to maintain lysogeny.

8.2 TRANSCRIPTION IN EUKARYOTES

In eukaryotic cells, three different RNA polymerases exist. Each one is responsible for transcribing a set of genes and each recognizes a different kind of promoter. They are called as RNA polymerase I, II, and III. These three polymerases differ in their response to ionic strength and divalent metal ion condition. Polymerase I, located in nucleoli, is most active at low ionic strength. It is equally active in presence of Mg^{2+} and Mn^{2+} and is responsible for synthesis of rRNA. Polymerase II is most active at high ionic strength, and it prefers Mn^{2+} to Mg^{2+} for reaction. It is responsible for mRNA synthesis and resides in the nucleoplasm. RNA polymerase III is active over a broad range of ionic strength. It has a preference for Mn^{2+} and is involved in synthesis of tRNA, 5S rRNA and several other small cellular and viral RNAs. They show differential sensitivity towards α-aminitin. Polymerase II is inhibited at very low concentrations, whereas polymerase III requires 1000 fold higher concentration, and polymerase I is resistant to inhibition. All three polymerases are large enzymes containing 12 or more subunits. The two largest subunits of each polymerase have homology to each other and are similar to β' and β subunits of *E. coli* RNA polymerase.

Like bacterial RNA polymerase, each eukaryotic RNA polymerase catalyzes transcription in a 5'-3' direction and synthesizes RNA complementary to the antisense template strand. The reaction requires precursor nucleotides ATP, GTP, CTP, and UTP, and does not require a primer for initiation of transcription. These enzymes, unlike bacterial enzyme, require presence of additional initiation proteins before they are able to bind to promoters and initiate transcription. Requirement at each promoter is different.

8.2.1 Promoters of Eukaryotic Polymerases

Class II Promoters

Class II promoters have two distinct parts, namely *core promoter* and *upstream element*. Core promoter has three elements, namely a *TATA box* beginning at position -30, *an initiator* centered on the transcription start site, and a *downstream element* further downstream. Many natural promoters lack recognizable version of one or more of these elements (Fig. 8.16b). TATA box sequence is centered about 25 bp upstream of transcription start site. The consensus sequence of this element is

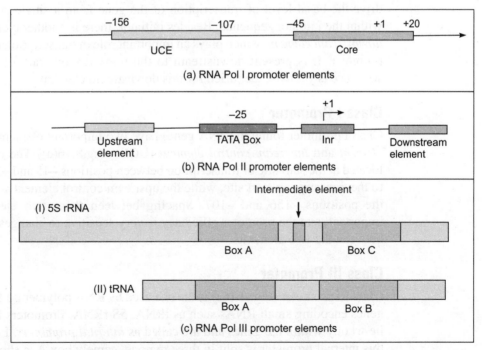

Figure 8.16   Promoters for eukaryotic RNA polymerases. RNA polymerase I, II, and III have distinct requirements of promoter for initiating transcription.

TATAAAA in the non-template strand. There are promoters, which lack a TATA box and therefore are referred to as *TATA less promoters*. These are found in many *housekeeping genes*, which are constitutively active in all cells because they control common biochemical pathways, e.g. nucleotide synthesis required to sustain cellular life. Thus, the TATA less promoters are found in genes for thymidylate synthetase, dihydrofolate reductase, etc. The second class of genes of TATA less promoters are developmentally regulated genes, e.g. homeotic genes. Deletion of TATA box sequence does not lead to a decrease in efficiency of transcription. However it leads to loss of specificity of initiation of transcription, thereby starting transcription at a different site.

Upstream element is present upstream to the TATA box and is rich in GC sequence, and therefore is referred to as GC box. The GC boxes are orientation-independent, and mutation in the GC box reduces the transcriptional activity of the promoter. Another upstream element found in a wide variety of class II promoters is CCAAT box (pronounced as *cat box*).

Some class II promoters have conserved sequences called *initiators* around their start site and are required for optimal transcription. They have consensus sequences PyPyANt/ApyPy (Py stands for pyrimidine C or T, N stands for any base, and the underlined A is a transcription start site). An example of a gene with an important initiator is the mammalian terminal deoxynucleotidyl transferase gene activated during development of B and T lymphocytes. This initiator is sufficient to drive the basal level of transcription of the gene from a single start site located within the initiator sequence. Besides initiator, there is another element called the *downstream element*, which plays an important role in transcriptional activity of the promoter. It is present downstream to the transcription start site. A consensus sequence is yet to be identified for this downstream element.

Class I Promoter

Class I promoter found in rRNA genes have two important elements, namely *core element* and *upstream control element* (UCE) (Fig.8.16(a)) The core element is located at the start site of transcription between positions –45 and +20 with respect to the transcription start site, while the upstream control element is located within the positions –156 and –107. Spacing between these two elements is very important, and the promoter efficiency is very sensitive to changes in the spacing between the two elements.

Class III Promoter

Class III promoter (Fig. 8.16c) is the one used by RNA polymerase III to transcribe genes encoding small RNAs such as tRNA, 5S rRNA. Promoters for these genes lie wholly within the genes and are called as *internal promoters*. For some genes, this internal promoter is split in three regions, namely box A, a short intermediate element, and box C; while for other genes, it is split in two parts, box A and box B.

There are some class III genes like the 7SL RNA gene that contains a weak internal promoter as well as a sequence in the 5′ flanking region of the gene required for a high level of transcription. On the other hand, 7SK and U6 RNA genes lack internal promoters altogether and have promoters that strongly resemble class II promoters which lie in the 5′ flanking region and contain a TATA box.

Other Regulatory Sequences

Many eukaryotic class II genes contain other *cis-acting DNA elements* that are not strictly part of the promoter, yet strongly influence transcription. *Enhancers* are elements, which stimulate transcription, whereas *silencers* are elements that repress transcription. Enhancers activate transcription by binding to enhancer binding proteins or activators and stimulate transcription by interacting with other proteins called general transcription factors at the promoter. This interaction promotes formation of a pre-initiation complex necessary for transcription. Enhancers are also position- and orientation-independent, which means that enhancers are present upstream, downstream, or within the gene, and in either orientation. They are also found to be tissue-specific.

Silencers inhibit rather than stimulate transcription. They cause chromatin to coil up into a condensed inaccessible form, thereby preventing transcription of neighbouring genes. Same DNA element can have both enhancer and silencer activity, depending on the protein they are bound to. For example, thyroid hormone response element acts as a silencer when thyroid hormone receptor binds to it without its ligand thyroid hormone. The same element acts as an enhancer when thyroid hormone receptor binds along with the thyroid hormone.

8.2.2 Transcription Factors in Eukaryotes

Eukaryotic RNA polymerases are incapable of binding to their respective promoters themselves. They are dependent on several proteins called *transcription factors* to guide them to the promoters. These factors are grouped into two classes, *general transcription factors* and *gene-specific transcription factors*. The general transcription factors by themselves can attract the RNA polymerases to their respective promoters and stimulate a basal level of transcription. However for fine control of transcription, gene-specific transcription factors are necessary.

Class II Factors

The class II pre-initiation complex (Fig. 8.17) has polymerase II and six general transcription factors (TFs), viz. TFIIA (TFII for polymerase II), TFIIB, TFIID, TFIIF, and TFIIH. These factors bind in a specific order to form pre-initiation complex. TFIID with help from TFIIA binds to TATA box forming a DA complex. TFIIA increases the affinity of TFIID for TATA box, though under certain conditions, TFIID can bind without the help of TFIIA. TFIIB then binds to this

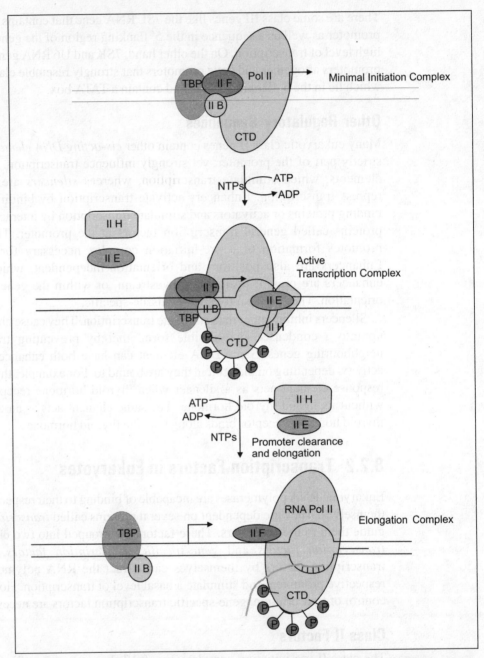

Figure 8.17 Transcription initiation by RNA polymerase II. Initiation begins with assembly of TF at the promoter which then recruits RNA polymerase II to initiate transcription. On initiation, CTD of RNA polymerase is phosphorylated and it leaves the promoter.

complex leading to the formation of *DAB complex*. TFIID itself is a complex protein containing TATA box binding protein (TBP) and 8-10 TBP associated factors (TAF$_S$ or TAF$_{IIS}$). TBPs are extremely important for transcription, and the cells mutant for TBP have defective transcription of class I, II, and III genes.

The TAF$_{IIS}$ of TFIID serve several functions, more important among those are: interaction with core promoter element and with gene-specific transcription factors. TAF$_{II}$250 and TAF$_{II}$150 help TFIID bind to the initiator and downstream elements of promoters, whereas TAF$_{II}$150 and TAF$_{II}$110 help TFIID interact with SP1 bound at GC boxes upstream to transcription start site. These TAF$_{IIS}$ therefore ensure that TBPs can bind to TATA less promoters having GC boxes. Various combinations of TAF$_{IIS}$ are required to interact with various transcription activators. TAF$_{II}$250 additionally has two enzyme activities, namely *histone acetyl transferase* and *protein kinase*. Some promoters in eukaryotes respond to alternative proteins like TRF 1 (TBP related factor 1 in *Drosophila melanogaster*), and not to TBP.

TFIIA has two subunits, which together form a functional unit. They bind to TBP and stabilize binding between TFIID and the promoter. TFIIB serves as a linker between TFIID and TFIIF-polymerase II. TFIIB has two domains, one of which binds to TFIID and the other is responsible for the assembly of pre-initiation complex.

Binding of RNA polymerase II to the DAB complex already bound to the promoter requires prior interaction with TFIIF which is composed of two subunits, RAP30 and RA70. RAP30 subunit shows the way to RNA polymerase into the growing complex. TFIIE is composed of two subunits (56 and 34 kDa), and it binds to DAB-Pol F complex; both subunits are required for binding and stabilization.

TFIIH is the last to join the pre-initiation complex. It contains nine subunits and can be separated into two complexes, a protein kinase complex composed of core subunit and a 5-subunit core TFIIH complex with DNA helicase and ATPase activity. The helicase activity of TFIIH is absolutely essential for transcription and it causes full melting of DNA at the promoter and facilitates *promoter clearance*. The protein kinase complex of TFIIH phosphorylates carboxy terminal domain (CTD) of the largest subunit of RNA polymerase II. This leads to conversion of RNA polymerase II from unphosphrylated state to phosphorylated state marking the beginning of transcription. Thus, TFIID along with TFIIB, TFIIA, RNA polymerase, and TFIIF form a minimal initiation complex.

Addition of TFIIH, TFIIE, and ATP allows DNA melting at the initiator region, and phosphorylation of the CTD of the largest subunit of RNA polymerase initiates transcription producing a small transcript called *abortive transcript*. With energy provided by ATP, the DNA helicase activity of TFIIH causes further unwinding of DNA expanding the transcription bubble. This expansion allows RNA polymerase II to clear the promoter and continue addition of NTPs.

The elongation complex continues elongating the RNA. TBP and TFIIB remain at the promoter. TFIIE and TFIIH are not required for elongation and dissociate from the elongation complex. One factor, TFIIS, has been identified to stimulate transcription elongation by limiting pauses by RNA polymerase II at discrete sites. The initiation factor TFIIF which remains associated with the elongation complex is also reported to play a role in elongation by limiting transient pausing of RNA polymerase II at random DNA sites.

Additionally, TFIIS contributes to proofreading of transcripts by stimulating inherent RNase in the RNA polymerase to remove mis-incorporated nucleotides. On addition of a wrong base, polymerase backtracks extruding the 3′ end of growing RNA chain with its mis-incorporated nucleotides out of the active site of the enzyme. The incorrect nucleotide is then clipped of the 3′ end by RNase activity of the RNA polymerase, and transcription resumes.

Class I Factors

The pre-initiation complex formed at rRNA promoters (Fig. 8.18) is much simpler and contains a TF called SL1 and an upstream binding factor (UBF) in addition to RNA polymerase I. SL1 by itself does not bind to the promoter, instead it interacts with polymerase I to strengthen its binding to UCE and perturbs the DNA in the core element. SL1 also determines *species-specificity* as does the core promoter element.

SL1 is a complex protein composed of TBP and three associated factors, namely TAF_1110, TAF_163, and TAF_148. UBF is composed of two polypeptides (94 and 97 kDa) and it stimulates transcription by polymerase I. UBF activates inactive promoter or the core element alone and mediates activation by UCE. Thus, both polymerases I and II rely on TF SL1 and TFIID, respectively, composed of TBP and several TAFs. The TBP is identical in the two factors, but TAFs are completely different.

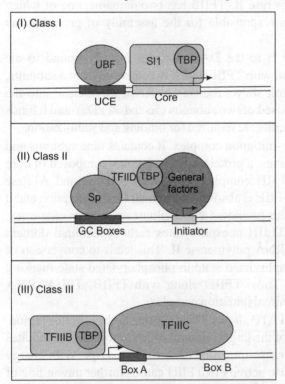

Figure 8.18 Pre-initiation complexes on TATA-less promoters. For all three polymerases, transcription factors bind first which then attract other factors and finally recruit the RNA polymerase to initiate transcription.

Class III Factors

Transcription of all Class III genes requires TFIIIB and TFIIIC and transcription (Fig. 8.18) of 5S rRNA genes requires TFIIIA in addition to

TFIIIB and TFIIIC. TFIIIB and TFIIIC depend on each other for their activities. TFIIIB by itself does not bind to the promoter region and it totally depends on TFIIIC. TFIIIC binds first to the internal promoter and serves as a center around which other components of pre-initiation complex can organize. On binding of TFIIIC, assembly factors allow TFIIIB to bind to upstream region, and TFIIIB then helps polymerase III bind at the transcription start site and initiate transcription. TFIIIC is removed in the process, but TFIIIB remains bound so that it can continue to promote further rounds of transcription. TFIIIC serves one of the same functions as TBP in TATA containing class II genes; that means it is the first factor to bind to the internal promoter and serves as a center for building up pre-initiation complex.

Thus, the assembly of pre-initiation complex on each kind of eukaryotic promoter begins with the binding of the assembly factors of promoter. With TATA containing promoters, this factor is usually TBP, while other promoters have their own assembly factors. Though TBP is not the first assembly factor at a given promoter, it becomes a part of growing pre-initiation complex on most known promoters and serves an organizing function in building the complex. There are several TAFs specific for each of the promoter complexes. Thus, the general transcription factors specific for each promoter, by themselves, dictate the starting point and direction for transcription and are capable of sponsoring basal level of transcription. To increase transcription substantially, eukaryotic cells additionally have gene-specific transcription factors or transcription activators. These factors activate and regulate the transcription.

8.2.3 Transcription Activators

Activators which are composed of at least two functional domains, a *DNA binding domain* and a *transcription activation domain* (Fig. 8.19), can either stimulate or inhibit transcription. Many of them have a dimerization domain that allows the activators to bind to each other forming *homodimers* (two identical monomers bound together), *heterodimers* (two different monomers bound together), or even higher multimers like *tetramers*. Some would even have binding site for effector molecules like steroid hormones.

DNA Binding Domain

A protein domain is an independently folded region and a DNA binding domain has a DNA binding motif. Most DNA binding motifs belong to:

   (a) Zinc containing modules,

   (b) homeodomains, and

   (c) b-zip and b-HLH motifs.

Zinc containing modules use one or more Zn ions to create a proper shape such that the α-helix within the motif can fit into the DNA major groove and make

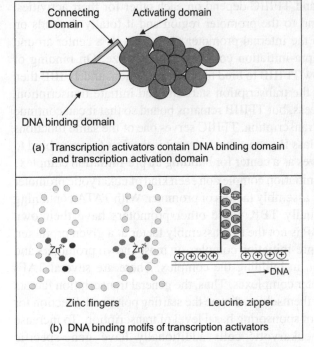

(a) Transcription activators contain DNA binding domain and transcription activation domain

(b) DNA binding motifs of transcription activators

Figure 8.19 Transcription activators contain two domains **(a)** DNA binding domain and transcription activation domain which can function independently. Among the DNA binding domains are **(b)** Zinc fingers and Leucine zippers.

specific contact. The Zn containing modules include Zn fingers, such as those found in TFIIIA and SP1, Zn modules found in nuclear receptors, and modules containing two Zn ions and 6 cysteines as found in yeast transcription activator GAL4.

Homeodomain motifs contain about 60 amino acids and resemble in structure and function to helix-turn-helix DNA binding domain of prokaryotic protein such as λ phage repressor. They are found in activators called *homeobox proteins* that regulate development.

b-zip and b-HLH motifs have a highly basic DNA binding motif linked to one or both of the protein dimerization motifs known as *leucine zippers* and *helix-loop-helix motifs*.

Most DNA binding domains have well-defined structures, and from the X-ray crystallography data, it has been possible to understand their interaction. Most of these proteins are incapable of binding to DNA in a monomer form and must at least form a dimer.

Zn fingers are composed of an anti-parallel beta strand followed by an alpha helix. Beta strand contains two cysteines followed by two histidines that are co-ordinated to Zn ions. This co-ordination of amino acid to ions helps to form a finger-shaped structure. The specific recognition between the finger and its DNA target occurs in the major groove. The GAL4 protein is a member of Zn finger family but does not have a Zn finger, instead it contains DNA binding motifs with six cysteines which co-ordinate with two Zn ions in a bi-metal thiolate cluster. Each of these motifs also has a short alpha helix that protrudes into the major groove of DNA and makes short contacts. The other end of each monomer is a alpha helix that serves a dimerization function. The third class of Zn modules is found in nuclear receptors, which interact with a variety of steroid and other hormones that diffuse through the cell membrane. They form hormone receptor complexes that function as activators by binding to enhancers or hormone response elements and stimulate transcription of their associated genes.

Thus, these activators must first bind to an effector molecule to function as an activator. This means that they must have an extra important domain for binding to an effector. Some of the hormones which function in this way are androgens, estrogens, progesterone, glucocorticoids such as cortisol, vitamin E that regulate calcium metabolism, and thyroid hormones and retinoic acid which regulate gene expression during development.

Some nuclear receptor, e.g. glucocorticoid receptor exists in cytoplasm complexed with other proteins like heat shock protein 90 (Hsp 90). In presence of glucocorticoid, it dissociates from Hsp90 and binds to the hormone. This hormone receptor complex then moves to the nucleus to activate transcription. On the other hand, other nuclear receptors like thyroid hormone receptor, in absence of the hormone, bind to the respective enhancer and represses transcription. In the presence of thyroid hormone, the complex formed functions as activator by binding to the same enhancers.

The DNA binding motifs of these nuclear receptors contain a zinc module with two Zn ions each of which is complexed to four cysteines to form a finger-like shape. The amino terminal finger in each binding domain is engaged in most of the interaction with DNA target. Some members of the steroid hormone receptor family bind to their respective ligands.

Homeodomains are DNA binding domains found in large family of activators. They are encoded by homeoboxes, first discovered in regulatory genes of Drosophila called homoeotic genes. Homeodomain proteins are the members of helix-turn helix family. Each homeodomain contains three α-helices, two of which form helix-turn helix motif and the third serves as recognition helix. In addition, the N-terminus of homeodomain proteins forms an arm that inserts into the minor groove of DNA. Most of these proteins rely on other proteins to help them bind effectively to DNA targets.

b-Zip and b-HLH domains combine two functions of DNA binding and dimerization. The Zip and HLH part refers to the leucine zipper and helix-loop-helix, respectively, of the domains, which are the dimerization motifs. The b in the name refers to the basic region of each domain that forms majority of DNA binding motifs. b-Zip proteins dimerize through a leucine zipper, which puts the adjacent basic regions of each monomer in position to embrace the DNA target site like a pair of tongs. Similarly, the b-HLH proteins dimerize through helix-loop helix, allowing the basic part of each long helix to grasp the DNA target site. Some proteins like the oncogene products (*Myc* and *Max*) have b-HLH–Zip domains with both HLH and Zip motifs adjacent to a basic motif.

Transcription Activation Domain

The DNA binding and transcription activating domains of activator proteins are independent modules. One can prepare a hybrid protein with a DNA binding domain of one protein and a transcription activation domain of another, and the hybrid protein still functions as an activator.

Most of the transcription activation domains fall into the following three classes:

(a) acidic domain, e.g. yeast activator GAL4 whose 49 amino acid domain has 11 acidic amino acids,

(b) glutamine rich domains, which are rich in glutamine content, e.g. SP1 has 39 glutamines in a span of 143 amino acids, and

(c) proline rich domains having a high proportion of prolines, e.g. activator CTF has 19 prolines in a domain of 94 amino acids.

The transcription activators of eukaryotes stimulate binding of general transcription factors and RNA polymerase II to the promoter and help build the pre-initiation complex. The acidic transcription activating domain of the Herpes virus TF VP16 binds to TFIID and recruits it to the pre-initiation complex, whereas acidic activation domain of GAL4 stimulates transcription by facilitating binding of TFIIB to the pre-initiation complex. Activators usually interact with one another for transcription activation, and this can occur in two ways. Individual factors can interact with one another to form protein dimer to facilitate binding to a single DNA target site or specific factors bound to DNA target sites can collaborate in activating a gene.

Dimerization increases the affinity between the activator and the DNA targets. Some activators form homodimers but others such as *jun* and *fos* function better as heterodimers. *Jun* and *fos* are the products of proto-oncogens *jun* and *fos* having a role in development of tumors. Both proteins are b-Zip family DNA binding proteins and *jun* and *fos* heterodimner binds tightly and specifically to DNA.

Protein-protein interaction between activators bound to the enhancers and general transcription factors and RNA polymerase bound to the promoter is in many cases mediated by looping out of DNA in between. DNA looping brings the activators bound at enhancers close to the promoter where they stimulate transcription in a co-operative way. Many genes have more than one enhancer, so they can respond to more than one stimulus, e.g. metallothionine gene.

Some transcription factors are referred to as architectural TF, which change the shape of DNA control region so that other proteins can interact successfully to stimulate transcription. One of the examples is human T cell receptor alpha chain gene control region which contains three enhancers, binding sites for three activators E-ts-1, LEF-1, and CREB within just 112 bp of transcription start site. Among these, the activator LEF-1 binds to DNA and induces bending. This bending helps other activators to bind and stimulate transcription. Thus, LEF-1 is an architectural factor.

Insulators

Enhancers can act at a great distance from the promoters that they activate. With such a large distance, some enhancers are likely to be close enough with other

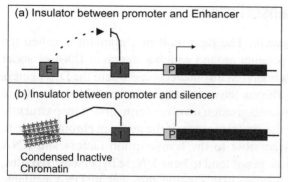

(a) Insulator between promoter and Enhancer

(b) Insulator between promoter and silencer

Condensed Inactive Chromatin

Figure 8.20 Insulator sequences: Insulator between a promoter and enhancer prevents the promoter from being activated from the enhancer. Insulator between a promoter and a silencer prevents promoter from the repressive effect of silencer.

unrelated genes to activate them. This inappropriate activation needs to be prevented. The DNA elements called *insulators* (Fig. 8.20) are used to block activation of unrelated genes by nearby enhancers. It appears that insulators create a boundary between the domain of the gene and that of the enhancer (or silencer) so that the gene can no longer feel the activating (or repressing) effects. The function of an insulator depends on the protein binding to it, e.g. Drosophila insulators contain the sequence GAGA and are known as GAGA boxes. These require GAGA binding protein Trl for insulator activity.

Mediators

Some class II activators can recruit basal transcription complex by themselves by contacting one or more general transcription factors or RNA polymerases, but many activators cannot do so and require proteins called as *mediators*. They mediate the effect of an activator by stimulating activation of transcription, but not basal transcription. CREB-binding protein (CBP) is an example of such mediators (Fig. 8.21).

When the level of cAMP rises in eukaryotic cells, it stimulates the activity of protein kinase A. This in turn phosphorylates an activator called the cAMP response element-binding protein (CREB), which consequently moves to the nucleus. In the nucleus, protein kinase A phosphorylates CREB which binds to the cAMP responses elements (CRE) and activates associated genes. CBP binds with a great affinity to phosphorylated CREB and CBP then contacts and recruits elements of the basal transcription apparatus. Thus, by coupling CREB to the transcription apparatus, CBP acts as a mediator of activation.

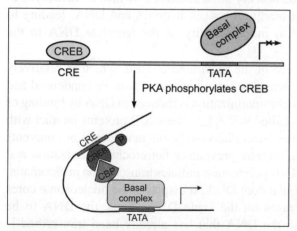

CREB

CRE

Basal complex

TATA

PKA phosphorylates CREB

CRE

CREB

CBP

Basal complex

TATA

Figure 8.21 Activation of cAMP responsive element (CRE) linked genes: unphosphorylated cAMP response element binding protein (CREB) binds to CRE. When CREB is phosphorylated, it binds to CREB-binding protein (CBP). CBP binds to the basal transcription complex and recruits it to the promoter thus activating transcription.

8.2.4 Chromatin and Transcription

Eukaryotic DNA exists as *chromatin*. The basic unit of chromatin is called the *nucleosome*. It contains histone octamer around which a stretch of DNA of about 146 bp is wrapped. Histone H1 is bound to this nucleosome where the DNA enters into and exits from nucleosome. The nucleosome fiber is further folded into 30 nm fiber. The third order of chromatin condensation involves formation of loop structure anchored to the central matrix of the chromosomes. If the genes in chromatin are to be transcribed, they must be accessible to the transcription factors and RNA polymerases. It is known that active genes tend to have DNase hypersensitive sites, indicating that there is absence of histones in this region, thus making DNA available for DNase activity.

Nucleosome remodeling agents such as SWI/SNF (switch–sniff) and ISWI (imitation switch) are required for activation of some genes. These proteins disrupt the core histones in nucleosome or move the nucleosome on DNA. This remodeling combined with core histone acetylation creates nucleosome free region in the DNA, which can bind readily to transcription activators. Acetylation is catalyzed by histone acetyl transferase (HAT A), an enzyme that transfers acetyl group from a donor acetyl CoA to core histones. This acetylation by neutralizing positive charge on histones leads to loosening the contact between histones and DNA, thus making the remodeling of the chromatin easier, which is required for transcription activation. Similarly, core histone deacetylation catalyzed by histone deacetylases (HDAC 1 and 2) leads to tight binding between histones and DNA, leading to inhibition of transcription due to inaccessibility of the template DNA to the transcription apparatus.

Some regions of chromatin are in the *euchromatin* form which is relatively extended and potentially active, whereas *heterochromatin* is more condensed and transcriptionally inactive. Heterochromatinization is induced in DNA by binding of silencing information regulators called SIR 3,4,2. These SIR proteins interact with histone H3 and H4 in nulceosomes. Acetylation of lysine in nulceosomes prevents its interaction with SIR protein, thereby preventing hetrerochromatinization and promoting transcription. When RNA polymerase initiates transcription in chromatin, it has to move through nucleosomal core DNA. It displaces the nucleosome cores behind the advancing polymerases on the same DNA. Thus, the DNA to be transcribed is unspooled, while the DNA that has already been transcribed is spooled around the core histone, and RNA polymerase continues the transcription between these spooling and unspooling regions.

8.3 POST-TRANSCRIPTIONAL EVENTS

In prokaryotes, once the RNA is synthesized during transcription, ribosomes bind to mRNA and translate it to make protein. Thus, transcription and translation are coupled. On the other hand, in eukaryotes, transcription and translation occur in different compartments. On finishing the transcription, RNA has to move to the cytoplasm where it is translated. Before translation, RNA undergoes extensive processing.

8.3.1 Splicing

Most eukaryotic genes compared to typical prokaryotic genes are interrupted by non-coding DNA. During transcription, RNA polymerases transcribe entire genes without distinguishing between coding and non-coding regions. Coding regions are referred to as *exons*, while non-coding regions are called as *introns*. These transcripts are referred to as pre-messenger RNA (Pre-mRNA). From this, the non-coding region is removed by a process called *splicing*. 5′ and 3′ ends of these mRNA are also processed. The 5′ end is capped, while a string of As is added to the 3′ end. All these events occur before the mRNA migrates to the cytoplasm for translation.

The Mechanism of Splicing

The splicing signals in nuclear mRNA precursors are uniform. The first two bases of the introns are almost always GU and the last two are AG. In addition to these two sequences, at intron-exon boundaries, another sequence called *branch point* sequence upstream to the 3′ end of intron is important. The splicing reaction is catalyzed by *spliceosomes* composed of many proteins as well as *small nuclear ribonuclear proteins* (snRNPs). Figure 8.22 shows the mechanism of splicing in a pre-mRNA.

A typical mammalian pre-mRNA has many introns, therefore, the splicing process must ensure that the exons are spliced in correct order. Every precursor mRNA has a unique pathway for removal of introns during splicing, dictated by the conformation of RNA that influences the accessibility of the splice site. The 5′ and 3′ splice site and the branch sequences are recognized by components of the splicing apparatus that assemble to form large complexes called *spliceosomes*.

The snRNPs involved in splicing are U1, U2, U3, U4, U5, and U6. Each snRNP contains a single small nuclear RNA (snRNA) and several proteins. The U4 and U6 snRNPs are usually found as single particles. Common structural core proteins are found to be associated with all snRNPs. In addition, each snRNP has a unique protein associated with it. The core proteins are called as *sm proteins* and they bind to the conserved sequence in snRNA. In addition to these proteins associated with snRNA, there are 70 other proteins found in spliceosomes called *splicing factors*. They include proteins required for assembly of spliceosomes and proteins involved in catalytic processes.

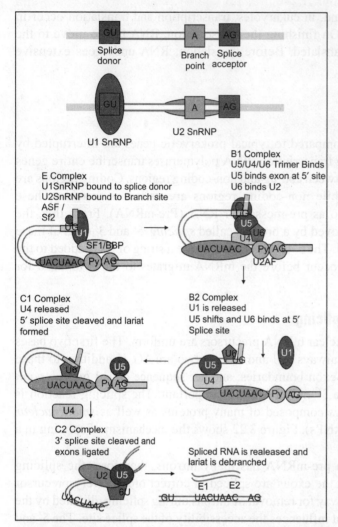

Figure 8.22 Splicing of pre-mRNA: In the nucleus, pre-mRNA is spliced using spliceosome, which consists of several snRNAs and associated proteins. Introns are removed in a specific order and exons are spliced together.

There are two distinct steps of splicing reaction. First, the consensus sequences at the 5′ splice site, the branch sequence, and the adjacent pyrimidine tract are recognized when a complex assembles containing all of the splicing components. Second, the cleavage and ligation reactions which change the structure of the substrate RNA take place. Components of the complex are released or reorganized during the splicing reaction.

The first complex formed during splicing reaction is complex E which contains U1snRNPS, splicing factor U2AF, and a member of family called SR proteins which are splicing factors and regulators. They are essential components of spliceosomes and they connect U2AF to U1. The E complex is also called as the *commitment complex*, since it identifies pre-mRNA as substrate for formation of splicing complex. In the E complex, U2AF is bound to the region between 3′ splice site and the branch site. U2AF has the large subunit (U2AF65), which contacts the pyrimidine tract downstream of the branch site, and a small subunit (U2AF35), which contacts dinucleotide AG at the 3′ splice site.

Another splicing factor called SF1 also binds to the E complex and connects U2AF to the U1 snRNP bound at the 5′ splice site. The E complex is converted to A complex when U2 snRNP binds to the branch site. Both U1 snRNP and U2AF are needed for U2 binding. U2 snRNA has a sequence complimentary to the branch sequence. The binding of U2 snRNP requires ATP hydrolysis and commits a pre-mRNA to the splicing pathway. Following the complex E, complex B1 is formed when a trimer containing U5 and U4/6 snRNPs binds. This complex is regarded as spliceo-somes. This is converted to B2 complex after U1 is released.

The dissociation of U1 is required for other components to come into juxtaposition with the 5′ splice position, most notably U6 snRNA. At this point, U5 snRNA changes its position. Initially, it is close to exon sequences at 5′ splice site, but in B2 complex, it shifts closer to the intron sequences. The catalytic reaction is triggered by the release of U4, and it requires ATP hydrolysis. The role of U4 snRNA is to sequester U6 snRNA until it is required; when U4 dissociates, U6 is free to interact with U2.

Several pairing reactions between snRNA and the substrate RNA occur in the course of splicing. The base pairing between branch point and U2 and between U2 and U6 resembles the active center of group II self-splicing introns. After the 5′ splice site is cleaved, intron forms a lariat with its free 3′ hydroxyl end and a branch site sequence. This *lariat formation* is responsible for determining the use of 3′ splice site because the 3′ consensus sequence (AG dinucleotide) nearest to the 3′ side of the branch becomes the target for second transesterification reaction. The bond is made between the first exon and the second exon, releasing the intron in the lariat formed. Thus, during splicing reaction, two bonds are broken (between exon and intron and intron and exon) and two bonds are made (a bond between two exons and a bond between 3′ hydroxyl end of the intron with the branch point sequence forming a lariat). Thus, no net energy is required for splicing reaction *per se*, however extensive amount of ATP is necessary for spliceosome assembly.

In alternative splicing, mRNA is produced from percursor pre-mRNA by skipping some exons. An alternative splicing pathway uses U12 spliceosome, which consists of U11 and U12, a U5 variant and U4$_{atac}$ and U6$_{atac}$ snRNAs. The splicing reaction essentially remains the same. The target introns are defined by longer consensus sequences but usually include the same GU/AG sequences.

Splicing can also occur between two different RNA molecules where exons from two different RNA molecules are spliced together and this is called as *trans-splicing*.

tRNA Splicing

tRNA splicing occurs by successive cleavage and splicing reaction (Fig. 8.23). Splicing depends on short consensus sequences. All the introns include a sequence complementary to the anticodon of the tRNA. This creates an alternative conformation where anticodon is base paired to form an extension of the usual arm. Splicing of tRNA depends principally on recognition of common secondary structure in tRNA rather than a common sequence of the intron.

The first step of the reaction involves phosphodiester bond cleavage by an atypical nuclease reaction catalyzed by an endonuclease. On cleavage, the termini formed are 5′ hydroxyl and 2′-3′ cyclic phosphate. The intron is released and two half tRNAs are formed which pair to form the native structure. For ligation reaction to occur, these ends need to be polished. The 5′ hydroxyl end is phosphorylated by a

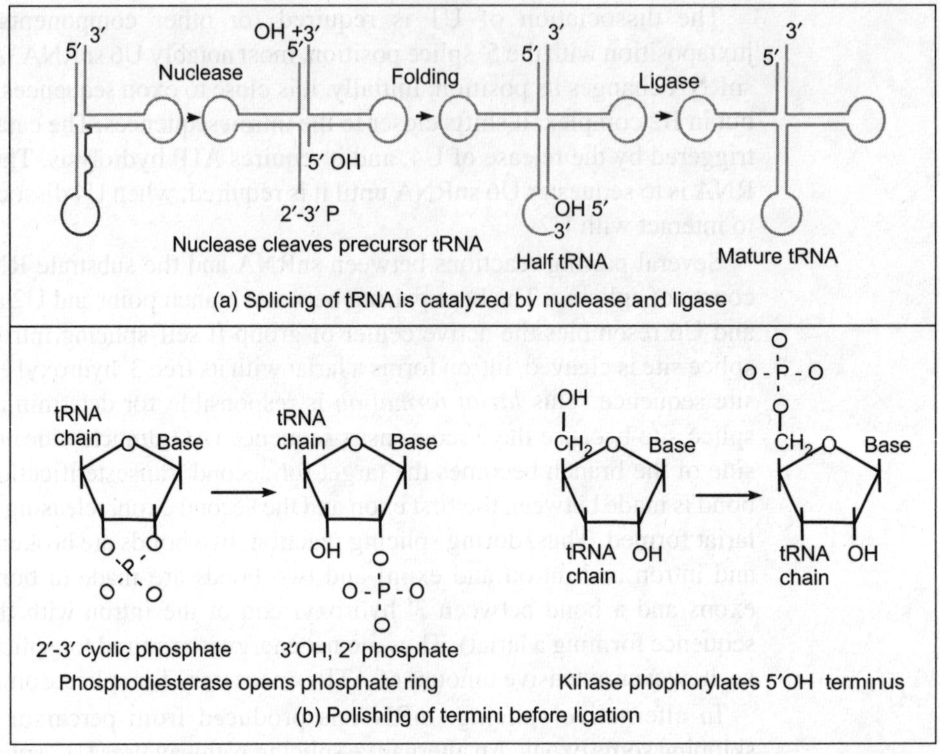

Figure 8.23 Splicing of tRNA: **(a)** precursor tRNA is spliced to obtain a mature tRNA. Splicing is catalyzed by endonuclease and ligase. **(b)** Before ligation, the ends generated by nuclease need to be polished and converted to 3'OH and 5'PO₄ termini.

polynucleotide kinase. The 2'-3' cyclic phosphate group is opened by phosphodiesterase to generate 2'phosphate and 3' hydroxy group. The 2' phosphate is then removed by phosphatase. The two termini, 3' hydroxy and 5' phosphates, are joined by ligase to restore the structure of tRNA. All three activities phosphodiesterase, 5' nucleotide kinase and ligase are arranged in different functional domains on a single protein and act sequentially to join the two tRNA halves.

Splicing of group I and group II introns

The third group of splicing reaction is called *self-splicing/autosplicing*, which is found in group I or group II introns. Group I introns (Fig. 8.24(A)) are found in genes coding for rRNA and also in some mitochondrial genes. They are present in genes of phage T4 and also in bacteria. Group I introns form a secondary structure with nine duplex regions named as P1, P2,.....,P9. Some of them are highly conserved. P1 includes 3' end of the left exon. The sequence within the intron that pairs with the exon is called as *internal guide sequence* (IGS). A very short sequence between P7 and P9 base pairs with the sequence that immediately precedes reactive G at the 3' end of the intron.

The splicing of group I intron depends on the secondary structure between pairs of consensus sequences within the intron. Group I intron splicing reaction requires a guanine dinucleotide factor. No base can be substituted for guanine and guanine can be used in any form, i.e. a base, a nucleoside, a nucleotide, mono, di, tri, phosphate form. Each stage of the self-splicing reaction occurs by a *trans-esterification reaction* in which one phosphate ester is converted directly into another without hydrolysis. The RNA forms a secondary structure in which the relevant groups are brought close to each other, a guanine nucleotide can be bound to a specific site, and then the bond breakage and reunion reactions follow (Fig. 8.24B).

The group I intron has a guanosine binding site and a substrate binding site and the excision of this intron occurs by successive reactions between the occupants of a guanosine binding site and substrate binding site (Fig. 8.24C). The substrate binding site is formed from the P1 helix in which 3′ end of the first intron base pairs with the IGS in an intermolecular reaction. A guanosine binding site is formed by sequences in P7, which can be occupied by a free guanosine nucleotide or G residue within the intron itself.

The first event to occur is the binding of guanine to a guanosine binding site present in the intron, while 5′ exon occupies the substrate binding site. The guanine attacks the exon-intron junction, the phophodiester bond is broken, and guanine is transferred on the 3′ end of the intron. In the second reaction, 3′ end of the exon attacks the next intron-exon junction, splicing the two exons together. The 5′ end of intron now occupies the substrate binding site which was earlier occupied by the first exon. The intron undergoes further cyclization and 3′ end of the intron attacks a phospodiester bond within the intron releasing a small 15 nucleotide long linear RNA and cyclized intron. The catalytic activity of group I intron is conferred by its ability to generate a particular secondary and tertiary structure that creates an active site equivalent to the active site of a conventional enzyme.

Certain introns of group I and group II classes contain open reading frames, which are translated into proteins. These proteins allow the intron to be mobile, which means introns can insert itself into a new genomic site. The intron perpetuates itself only by insertion into a single target site and not elsewhere in the genome. This is called *intron homing*.

Group II intron (Fig. 8.24A) can autosplice but may be assisted by proteins coded within the intron. Some of the group II introns code for protein with its N-terminal having reverse transcriptase activity, whereas the central domain is associated with an ancillary activity and assists in folding of an intron with its active structure called *maturase*. This reverse transcriptase activity is specific for the intron and generates a DNA copy of the intron, which is then inserted into the target site as a duplex DNA. Some autosplicing introns may require maturase activity encoded by the intron for splicing reaction. The maturase is in effect a splicing factor required for splicing of sequences that code for it. It functions to assist the folding of catalytic core to form an active site.

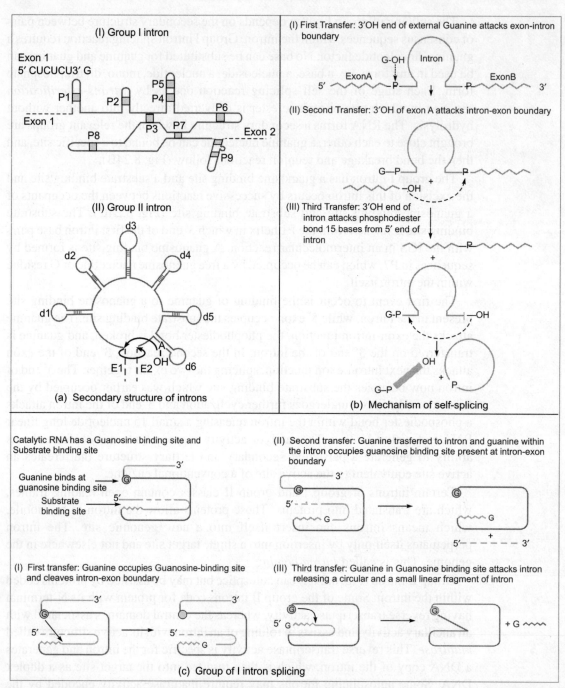

Figure 8.24 **(A)** Secondary structure of Group I intron as found in precursor rRNA and of group II intron as found in mitochondrial RNA. **(B)** Self-splicing by group I intron: Guanine nucleotide is required to initiate the splicing in which the exons are spliced together removing an intron. **(C)** Group I introns are spliced by successive occupation by guanosine binding site and substrate binding site within an intron.

Group II introns are also found in some mitochondrial genes and chloroplast genes and the mechanism used for autosplicing is different than that used for group I introns. Splicing of group II introns resemble nuclear splicing reactions. The important sequences for splicing reaction are splice junction GU at the 5′ end of intron and AU at the 3′ end of the intron. The third nucleotide important for splicing is A present upstream to the 3′ end of the intron. The first event in group II splicing involves intramolecular attack by an 'A' residue in the intron to form a lariat. The lariat formation by group II is similar to the situation in spliceosomal splicing of nuclear mRNA precursor. It has been proposed that nuclear mRNA introns evolved from these group II introns. The catalytic activity found in the RNA of group I and group II is referred to as *ribozyme*.

The catalytic activity of RNA was first detected in the enzyme RNase P involved in tRNA processing in *E. coli*. RNase P can be dissociated in two components, the 375 base RNA and the 20 kDa polypeptide; and the enzyme loses its catalytic activity if treated with RNase. Surprisingly, in RNase P, the catalytic activity resides in the RNA, while the protein component increases the speed of the reaction.

8.3.2 RNA Editing

The central dogma states that DNA is transcribed into a sequence of RNA that is in turn directly translated into protein. RNA splicing adds an additional step into the process of gene expression, wherein the interrupted sequences are removed and coding sequences are joined. This coding sequence in DNA remains unchanged. Changes in the information coded in DNA occur in few cases that lead to generation of new coding sequences, which means the information changes at the level of mRNA (Fig. 8.25A). Therefore, coding sequence in RNA differs from the sequence of DNA from which it was transcribed. This process is known as *RNA editing*.

During RNA editing, a base can be added, deleted, or substituted. An example of RNA editing where the base is substituted is apolipoprotein gene mRNA. This RNA is translated into a 512 kDa protein representing a full coding sequence in the liver, whereas a shorter form of protein of 250 kDa is synthesized in the intestine. This protein is translated from an mRNA whose sequence is identical to that of the liver except for a change from C to U at codon 2153. This substitution changes the codon CAA for glutamine into the stop codon UAA. Thus, the editing event in apo B mRNA causes single cytosine at a specific position to uridine and thereby produces a smaller protein. This substitution is achieved by *deamination*, whereby an amino group of the nucleotide ring is removed, and it is catalyzed by an enzyme called *cytidine deaminase*. This cytidine deaminase has a catalytic subunit, which is related to bacterial cytidine deaminase but has an additional sequence recognition domain that helps to recognize the specific target site for editing.

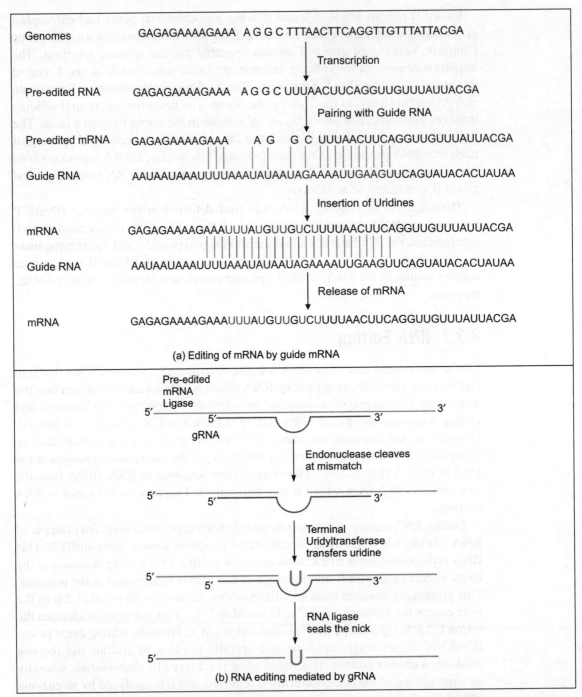

Figure 8.25 Editing of mRNA **(a)** Editing leads to insertion of U in mRNA as a post-transcriptional event. **(b)** Edition is catalyzed by gRNA along with endonuclease TUTase and RNA ligase.

Extensive RNA editing in trypanosome mitochondrial RNA occurs by insertion or deletions of uridine, and the entire process is driven by guide RNA (gRNA). The guide RNA contains a sequence that is complementary to the correctly edited mRNA. Editing reaction begins with base pairing of gRNA with the mRNA to be edited. Significant complementarity is found in gRNA and mRNA surrounding the edited region. Complementarity is more extensive at the 3′ side of the edited region compared to the 5′ side.

Pairing between the gRNA and the pre-edited RNA leaves gaps where unpaired A residues in the gRNA do not find complements in the pre-edited RNA. The gRNA provides a template that allows missing U residues to be inserted at these positions. When the reaction is completed, gRNA separates from the mRNA, which becomes available for translation. The gRNA is encoded as an independent transcription unit. In addition to gRNA, editing of uridines is catalyzed by an enzyme complex that contains an endonuclease, a terminal uridyl transferase, and an RNA ligase (Fig. 8.25B).

On base pairing of gRNA with the pre-edited mRNA, the substrate RNA is cleaved at a site identified by the absence of pairing with gRNA. A uridine is inserted into or deleted from the base pair with the gRNA and then the substrate RNA is ligated. Uridine is added by TUTase activity using UTP. An exonuclease is believed to be responsible for deletion of U. The genes whose mRNA is edited are referred to as *cryptogenes*, and the mRNA can be edited either at 3′ end, 5′ end, or along the length (pan edited). The direction of editing is always 5′ to 3′.

8.3.3 Processing of mRNA at 3′ End

The 3′ ends of mRNAs are generated by cleavage followed by endonucleolytic cleavage followed by polyadenylation (Fig. 8.26). RNA polymerase II always transcribes the genes past the site corresponding to the 3′ end, and sequences in the RNA are recognized as target for polyadenylation. A single protein complex is involved in both cleavage and processing of mRNA.

Polyadenylation signal AAUAAA is required for 3′ processing of the RNA. AAUAAA is present 1-30 nucleotides upstream of the site of poly A addition, and this signal is required for both cleavage and polyadenyaltion. The cleavage event provides a trigger for termination of transcription by RNA polymerase II. The RNA beyond the AAUAAA signal is cleaved, while the synthesis still continues by RNA polymersae. This cleavage leads to generation of a transcript with 3′ end where poly A is to be added and an exposed 5′ end of the RNA that continues to be transcribed after cleavage. Exonuclease degrades the RNA in 5′-3′ direction faster than it is synthesized so that it catches up with RNA polymerase. Then, it interacts with proteins bound to carboxyl terminal domain of the polymerase, and this interaction triggers the release of RNA polymerase from DNA causing transcription to terminate.

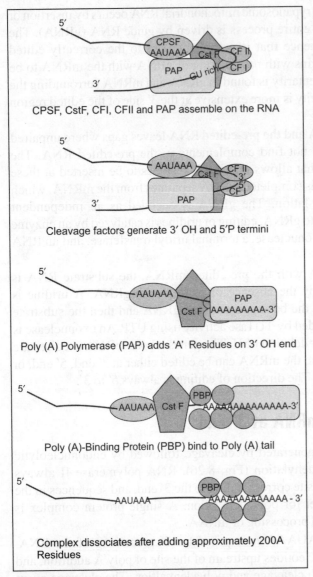

CPSF, CstF, CFI, CFII and PAP assemble on the RNA

Cleavage factors generate 3′ OH and 5′P termini

Poly (A) Polymerase (PAP) adds 'A' Residues on 3′ OH end

Poly (A)-Binding Protein (PBP) bind to Poly (A) tail

Complex dissociates after adding approximately 200A Residues

Figure 8.26 3′ end processing of mRNA: 3′end of mRNA is polyadenylated. The enzyme Poly (A) polymerase is involved in adding a stretch of A's, along with other proteins.

The 3′ processing complex consists of several activities. Generation of 3′ terminal structure requires an endonuclease consisting of cleavage factor 1 and 2 (CF I and CF II) to cleave the RNA, a poly A polymerase (PAP) to synthesize a poly A tail, and a cleavage and polyadenylation specificity factor (CPSF) that recognizes the AAUAAA sequence and directs the other activities. A stimulatory factor CstF binds to a GU rich sequence downstream to the polyadenylation signal.

Polyadenylation occurs in two phases. First phase is absolutely dependent on the AAAUAAA sequence, and Poly A polymerase performs it under the direction of specificity factor. In the second phase, oligo A tail is extended to the full 200 residue end. This reaction is AAUAAA-independent and requires a stimulatory factor that recognizes the oligo A tail and directs poly A polymerase to extended 3′ end. Intrinsically, Poly A polymerase is a distributive enzyme and would dissociate after each nucleotide is added. In presence of CPSF and poly A binding protein (PABP), it functions processively to extend poly A chain. PABP is a 33 kDa protein and it controls the length of poly A. PABP binds to the translation initiation factor eIF-4G, thus generating a closed loop in which a protein complex contains both the 5′ and 3′ end of the RNA.

Polyadenylation is an important determinant of mRNA function. It affects both stability and initiation of translation. Polyadenylated mRNAs are shown to have longer half-life compared to their deadenylated counterparts. In addition to polyadenylation, the regions in the 3′ untranslated regions (3′ UTR) and 5′ untranslated regions (5′UTR) also contribute to the stability of RNA. Some mRNAs are not polyadenylated, and therefore, their 3′ end is formed differently. A good example is histone mRNA, wherein the formation of 3′ end depends upon the secondary structure. The structure at the 3′

terminus is a highly conserved stem loop structure, and cleavage occurs 4-5 bp downstream of the stem loop. Two factors are required for cleavage reaction. Stem loop binding protein (SLBP) recognizes the structure and U7 snRNA pairs with the sequence located 10 nucleotides downstream of the cleavage site. Cleavage to generate the 3′ terminus occurs at a fixed distance from the site recognized by U7 snRNA.

Unlike RNA polymerase II, polymerase I and III do not transcribe RNA much beyond the 3′ end of the gene. For RNA polymerase I, termination occurs at a discrete site IkB downstream of the matured 3′ end which is generated by cleavage. Termination involves recognition of an 18 base terminator sequence by an ancillary factor. With RNA polymerase III transcription, the termination reaction resembles intrinsic termination by bacterial RNA polymerase. Like in prokaryotic terminators, there is a run of U embedded in a GC rich region. The critical feature in termination is therefore recognition of U_4 sequence in the context of a GC rich sequence.

8.3.4 Processing of mRNA at 5′ End

Transcription begins with a 5′ triphosphate, usually with a purine A or G. The first nucleotide retains the 5′ phosphate group and then a polynucleotide chain is synthesized by formation of a phosphodiester bond with incoming nucleotides. The 5′ end G is added by the enzyme guanilyl transferase, and it is linked to the first nucleotide by unusual 5′-5′ triphosphate linkage. This structure is called a *cap* and it is a substrate for several methylation events (Fig. 8.27). This guanine is then methylated.

The types of caps are distinguished by the number of methylations that have occurred. The first methylation involves the addition of first methyl group to the 7 position of terminal guanine and is called cap0. The next step is to add another methyl group to the 2′-O position of the penultimate base, and the reaction is catalyzed by 2′-O methyl transferase. A cap with 2 methyl groups is called cap1. In some cases, another methyl group is added to the second base, especially when the position is occupied by adenine and is methylated at N^6 position. In some RNAs, one more methyl group is added in the cap1 RNA, the third base is methylated by 2′-O ribose methylation. This creates a cap2 mRNA. In all these reactions, S-adenosyl methionine is the methyl donor, which is converted to S-adenosyl homocysteine. Some proteins are associated with the cap called the *cap binding proteins*. The cap protects the mRNA from attack by nucleases, helps translation of mRNA, and is also a signal for transport of mRNA from nucleus to cytoplasm.

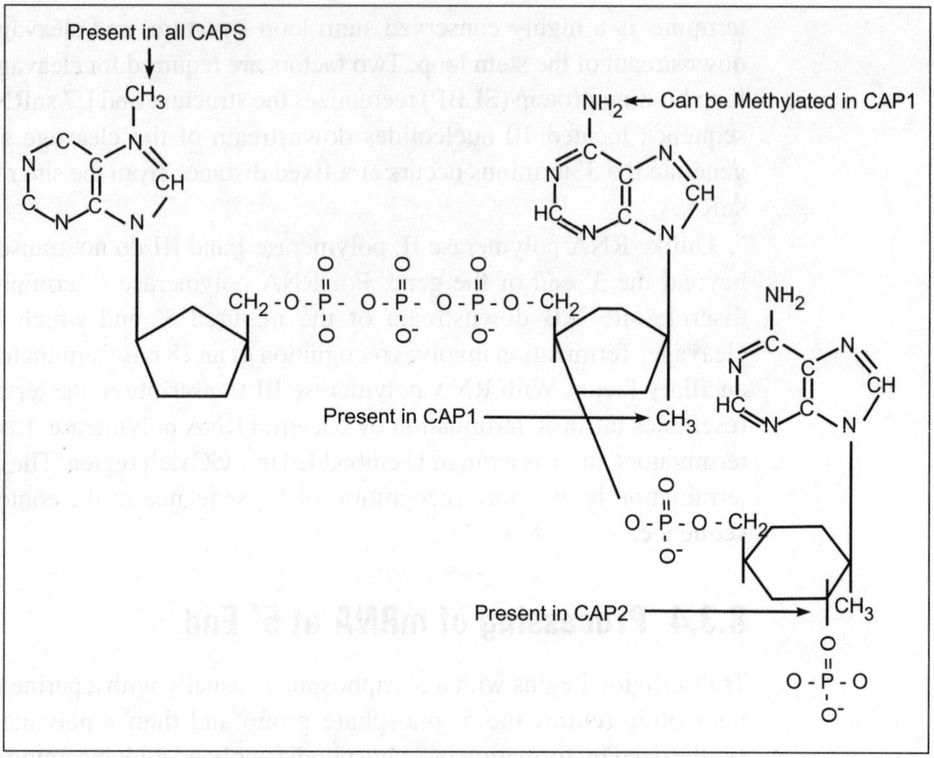

Figure 8.27   5′ end processing of mRNA: mRNA is capped by transferring a 7Methyl guanosine onto the 5′ end.

8.3.5 Production of Mature rRNA

Both in prokaryotes and eukaryotes, rRNA precursor is processed to generate mature rRNA. An eukaryotic precursor rRNA contains the sequences of 18S, 5.8S, and 28S rRNA based on its sedimentation coefficient, and it is called as 45S rRNA (Fig. 8.28). The mature rRNAs are released by a combination of cleavage and trimming reaction. Many ribonucleases are implicated in the processing. 45S precursor is ultimately processed to get 18S, 5.8S, and 28S rRNA. 5S RNA is transcribed from separate genes by RNA polymerase III. In prokaryotes, precursor rRNA contains 16S rRNA, 23S rRNA, 5S rRNA, and one or two tRNA. Each unit is referred to as *rrn operon*. Processing is required to release the products, and then the maturases are involved in producing mature 16S, 23S, and 5S rRNA.

Processing and modification of rRNA, especially in eukaryotes, requires a class of small RNAs called *sno RNAs* (small nucleolar RNA). Two groups of sno RNAs are required; one group is required for the addition of methyl group to the 2′position of ribose. The sno RNA base pairs with the rRNA to create a duplex region that is recognized as a substrate for methylation. Another group of sno RNAs is involved in synthesis of pseudouridine.

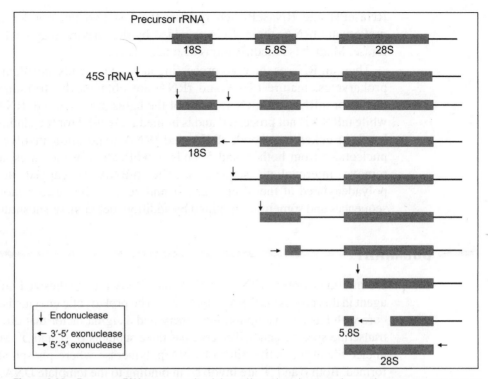

Figure 8.28 Precursor rRNA is processed and cleaved by endonucleases and exonucleases to generate mature 18S, 28S, 5.8S rRNAs.

8.3.6 Production of Mature tRNA

tRNAs are made as long precursors that are processed by removing excess RNA at both ends. In prokaryotes, as explained earlier, it is synthesized along with rRNA. The precursor is cut to release tRNA (Fig. 8.29) after which the extra nucleotides at both the 3′ and 5′ ends are removed; the enzyme involved is RNase P which has an RNA as the catalytic sub-unit. The 3′ end maturation is more complex and many RNases are involved in this, namely polynucleotide-phosphory-lase (PNPase), RNaseD, RNaseBN, RNaseT, RNasePH, and RNaseII. RNaseII and PNPase cooperate to re-move most of the extra nucleotides at the end of the tRNA precursor but they stop when two extra nucleotides remain.

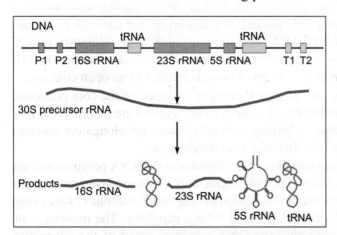

Figure 8.29 In prokaryotes, a precursor RNA is synthesized which includes transcripts for rRNA an tRNA and this is then processed to release mature rRNA and tRNA molecules.

RNasePH and RNaseP then remove the last two nucleotides. In contrast to prokaryotic tRNA, the 3' CCA is added by the enzyme tRNA nucleotidyl transferase. Many base modifications also take place.

Thus, all RNases undergo processing at 5' and 3' ends and along the length. In prokaryotes, mature tRNA and rRNA are obtained by trimming the ends of precursor mRNA and tRNA. Some of the bases in rRNA and tRNA are modified while mRNA is not processed and is immediately used for translation. On the other hand, in eukaryotes, both rRNA and tRNA maturation involves trimming of nucleotides from both 3' and 5' ends, modification of the bases, and splicing to remove intermediate sequences. The mRNA is capped at the 5' end, polyadenylated at the 3' end, and it undergoes splicing to remove intervening sequences and sometimes is edited by addition, deletion, or substitution of a base.

■ SUMMARY ■

During transcription rRNA, mRNA and tRNA are synthesized and the catalytic agent in this process is RNA polymerase. The prokaryotic enzyme is composed of a core with basic transcription machinery and a sigma factor that directs the core to transcribe specific genes. The core has three subunits α, β, and β'. Subunit β binds nucleotide at the active sites of RNA polymerase where phosphodiester bond is formed. Both β and β' are involved in binding to the template DNA, and α subunit serves to assemble the RNA polymerase holoenzyme. The α subunit is involved in binding to promoters UP elements allowing tight binding between polymerase and promoter.

Sigma factor allows RNA polymerase holoenzyme to bind tightly to promoter sequence from where the transcription initiates and loosens non-specific interaction between the DNA and the polymerase. Prokaryotic promoter contains two important regions at –10 and –35 base pair at upstream of start site with consensus sequences TATAAT and TTGACA, respectively, in *E. coli*. Mutation in –35 sequence affects initial binding of RNA polymerase, whereas mutation in –10 affects the melting reaction that converts a closed complex to an open complex.

Some strong promoters contain an UP element upstream of the core promoter. The initiation process continues until around 10 nucleotides are incorporated into RNA after which σ dissociates leaving core polymerase for elongation process. The σ factor can be reused for different core polymerase.

Elongation involves polymerization of nucleotides as the RNA polymerase core travels along the template DNA. The polymerase during its movement maintains a short melted region referred to as transcription bubble which is about 17 bases long and contains a RNA-DNA hybrid of about 9 base pair long. The movement of transcription bubble requires that the DNA unwinds ahead of the advancing polymerase and winds again behind it. This generates strain into the template strand, which is relaxed by topoisomerase.

Termination of transcription occurs by rho-dependent or rho-independent pathway. Rho-independent termination uses intrinsic terminator with an inverted repeat that allows a hairpin to be formed at the end of the transcript followed by a string of U. Together, these elements cause polymerase to pause and the transcript to be released. Rho-dependent terminators consist of an inverted repeat which can cause a hairpin to be formed in the transcript and a rho factor which binds to the growing transcript. This rho factor catches up to RNA polymerase; then it stalls at the hairpin, and the RNA-DNA helicase activity of rho releases the transcript, terminating the transcription.

Transcription is strictly regulated in prokaryotic cells. The genes are turned on only when their products are required by the cell. One of the strategies required to control the expression of genes is by grouping functionally related genes together, as they can be regulated together. Such a group of contiguous, coordinately controlled genes is called an operon. Two types of operons are functional in *E. coli*: (i) inducible operons and (ii) repressible operons. Inducible operons are functional when the substance that their gene product metabolizes is present in the medium, whereas repressible operons are those that are functional only when the substance that their gene product synthesizes is absent in the medium, e.g. tryptophan operon.

Lactose operon is controlled by both positive and negative regulation. Negative regulation is through a repressor, which normally keeps the operon turned off by binding to the operator in presence of lactose. The repressor is inactivated and dissociates from the operator allowing RNA polymerase to transcribe the genes. Positive control is mediated by catabolite activator protein which in conjunction with cyclic-AMP stimulate transcription. This positive regulation ensures that *lac* operon is activated only when easily metabolizable glucose is exhausted and there is a need to metabolize an alternative energy source. Both these regulations control transcription of a limited number of genes at a time. At other times, radical shift in gene expression takes place in a bacterial cell, e.g. when a phage infects a bacterial cell or when a bacterial cell undergoes sporulation.

A massive shift in transcriptional activity takes place making fundamental changes in transcriptional machinery. This is achieved by specific sigma factors that associate with core RNA polymerase and change the specificity of transcription. This strategy is commonly used during phage infection. During sporulation, a whole new set of sporulation genes are turned on and vegetative genes are turned off. This is achieved by synthesizing several new sigma factors that displace the vegetative sigma factor from the core RNA polymerase.

In case of bacteriophage λ, the transcriptional switching during lytic cycle is controlled by anti-terminators N and Q. These anti-terminators allow RNA polymerase to ignore the terminators and continue transcription. Phage lambda can establish lysogeny, and whether a given cell is lytically or lysogenically infected depends on the outcome of race between the products of cI and *cro* genes. Both these gene products control each other's transcription. Whichever gene product occurs first in high concentration to block its competitor's synthesis wins the race and determines the cell's fate.

In eukaryotes, three kinds of RNAs are synthesized by three different polymerases. Each of these recognizes different promoters. RNA polymerase I makes a large precursor to the major rRNAs, polymerase II synthesizes pre-mRNAs which are precursors to mRNAs, and polymerase III makes precursor to tRNAs and several other small cellular and viral RNAs. All three polymerases have some common subunits and some polymerase-specific subunits. The general architecture of class I promoter is well conserved and consists of two elements, a core element surrounding the transcription site and an upstream control element about 100 base pair farther upstream. The spacing between these two elements is important. Class II promoters may contain any of the following four elements: an initiator, a TATA box, an upstream element and a downstream element. At least one of these is missing in many promoters. The classical polymerase III transcribed genes have promoters that lie within the genes. The internal promoter is split into two or three regions. Some class III genes contain a promoter that strictly resembles class II promoters. In addition, elements such as enhancers and silencers, which are position- and orientation-independent DNA elements, stimulate or repress, respectively, transcription of associated genes.

All three polymerases depend on transcription factors for their binding to respective promoters. These transcription factors bind to promoter in a specific order and finally recruit RNA polymerase after which transcription initiates. These are basal transcription factors necessary for bringing RNA polymerase to their respective promoters and dictate the starting point and direction of transcription. Thus, they are capable of sponsoring basal level of transcription. When the transcription is to be enhanced above the basal level, additional gene-specific transcription factors are required.

Activators can either activate or stimulate transcription by RNA polymerase II and are composed of at least two functional domains, a DNA binding domain and a transcription activating domain. DNA binding domains include motifs such as a zinc module, a homeodomain, b-ZIP, and b-HLH motif. Transcriptional activation domain can be acidic, and glutamine or proline rich. The DNA binding and transcription activation domains of activator proteins are independent modules. Activators function by contacting general transcription factors and stimulating the assembly of pre-initiation complex at promoters.

Most of the activators work as homodimers or heterodimers. Dimerization increases the affinity between the activator and its DNA target, thereby it is advantageous to an activator. The stimulation of transcription by enhancers is also achieved by binding of these activators to the enhancers, which then interact with the general transcription factor and the RNA polymerase bound at the promoter. Some activators are called as architectural transcription factors. These do not enhance transcription by themselves but induce bending in the DNA, which help binding of activators. Insulators are DNA elements that shield genes from activation or repression by enhancers or silencers. Many activators do not activate transcription by contacting the basal transcription apparatus directly, instead they contact a protein called mediator, which in turn contacts the basal transcription factor and recruits it to the promoter.

Eukaryotic DNA is in the form of chromatin whose basic unit is a nucleosome, wherein the DNA is wound around the histone octamer and is condensed by a factor of 6–7. For transcription of the genes, the chromatin has to undergo remodeling whereby the contacts between DNA and protein are loosened, making the DNA accessible for transcriptional machinery. Nucleosome remodeling proteins disrupt the core histones in nucleosomes. This combined with core histone acetylation creates nucleosome-free regions in the DNA which can readily bind to transcription activators to begin transcription. When RNA polymerase moves through nucleosomal DNA, it keeps displacing the core histones to a new location, apparently behind the advancing polymerase on the same DNA molecule. This is accomplished by a spooling mechanism in which upstream DNA spools around core histones as downstream DNA unspools with the RNA polymerase transcribing alone between these spooling and unspooling DNAs.

Precursor RNA molecules formed by three different RNA polymerases are extensively processed. Nuclear mRNA precursors are capped at the 5′ end, polyadenylated at the 3′ end. Both these are responsible for the stability of the mRNA. These precursor RNAs are also spliced to remove introns and splicing is catalyzed by spliceosome containing several sn-RNAs and proteins. Some mRNAs are also edited which involves addition, deletion, or substitution of base. Type I editing is catalyzed by gRNA whereas type II is catalyzed by editosome. Precursor rRNA is cut sequentially into individual rRNA subunits and it undergoes autocatalytic splicing, whereas tRNA precursor is trimmed from both 5′ and 3′ ends to remove excess nucleotides and is spliced by enzyme catalyzed reaction to remove the intron.

SHORT ANSWER QUESTIONS

1. What does coding strand mean? Which strand is used as a template during transcription?
2. Define a promoter. Why is a promoter sequence important for transcription? Is promoter sequence gene-specific in prokaryotic system?
3. What is sigma factor? Why is it necessary for initiation of transcription?
4. Differentiate rho-dependent and rho-independent termination of transcription.
5. Justify the statement, 'In prokaryotes transcription and translation is coupled.'
6. How is it ensured that E. coli will use lactose only after glucose is exhausted from the medium?

7. How is termination of transcription by attenuation is different from regular termination of transcription?
8. Justify the statement, 'CI protein of bacteriophage λ is both positive and negative regulator of transcription.'
9. Compare the basic promoter structures used by RNA polymerase I , II, and III for transcription in eukaryotic cells.
10. Why are transcription factors necessary for transcription in eukaryotes? Can RNA polymerases function in absence of transcription factors?
11. What is meant by transcription activators? How do they function?
12. How do the architectural transcription factors differ from other transcription factors?
13. Why are chromatin remodeling proteins necessary during eukaryotic transcription?
14. How does splicing reaction of pre-mRNA differ from that of pre-rRNA?
15. What will happen if hnRNA is not spliced in a cell?
16. Explain the process of polyadenylation of mRNA. Why does transcrptionally active chromatin become hypersensitive to DNase?
17. What is meant by RNA editing? What is a cryptogene?
18. What are ribozymes? Give examples.
19. What are non-coding RNAs? Give examples.

FURTHER READING

- Helmann, J.D. and Chamberlain, N. (1988). Structure and function of bacterial sigma factors. *Annul. Rev. Biochem.* **57,** 839–872.
- Das, A. (1993). Control of transcription termination by RNA binding protein. Annu. Rev. Biochem. **62,** 893–930.
- Greenblatt, J., Noddwell, J.R., and Mason, S.W. (1993). Transcriptional anti-termination. *Nature* **363,** 401–406.
- Cramer, P. (2002). Multisubunit RNA polymerase. *Curr. Opin. Struct.Biol.***12,** 89–97.
- Darst, S.A. (2001). Bacterial RNA polymerases. *Curr. Opin. Struct.Biol.* **11,** 155–162.
- Marakami, K.S. and Darst, S.A. (2003). Bacterial RNA polymerases : The Whole Story. *Curr. Opin. Struct.Biol* **13,** 31–39.
- Richardson, J.P. (1993). Transcription termination. *Crit.Rev.Biochem.Mol.Biol.* **28,** 1–30.
- Dieci, G. and Sentenac, A. (2003). Detours and shortcuts to transcription intiation. *Trends Biochem. Sci.* **28,** 202–209.

- Latchman, D.S. (2001). Transcription factors bound to activate or repress. *Trends Biochem. Sci.* **26,** 211–213.

- Klug, A. (2001). A marvelous machine for making messages. *Science* **292,** 1844–1846.

- Studitsky, V.N., Walter, W., Kireeva M, Kashlevm, and Felsonfeld, G. (2004). Chromatin remodeling by RNA polmymerases. *Trends Biochem. Sci.* **29**, 127–132.

- Lamond, A.I. and Travers, A.A. (1985). Stringent control of bacterial transcription. *Cell* **41**, 6–8.

- Gott, J.M. and Emeson, R.B. (2000). Functions and mechanisms of RNA editing. *Annu. Rev. Genet.* **34**, 499–531.

- Cech, T.R. (1990). Self splicing of group I intron. *Annu.Rev.Biochem.* **59**, 543–568.

- Kramer, A. (1996). The structure and functions of proteins involved in mammalian mRNA splicing. *Annu.Rev.Biochem.* **65**, 367–409.

- Proudfoot, N.J., Furger, A., and Dye, M.J. (2002). Integrating mRNA processing with transcription. *Cell* **108,** 501–512.

- Tanaka Hall, T.M. (2002). Poly (A) tail synthesis and regulation: recent structural insights. *Curr.Opin.Struct.Biol.* **22**, 132–137.

9

Protein Synthesis

OVERVIEW

- Protein synthesis is the second most important step of gene expression.

- We first try to understand various characteristics of the major components of protein synthesizing machinery including the recent update on the structure-function relationship of ribosomes.

- The next step is to understand the mechanism of the act of translation, an orchestrated organization and functions of various components.

- The most important part of protein synthesis is its regulation by the cell. We discuss various cell strategy of global and mRNA-specific regulation, keeping in mind that most of these regulatory steps are linked to the initiation of protein synthesis.

- *In vitro* translation systems as a major tool for understanding regulation of gene expression and for various research in molecular biology have been elaborated.

Protein synthesis, a process in which messenger RNAs are translated into proteins, constitutes the second major step of gene expression. Protein synthesis, being a continuous process for generating a large number of proteins required by a growing cell, consumes a substantial amount of a cell's energy. In fact, in prokaryotes, about 80% of a cell's energy is spent for this vital process. The rate of protein synthesis appears to be tightly linked to the growth rate, and thus, the regulation of protein synthesis is of high significance to the cell. Further, it has recently been established that under a number of very important cellular processes, such as oogenesis, fertilization, embryogenesis, and differentiation including abnormal differentiation leading to cancer, and also during environmental stresses, regulation of gene expression takes place predominantly at the level of translation.

These observations have led to the concept and growth of an area called the *translational regulation of gene expression*. Indeed, any abnormality or failure of translation regulation of a variety of genes leads to a number of pathophysiological conditions in various organisms including humans. Thus, translational control and pathophysiology has been emerging as a new field of extensive research. All these put together, therefore, necessitate a clear understanding of the mechanism of protein synthesis *per se*, and also its regulation. This chapter, therefore, has been organized accordingly. We first discuss about the components involved in protein synthesis, thereafter the mechanism, and lastly the regulation of protein synthesis. It is worth mentioning here that most of our understanding on the mechanism of protein synthesis has been possible by studying *in vitro* protein synthesis systems constituted by cell-free extracts of various organisms. Therefore, this topic has been detailed as a sub-chapter.

9.1 COMPONENTS OF PROTEIN SYNTHESIS MACHINERY

Protein synthesis involves the following three major components in a cell: the mRNA, the tRNA, and the ribosome. These components are referred to as the *template*, the *adapter*, and the *moving factory*, respectively. In addition, a number of components, such as various cytoplasmic protein factors, enzymes, chemical energies, salts, etc. are also required. For an understanding of the mechanism of protein synthesis, let us familiarize ourselves with the major components first, and then the additional components will be dealt along with their involvement at appropriate steps as discussed subsequently. It is to be noted that these components, although universal in prokaryotes as well as eukaryotes, often differ in the details of their structure and mode of action.

9.1.1 Messenger RNA

mRNA, the acronym for messenger ribonucleic acid, acts as the template, and it carries genetic information from the DNA in the nucleus to the cytoplasm (in case of eukaryotes), through a process called *transcription*. It is in the cytoplasm where the message is interpreted with the help of the other two components, tRNA and ribosomes, in the form of a polypeptide through the process of *translation*. The template is generated by transcription in the form of mRNA that represents one strand of the DNA duplex (gene). With this understanding, the two strands are differently called: the strand that bears the identical sequence as the mRNA except for T instead of U is called the *coding strand*, while the other strand which directs the synthesis of mRNA is called the *anti-coding strand*.

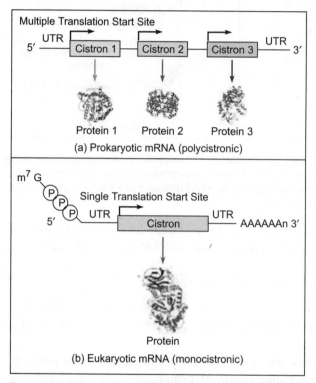

(a) Prokaryotic mRNA (polycistronic)

(b) Eukaryotic mRNA (monocistronic)

Figure 9.1 Differences between mRNAs of prokaryotes and eukaryotes. Prokaryotic mRNAs are polycistronic; proteins are individually translated from each cistron **(a)**. Eukaryotic mRNAs are mostly monocistronic that upon translation yield individual proteins **(b)**.

Nature has chosen this single-stranded nucleic acid molecule as the messenger, and accordingly, it has been attributed with other appropriate properties for the purpose for which it is needed. In general, although the function of mRNA is to be translated into proteins, there are differences in the details of the synthesis and structure of mRNA between the prokaryotes and eukaryotes. Let us now look at these differences and try to understand the relevance of such differences in the mRNAs of the two groups of organisms.

To start with, there is a basic organizational difference in the mRNAs between the prokaryotes and the eukaryotes. This relates to the structure and organization of genes in the DNA, as discussed in the previous chapter, and the origin of the mRNAs. In prokaryotes, the mRNAs are *polycistronic* (mRNAs that encode multiple polypeptides). For example, the lac operon in *E. coli* consists of three genes, and the respective protein products are translated from the same single polycistronic mRNA. Contrary to this, in eukaryotes, most of the mRNAs are *monocistronic* (Fig. 9.1).

Further, in eukaryotes, most of the nascent mRNAs in the nucleus are large precursors with introns, which are eliminated through splicing in the nucleus. Simultaneously, the mRNAs undergo maturation before they are transported to the cytoplasm. In prokaryotes, however, both the processes of transcription and translation occur simultaneously, and are coupled. Therefore, the process of gene expression in bacteria involves *coupled transcription-translation* (Fig. 9.2). Along the same lines, the stability of mRNAs differs in these two groups of organisms. For example, bacterial mRNAs are very short-lived with a half-life of

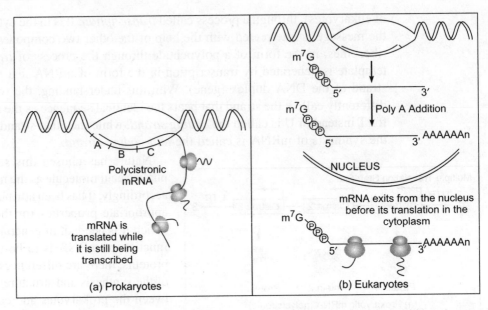

Figure 9.2 Difference in the strategy of gene expression between prokarotes and eukaryotes. **(a)** In prokaryotes, there is coupled transcription and translation, whereas in eukaryotes **(b)**, the two processes take place separately in nucleus and cytoplasm. Further, eukaryotic mRNAs undergo post-transcriptional processing in the nucleus prior to their transport to cytoplasm.

about 2 min, as opposed to eukaryotic mRNAs which are stable with a half-life of 6–24 h. However, in some exceptional cases, the mRNA half-life extends up to months (e.g., lens crystallin mRNAs).

A typical mRNA has the following characteristic features:

(a) It has a *polarity* with regard to translation: 5'-phosphate and 3'-hydroxyl, giving rise to amino- and carboxy terminus, respectively, of the translated polypeptide.

(b) It bears a *coding sequence* which codes for the amino acids of the polypeptide; it begins with the initiation codon AUG and ends with one of the three termination codons, UAG, UGA, and UAA (Fig. 9.3). This is often referred to as *reading frame* because the ribosome reads the code from the initiation codon AUG to the termination sequence in the form of trinucleotides in a linear manner, giving rise to the linear array of amino acids in the polypeptide.

(c) It has a *leader and a trailer sequence*: on both the sides of the coding sequence, there are untranslated regions (UTRs) as flanking sequences, called 5'-UTR and 3'-UTR. These are also referred to as the leader and trailer sequences, respectively. The mRNAs, although single-stranded, often forms double-stranded stem-loop structures at both the UTRs. These secondary structures confer stability to mRNAs, and are highly important for regulation of their translation.

(d) It has a 5'-cap, and a 3'-poly A tail.

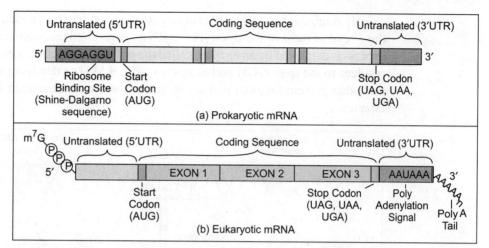

Figure 9.3 Structure of mRNAs in prokaryotes **(a)** and eukaryotes **(b)**. mRNAs contain coding sequence that begins with AUG and ends with a stop codon, and untranslated regions both at the 5' and 3' ends. Eukaryotic mRNAs contain a 5'cap and a 3' poly A tail and also introns that separate the exons.

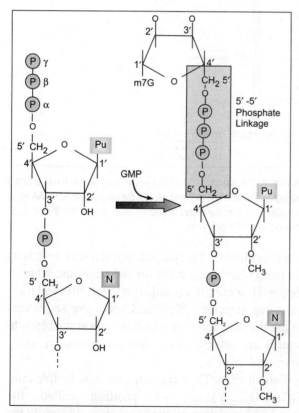

Figure 9.4 Capping of mRNA. 5' ends of mRNAs are added with a Guanine residue in a reverse orientation (5'-to-5' linkage). The 7th nitrogen of the Guanine is methylated.

The cap of mRNA refers to the modification of the 5'-end of mRNA: the terminal base is always guanine which is added to the original molecule of mRNA at the time of synthesis of mRNA. This process is catalyzed by a nuclear enzyme, *guanylyl transferase*. The reaction involves a condensation between GTP and the 5' triphosphate terminus of the original mRNA, forming an unconventional 5'-5'-phosphate linkage (Fig. 9.4). This structure is called a *cap*, and it is a substrate for several methylations. In fact, mRNA caps are classified as cap0, cap1, and cap2, depending on how many sites are methylated.

The first methylation (addition of a methyl group at the 7 nitrogen position of the terminal guanine) occurs in almost all eukaryotic mRNAs. A cap with this single methyl group is called cap0, whereas a cap with two methyl groups (second on the 2'-O position of the first base of the transcript catalyzed by 2'-O-methyl transferase) is called cap1. Cap2 mRNAs are those with a modification of the second base in the form of 2'-O ribose methylation. While most of the eukaryotic mRNAs are capped, the prokaryotic mRNAs are without caps, in general (Fig. 9.3).

Poly A tail due to addition of multiple adenine residues (not coded in the DNA) at the 3′-end of most of the eukaryotic mRNAs is resulted posttranscriptionally. This process is catalysed by an enzyme called the poly A polymerase which adds ~200 A residues to the free 3′-OH end of mRNA (Fig. 9.5). The details of the mechanism and other protein factors involved in this process were described in the previous chapter.

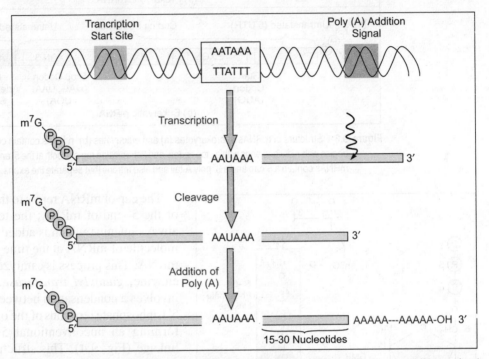

Figure 9.5 Polyadenylation of eukaryotic mRNA. The transcribed mRNA by RNA polymerase II has a signal sequence and downstream additional signals for cleavage at 3′ end. Mostly, the sequence is AAUAAA in the mRNA. Cleavage occurs at 15–30 nucleotides 3′ of the signal sequence. After this, at the 3′ end, AMP residues are added by poly A polymerase to form 3′ poly A tail.

Both the cap and poly A tail are important for the stability of mRNA as well as its translation. The cap structure is particularly important for the ribosome when it binds to mRNA during its translation. However, many eukaryotic mRNAs and most of the viral mRNAs do not have cap structure. Such mRNAs have sequences internal to the 5′UTR, and these are the sites through which the ribosome binds to mRNA internally. These structures are referred to as *internal ribosome entry sites* (IRES) or *ribosome landing pad* (RLP).

The prokaryotic mRNAs are without caps. They contain a hexanucleotide consensus sequence (5′.....AGGAGG.....3′) at −7 position called the *Shine-Dalgarno (S.D.) sequence* (AUG initiation codon is taken as +1, while the positions of any nucleotide sequence upstream and downstream to it are referred as

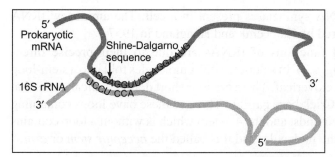

Figure 9.6 Signals in the mRNAs for ribosome binding. A hepta nucleotide sequence at −7 position in the mRNA called the S. D. sequence base pairs with the complementary sequence present towards the 3' end of 16S ribosomal RNA during translation initiation in prokaryotes.

– and + followed by the position number, respectively). The ribosome binds to these sequences through complementary sequences present at the 3'-end of the 16S ribosomal RNAs (Fig. 9.6). Recent evidences demonstrated that the efficiency with which a given S.D. sequence works can be modulated by proteins that bind to it and block its availability for ribosome binding. For example, in *E. coli.*, when the rate of synthesis of ribosomal proteins exceeds that of the ribosomal RNAs, free ribosomal proteins accumulate and some of them bind to the S.D. sequences on the mRNAs of ribosomal proteins, thereby inhibiting their translation.

9.1.2 Transfer RNA

The trinucleotide codons present in the mRNA are structurally so different from the amino acids present in the polypeptides, the immediate question is how this matching is done. In this context, Crick's initial proposition of the presence of an adapter molecule for the decoding process turned out to be true. Hoagland (1957) discovered tRNA as the adaptor molecule. Thus, an RNA molecule has been attributed by nature to carry out the role of a protein. Transfer RNA therefore is a very important molecule for the flow of genetic information from mRNA to protein.

Transfer RNAs are small molecules of 75-85 nucleotides. Alanine tRNA was the first tRNA sequenced from *Saccharomyces cerevisiae* by Holley (1966). Subsequently, as many as 600 molecules of tRNAs from a number of species have been purified and sequenced. In bacterial cells, there are about 40 different tRNAs that have been identified, while the number in eukaryotic cells (animals and plants) is close to 100. The discrepancy in number between the tRNAs and amino acids is accounted for by the fact that many amino acids can bind to more than one tRNA, and many tRNAs can recognize and attach to more than one codon. These tRNAs therefore are called *isoacceptor tRNAs*.

The adapter function of a tRNA molecule lies in its ability to chemically bind to a particular amino acid on one hand, and to base pair with a codon in the mRNA on the other, so that the amino acids as per the respective codons can be added to a growing peptide chain. The specific binding of tRNA to its cognate amino acid is catalyzed by a specific enzyme called the aminoacyl-tRNA synthetase, resulting in the formation of an aminoacyl-tRNA. Such a correctly assembled aminoacyl-tRNA then recognizes its codon through anticodon sequence present in it. Thus, there are exactly 20 aminoacyl-tRNA synthetases present in a cell. The aminoacyl-tRNA synthetases were discovered by Zamecnik and Hoagland in 1957.

The above-mentioned functions of tRNAs are due to their precise three-dimensional structures (Fig. 9.7). In solution, tRNA molecules fold into a stem-loop structure that resembles a cloverleaf. There are four short double-helical stems that are stabilized by Watson-Crick base pairing. Three of these have loops containing seven or eight bases at their ends, and the 4th stem which is without a loop contains the 3′ and the 5′ ends of the molecule, and it is called the *acceptor stem* or *arm*.

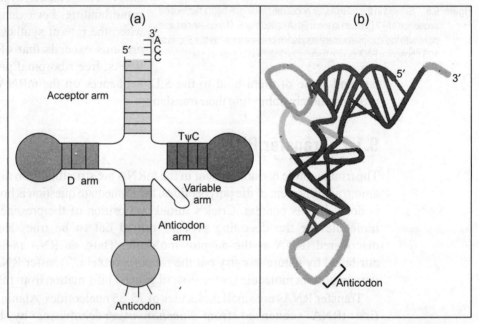

Figure 9.7 **(a)** tRNA clover leaf structure and **(b)** 3-dimensional folded structure. 3′CCA is the site for amino acid attachment, while the anticodon attaches to the mRNA codon through complementary base pairing.

The other arms are: the TψC arm named for the presence of this triplet sequence, the anticodon arm containing the anticodon triplet, the D arm which contains the base dihydrouridine, and the extra arm that lies between the TψC and the anticodon arms. The extra arm is highly variable. Based on the nature of this arm, tRNAs can be classified as: (i) class 1 tRNAs with a small extra arm, and (ii) class 2 tRNAs with a large extra arm.

It turns out that specific aminoacyl-tRNA synthetases recognize the surface structure of each tRNA for a specific amino acid and attach the correct amino acid covalently to the acceptor stem. The 3′ end of all tRNAs has the sequence CCA, and this sequence is added after synthesis of RNA. The three dimensional structure of tRNA has an L shape with anticodon loop and the acceptor stem forming the ends of the two arms (Fig. 9.7).

There are additional postsynthetic modifications of several bases in the tRNAs: most tRNAs are synthesized with a four base sequence of UUCG near the middle of the molecule. The first uridylate is methylated to become a thymidylate; the second is rearranged into a pseudouridylate, in which the ribose is attached to carbon 5 instead of nitrogen1 of uracil. These modifications produce a characteristic TΨCG loop in an unpaired region.

Activation of amino acids (charging of tRNA)

As mentioned above, it precedes the codon-anticodon recognition of the mRNA and tRNA on the ribosomes. The binding of the correct amino acid to its cognate tRNA (called charging of tRNA) is catalyzed by a specific aminoacyl-tRNA synthetase (Fig. 9.8). Thus, there are exactly 20 aminoacyl-tRNA synthetases, each specific for an amino acid, present in a cell. These enzymes couple amino acids to the cognate tRNAs at their free 2′- or 3′-OH of the adenosine at their 3′ terminus.

When an enzyme transfers the amino acid to the 2′-OH position, it is referred to as the class I enzyme, while the class II enzymes transfer the amino acid to the

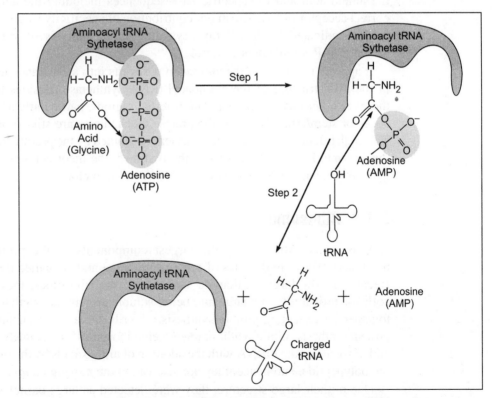

Figure 9.8 Charging of tRNA (attachment with amino acid) by aminoacyl-tRNA synthetases, which is a two-step procedure. The binding of the correct amino acid to its cognate tRNA is catalyzed by a specific aminoacyl-tRNA synthetase. In the first step, amino acyl AMP is formed with the hydrolysis of ATP. In the second step, amino-acyl tRNA is formed and AMP is released.

3′-OH position of the adenosine. This reaction process of linking the amino acid to the tRNA is through a high energy bond, and therefore it is also called as the *activation of amino acids*. Further, energy generated through hydrolysis of this bond drives the formation of the peptide bonds during elongation of the polypeptide chain on ribosomes. The charging of tRNA takes place in two steps:

(a) First, the amino acid is activated with ATP to form an aminoacyl-AMP-tRNA synthetase complex.

(b) In the second step, the activated amino acid is joined to the 3′ terminus of the tRNA with the simultaneous release of the aminoacyl-tRNA synthetase and AMP (Fig. 9.8).

Specific recognition between tRNA and cognate amino acid (Error correction/proof reading)

The specificity of the tRNA charging with the cognate amino acid is decided by the enzyme, aminoacyl-tRNA synthetase. This enzyme is highly selective in recognizing the amino acid and the specific base sequences including the anticodon of the correct acceptor tRNA. The major contribution of specificity by the anticodon was further indicated by the X-ray crystallographic analysis of the amino acid-aminoacyl-tRNA synthetase complex.

It appears that each of the anticodon bases fits into a separate specific 'pocket' in the 3-D structure of the aminoacyl-tRNA synthetase. Besides the anticodon, there are other identity elements in different regions, particularly at the end of the acceptor stem/arm. However, the precise mechanisms are still to be worked out. This high fidelity of recognition is further contributed by the proofreading activity of the enzyme. In case of an error in the first step, the incorrect aminoacyl-AMP is hydrolyzed, and not joined to the tRNA (second step block).

9.1.3 Ribosome

Ribosomes are structurally the largest components of the protein synthesis machinery. They are the sites of protein synthesis, and are called the *work bench*. Because of the large particulate structure of the ribosomes, they enhance the collision probability of a large number of soluble protein factors that need to come together for executing protein synthesis. The ribosomes, thus enhance the rate of protein synthesis. While synthesizing protein, ribosomes bind to mRNA at the 5′-end and move along the mRNA with the addition of amino acids by the tRNAs forming the polypeptide, and hence, they are also called the *moving factory*.

Due to their large structure, they were detected as large particles sedimenting with a high sedimentation coefficient by ultracentrifugation. They are designated with their sedimentation value S (Svedberg's unit). The sizes of ribosomes are 70S and 80S, respectively, for prokaryotes and eukaryotes. Both these types of ribosomes are composed to two subunits called the *large* and the *small subunits*, each containing characteristic proteins and ribosomal RNAs. Each type of cell contains a large number of ribosome molecules, reflecting the importance of ribosomes in protein synthesis. In a bacterial cell, there are about 20,000 ribosomes (25% of the dry weight), while there are about 10 million ribosomes in an eukaryotic cell.

The basic structure and composition of ribosomes are similar in prokaryotes and eukaryotes. There are minor variations in terms of the size of the subunits and molecular composition. For example, in *E. coli*, the small subunit (30S) is composed of 16S ribosomal RNA and 21 ribosomal proteins, while the large subunit (50S) contains 23S and 5S rRNAs and 34 proteins. Similarly, the eukaryotic small ribosomal subunit (40S) contains 18S rRNA and about 30 proteins, while the large subunit (60S) is composed of 28S, 5.8S, and 5S rRNAs, and about 40 proteins (Fig. 9.9).

Organelle ribosomes are distinct from the cytoplasmic ribosomes, and they are of varied forms. The large ones are about the size of the bacterial ribosomes and contain 70% rRNA; the smaller ones are only 60S and have less than 30% rRNA. The components of ribosomes have been purified and characterized. It has also

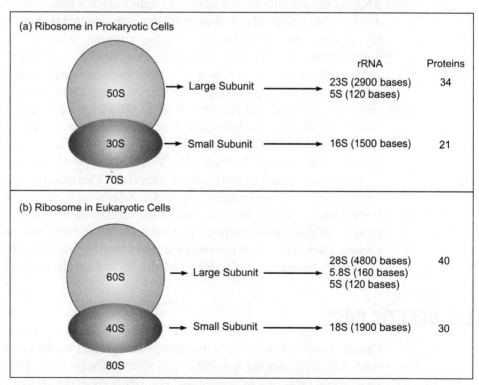

Figure 9.9 Ribosome subunits and their components. Both prokaryotes and eukaryotes contain similar structure and components; eukaryotic ribosomes are larger and more complex.

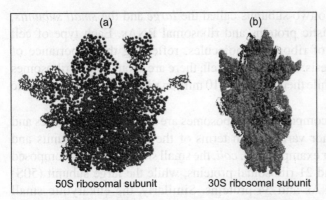

Figure 9.10 3-dimensional high resolution X-ray crystal structures of prokaryotic ribosomal subunits. **(a)** 50S ribosomal subunit. **(b)** 30S ribosomal subunit. **(See Colour Plate 13)**

been possible to reconstitute functional ribosomes *in vitro* from the purified components. Although such experiments have highlighted the importance of each of the components in terms of structure-function of ribosomes, the ribosome structure at the atomic level could not be understood until recently. Indeed, the high resolution structural analysis of ribosomes by X-ray crystallography (Fig. 9.10) in 2000 has enhanced our understanding of the structure-function relationship of ribosomes.

The complex nature of the structure of ribosome is due to the presence of extensive secondary structure of rRNAs which get further folded into a three-dimensional structure in association with the ribosomal proteins. Initially, the rRNAs were thought to play a structural role for the formation of ribosomes. However, with the discovery of enzyme activity of various RNA molecules, attention was paid to determine the functional activity of rRNA during *in vitro* assembly of ribosomes. It was discovered that for functional assembly of ribosomes, rRNAs, and not the proteins, were absolutely necessary. Subsequently, two landmark discoveries elucidated the catalytic role of rRNAs.

Harry Noller and his colleagues in 1992 demonstrated that the large ribosomal subunit is capable of catalyzing the formation of peptide bonds (peptidyl transferase reaction) even after the removal of up to 90% of the proteins. On the contrary, treatment with RNases abolishes this activity. However, the definitive evidence of the catalytic role of rRNAs came from the high resolution structural analysis of the 50S ribosomal subunit determined by Moore and co-workers in 2000 (Fig. 9.10).

The ribosome structure at the atomic level revealed that ribosomal proteins were strikingly absent from the site of peptidyl transferase activity, thereby suggesting the involvement of rRNA in the peptide bond formation. This also indicates that the proteins in the ribosomes perform predominantly a structural role. Besides this, various other activities in ribosomes are performed by various active centres which are constituted by precise assembly of rRNAs and specific proteins in ribosomes.

9.2 GENETIC CODE

Genetic code refers to the trinucleotide sets present in the coding sequences of mRNAs which code for specific amino acids in the proteins. The discovery of the genetic code was one of the seminal contributions in the area of gene expression.

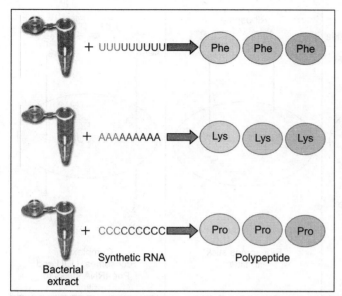

Figure 9.11 Deciphering the genetic code using synthetic mRNA (trinucleotide) translation in *E. coli* lysates *in vitro*. In this process, the bacterial extracts containing all translational machinery is incubated with a synthetic mRNA containing one trinucleotide sequence repeat. Thus, the synthetic mRNA gets translated into polypeptide having one type of amino acid. In this manner, a trinucleotide sequence coding for an amino acid is determined by analysing the amino acid in the polypeptide.

Considering the probability of any one of the four possible nucleotides occupying each of the three positions of the codon, there could be at least 4^3, i.e. 64 combinations of possible trinucleotide sequences. But, there are only 20 amino acids in a cell. To address this issue, various experiments were carried out, and it turned out that there are redundancies in the genetic code, i.e. one amino acid can be coded by more than one trinucleotide sequences. Therefore, there are so many codons. In reality, there are 61 codons coding for 20 amino acids, and the remaining three are *non-sense* or *termination codons*. They are so called because there are no tRNAs to recognize them in the cells.

Many scientists in the 1960s have contributed to the cracking of the genetic code, namely Ochoa, Nirenberg, and Khorana. Cracking of the genetic code refers to assigning amino acids to specific codons. These scientists developed two methods for this purpose using an *in vitro* cell-free protein synthesis system from *E. coli*. The first one was detection and characterization of the translated product of the synthetic polynucleotides in an *in vitro E. coli* lysate. The first success was Nirenberg's demonstration of synthesis of polyphenylalanine by polyuridylic acid (poly U) in 1961. This indicated that UUU must be a codon for phenylalanine (Fig. 9.11). Subsequently, many other synthetic polynucleotides were used by Khorana and others, and thus, almost half of the 64 codons were assigned to their respective amino acids.

The second approach involved codon assignment by ribosome binding assay (Nirenberg and Leder, 1964). A trinucleotide was used to mimic a codon, by causing the corresponding aminoacyl-tRNA to bind to ribosome. In such an assay, a triple complex of trinucleotide-aminoacyl-tRNA-ribosome can be isolated by taking advantage of the ability of ribosomes to bind to nitrocellulose filters. The retention of aminoacyl-tRNA is detected by means of a radioactive label (Fig. 9.12). In this manner, the meaning of each trinucleotide codon could be determined (Table 9.1).

These two techniques together assigned the meaning of 61 of the 64 codons *in vitro*. Since then, the sequencing of DNA has made it possible to compare corresponding nucleotide and amino acid sequences directly (Fig. 9.13).

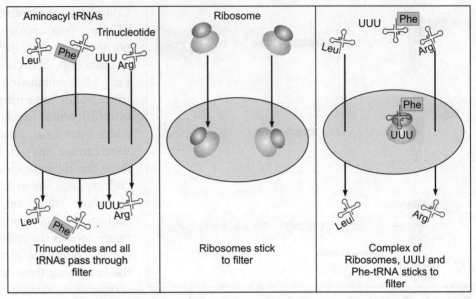

Figure 9.12 Deciphering the genetic code by filter-binding assay. Trinucleotides, aminoacyl tRNAs pass through the filter, while ribosomes do not pass through it. But when ribosomes, aminoacyl tRNAs, and RNAs (trinucleotide sequences) are incubated, ribosome binds to RNA and then tRNA binds to it, and this complex cannot pass through the filter and remains bound to it. Thus, by analyzing the bound ribosome, the genetic code can be deciphered.

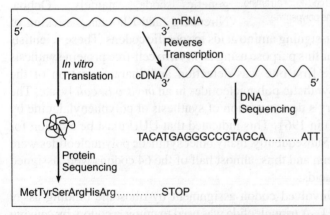

Figure 9.13 Protein sequencing vs DNA sequencing. Protein sequencing being more tedious than DNA sequencing, has become obsolete due to our knowledge of genetic code. By using these codes, the protein sequence can be easily deduced from the DNA sequence.

A striking feature of genetic code is its *degeneracy* or *redundancy*. The 61 codons represent 20 amino acids, i.e. almost every amino acid is represented by several codons. Codons that have the same meaning are called *synonyms*. Indeed, it was found that highly purified tRNA species of known sequence could recognize several different codons. The genetic codes and their corresponding amino acids are shown in Table 9.1. Further, it was also established that the anticodon bases of tRNAs did not always belong to the four regular ones: a fifth base, inosine (I), a derivative of adenine, was found.

These observations could subsequently be explained with the development of the *wobble concept* in 1966. According to this concept, the base at the 5′ end of the anticodon in the tRNA is not as spatially confined as the other two, thus allowing it to form hydrogen bonds with any of the several bases located at the 3′ end of the

Table 9.1 Genetic code and its redundancy

Second Position

	U	C	A	G	
U	UUU UUC } Phe UUA UUG } Leu	UCU UCC UCA UCG } Ser	UAU UAC } Tyr UAA UAG } STOP	UGU UGC } Cys UGA Stop UGG Trp	U C A G
C	CUU CUC CUA CUG } Leu	CCU CCC CCA CCG } Pro	CAU CAC } His CAA CAG } Gln	CGU CGC CGA CGG } Arg	U C A G
A	AUU AUC } Ile AUA AUG Met	ACU ACC ACA ACG } Thr	AAU AAC } Asn AAA AAG } Lys	AGU AGC } Ser AGA AGG } Arg	U C A G
G	GUU GUC GUA GUG } Val	GCU GCC GCA GCG } Ala	GAU GAC } Asp GAA GAG } Glu	GGU GGC GGA GGG } Gly	U C A G

First Position (5'-end) — Third Position (3'-end)

codon. For example, U at the wobble position can pair with either adenine or guanine, while I can pair with U, C, or A. The wobble rules, however, do not permit any single tRNA molecule to recognize four different codons. Three codons can be recognized only when inosine occupies the first (5′) position of the anticodon.

The next question was to ascertain if the genetic code is universal. When DNA sequences were compared with the protein sequences, it was observed that identical sets of codon assignments were used in bacteria and eukaryotes. This was further verified by the fact that mRNA from one species usually can be translated by the protein synthesis machinery of other species both *in vitro* and *in vivo*. Therefore, the codons present in the mRNA of one species have the same meaning and interpretation for the ribosomes and tRNAs of other species. Interestingly, however, there are rare exceptions, particularly regarding the termination codons. In mycoplasma, a prokaryote, UGA, normally a termination codon, codes for tryptophan. Similarly, in certain species of ciliates (protozoans), Tetrahymena and Paramecium, UAA and UAG code for glutamine. Further, systematic alterations of genetic code have occurred only in mitochondria. The organisms containing the genetic codes which code for unusual amino acids are shown in Table 9.2.

Table 9.2   Rare exceptions of the genetic code

Source	Codon	Usual Meaning	New Meaning
Fruit fly	UGA	Stop	Tryptophan
mitochondria	AGA & AGG	Arginine	Serine
	AUA	Isoleucine	methionine
Mammalian	AGA & AGG	Arginine	Stop
mitochondria	AUA	Isoleucine	Methionine
	UGA	Stop	Tryptophan
Mycoplasma	UGA	Stop	Tryptophan
Yeast	CUN*	Leucine	Threonine
mitochondria	AUA	Isoleucine	Methionine
	UGA	Stop	Tryptophan
Higher plant	UGA	Stop	Tryptophan
mitochondria	CGG	Arginine	Tryptophan

9.3 MECHANISM OF PROTEIN SYNTHESIS

Protein synthesis (translation of mRNA) can be divided into three steps: *initiation, elongation,* and *termination.* In the initiation step, binding of the initiator methionyl tRNA and the mRNA to the small subunit of ribosomes is followed by the joining of the large subunit of ribosome. Based on the formation of such a functional ribosome, elongation of the polypeptide chain occurs at a rapid rate. Once the ribosome reaches the termination codon in the mRNA, the translation process terminates with the completion of the polypeptide (Fig. 9.14).

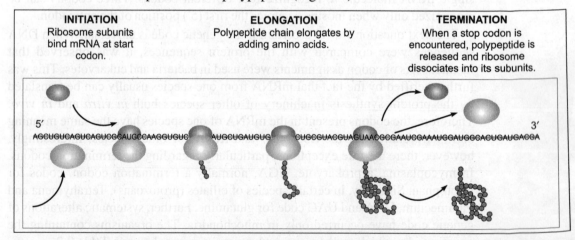

INITIATION	ELONGATION	TERMINATION
Ribosome subunits bind mRNA at start codon.	Polypeptide chain elongates by adding amino acids.	When a stop codon is encountered, polypeptide is released and ribosome dissociates into its subunits.

5′ AGCUGUCAGUCAGUCGGCAUGCCAAGGUGUC   AUGCUGAUGUG   CUGCCGUACGUAGUAACGCGAAUCGAAAUGGAUGGCACUGAUGACGA 3′

Figure   9.14   Overview of translation. The first step is initiation in which ribosomal subunits bind to the mRNA and translation starts at the initiation codon (AUG). The second step is elongation, in which the polypeptide is elongated by the addition of amino acids. In the third step, termination occurs at the termination codon (UAA, UAG, and UGA) by binding of release factor, and then the polypeptide is released and ribosomes dissociate. **(See Colour Plate 14)**

A number of cytoplasmic protein factors are essential to catalyze various steps during initiation, elongation, and termination. These are detailed in Table 9.3. Although the basic processes are similar in prokaryotes and eukaryotes, there are differences, particularly in terms of complexity. Accordingly, they have been described either separately for bacteria and eukaryotes, or together with highlights of the differences.

Table 9.3 Translation factors

Translation factors		
Role	**Prokaryotes**	**Eukaryotes**
Initiation	IF-1, IF-2, IF-3	eIF-1, eIF-2, eIF-2B, eIF-3, eIF-4A, eIF-4B, eIF-4C, eIF-4F, eIF-5, eIF-6
Elongation	EF-TU, EF-Ts, EF-G	eEF-1α, eEF-1$\beta\gamma$, eEF-2
Termination	RF-1, RF-2	eRF-1, eRF-3

The eukaryotic and prokaryotic translation factors involved in initiation, elongation, and termination of protein synthesis are shown in this table.

9.3.1 Initiation

In prokaryotes, in the first step, three initiation factors (IFs), IF-1, IF-2, and IF-3 bind to the 30S ribosomal subunit (Fig. 9.15) to form a pre-initiation complex. In the second step, the mRNA and the initiator N-formylmethionyl tRNA (fMet-tRNA$_i^{Met}$) join the pre-initiation complex, with IF-2 (bound to GTP) specifically recognizing the initiator tRNA. In the third step, IF-3 is released, allowing a 50S ribosomal subunit to associate with the complex. Such an association results in the hydrolysis of GTP bound to the IF-2 leading to the release of IF-1 and IF-2 (bound to GDP). Finally, the resultant 70S initiation complex, with mRNA and the initiator tRNA bound to it, is ready to begin the next step of protein synthesis, the elongation.

In eukaryotes, the initiation of protein synthesis is one among the most complex biochemical processes which they undertake. This involves the concerted action of dozens of polypeptides and RNAs. Incidentally, initiation is the crucial step, and it is at this level where most translational controls operate. Perhaps the least understood of this process is the entry of mRNA into the initiation cycle. Although the factors involved have been identified, there is little information about the biochemical reactions they carry out. Nonetheless, a series of recent findings have considerably clarified this area and made it possible to formulate models. These models are beginning to provide the basis for understanding various phenomena, such as the requirement of ATP in initiation, the influence of mRNA secondary structure on translational efficiency, and the regulation of translation at the mRNA entry step.

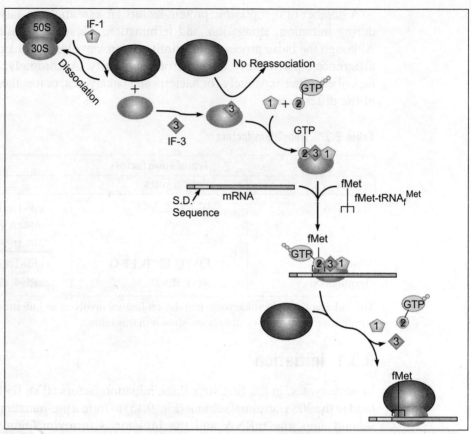

Figure 9.15 Initiation of translation in prokaryotes. In the first step, three initiation factors, IF-1, IF-2, and IF-3 bind to the 30S ribosomal subunit. In the second step, mRNA and the initiator tRNA (fMet-tRNA$_f$) bind, and thereafter, the large subunit of ribosome joins to form the initiation complex. The factors are released subsequently. **(See Colour Plate 14)**

Initiation in eukaryotes can be divided into three stages:

(a) formation of the 43S pre-initiation/ribosomal complex,

(b) formation of the 48S pre-initiation complex, consisting of mRNA binding to the 43S complex, followed by scanning for the AUG codon, and

(c) formation of the 80S initiation complex.

Formation of 43S Pre-initiation/Ribosomal Complex

Translation of all mRNAs begins with a common step, the binding of Met-tRNA$_i$ to the 40S ribosomal subunit. Eukaryotic initiation factor-2 (eIF-2) is the key protein and the ternary complex, eIF-2.GTP.Met-tRNA$_i$, is an obligatory intermediate in the binding of Met-tRNA$_i$ to the 40S subunit (Fig. 9.16). Under steady-state conditions, ternary complex formation involves the recycling of eIF-2 and therefore depends in part on the guanine nucleotide exchange reaction catalyzed by eIF-2B (also called

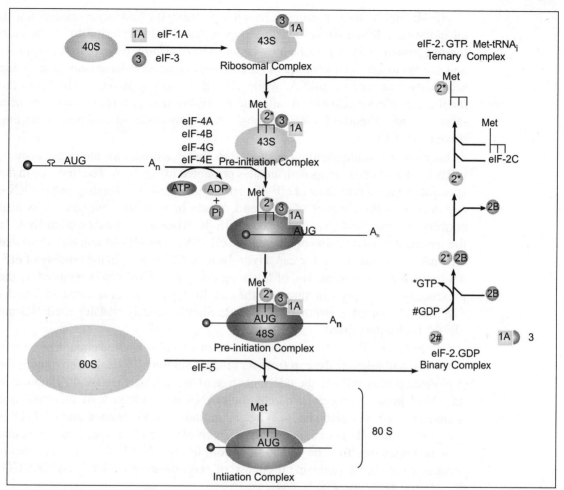

Figure 9.16   Initiation of translation in eukaryotic cells. The process is governed and catalyzed by several protein factors. 40S ribosomal subunit along with ternary complex binds at the 5′ m⁷G cap of mRNA to initiate the process. Formation of 80S initiation complex occurs at the initiation codon (AUG) with the introduction of 60S large ribosomal subunit to the 48S pre-initiation complex. After this, the peptide elongation starts.

the guanine nucleotide exchange factor, GEF or the reversing factor, RF). The formation and stability of the ternary complex are also influenced by other proteins such as co-eIF-2A and eIF-3.

Cap Recognition and the Entry of mRNA into Initiation

Four polypeptides cooperate to find the capped 5′-terminus of eukaryotic mRNA, begin the melting out of the 5′-terminal secondary structure at the expense of ATP, and transfer the mRNA to the 43S initiation complex. The polypeptides that carry out this step consists of the eIF-4 group of polypeptides: eIF-4A (46 kDa), eIF-4B (80 kDa), eIF-4E (25 kDa), and eIF-4G (p220, 220 kDa). The first three protein factors are also referred to as eIF-4F.

eIF-4E binds to the cap structure, and it is probably the first factor to interact with an initiating mRNA. Therefore, this polypeptide is also referred to as the *cap-binding protein* (CBP). It is the only factor to have affinity for mRNA caps *per se*, and it binds to the cap in absence of ATP. eIF-4A is the ATP-driven unwinder of the secondary structure of mRNA, while eIF-4B is the cap releaser. eIF-4G is the mediator of eIF-4A and eIF-4E alignment. eIF-4G also acts as an adaptor protein linking 5′- and 3′-ends of mRNA through its binding to eIF-4E and poly A binding protein (PABP).

Based on our understanding as discussed above, a model for the mechanism of mRNA entry has been proposed, and it is presented in Fig. 9.16. The first step in the speculative model is binding of eIF-4E to an initiating mRNA. Binding to the mRNA cap occurs in the absence of ATP and results in a 'weak complex'. It is also conceivable that eIF-4E is complexed to eIF-4G when it first binds with mRNA. In the second step, association of eIF-4G and eIF-4A aligns eIF-4A and eIF-4E so that unwinding can begin near the cap. Hydrolysis of ATP results in the binding of eIF-4A to the RNA and formation of a 'strong complex'. This step is retarded by the presence of secondary structure near the cap. In step 3, mRNA is transferred to the 43S initiation complex together with eIF-4E. Simultaneously, transfer of eIF-4G and eIF-4A is also postulated.

In step 4, eIF-4B associates with the initiation complex in close enough proximity to the cap and releases the cap from its binding site on eIF-4E. In step 5, with the cap released from eIF-4E, the mRNA is free to move relative to the 40S subunit in an ATP-dependent process, whereby the mRNA is simultaneously unwound and scanned for the initiation codon. When an initiation codon is encountered, GTP is hydrolyzed, eIF-4E is expelled together with eIF-2 and -3, and 60S ribosomal subunit joins (step 6). The initiation codon, usually 5′-AUG-3′ in eukayotes is recognized by the ribosome in the context of a consensus sequence 5′-ACCAUGG-3′, referred to as the *Kozak consensus*.

Scanning model of translation initiation

A scanning model has been proposed by Kozak (1989) to explain the mechanism by which eukaryotic mRNA initiates translation. As per this model, ribosome and associated factors bind at or near the 5′ end of the mRNA in a process that is facilitated by the presence of a cap structure, and scan the mRNA until the appropriate initiation codon is reached (Fig. 9.17). However, two modifications were subsequently introduced to this model to explain translation on polycistronic mRNAs.

It was postulated that ribosomes can terminate translation at an upstream ORF, then resume scanning and initiate translation at the

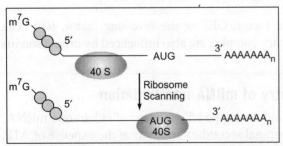

Figure 9.17 Signals for cap-dependent translation initiation in eukaryotes. The 40S ribosomal subunit first binds to the 5′ 7methyl guanosine cap of the mRNA and then the ribosome scans the mRNA in search of the initiation codon (AUG).

downstream ORF, and that ribosomes could skip an upstream AUG to initiate translation at a downstream AUG. The latter process was termed *leaky scanning*. Further, in some circumstances, ribosomes can skip a complex secondary structure of mRNA during cap-dependent initiation, which is called *ribosome shunt*. The details of ribosome scanning under different situations are presented in Fig. 9.18.

Cap-independent initiation of naturally uncapped mRNAs

It is known that viral mRNAs, in general, and also a number of eukaryotic mRNAs are naturally uncapped. Therefore, the above-described scanning model does not apply to translation initiation of these mRNAs. By using deletion mutagenesis in polio virus mRNAs, it has been shown that an internal sequence in the 5′ UTR of the RNAs is required for cap-independent translation. Subsequently, this sequence was found to confer cap-independent translation of heterologous mRNAs as well. This therefore raised the interesting possibility of binding of ribosomes internally at the 5′ UTR independent of the cap. The first such experimental evidence demonstrating internal ribosome entry in polio virus mRNA was reported by Sonenberg and co-workers in 1988. They termed the binding sites as *Ribosome Landing Pad* (RLP) or *Internal Ribosome Entry Sites* (IRES).

Now, the question is: what is the molecular mechanism by which ribosomes bind internally to the mRNAs, and what is the nature of the sequence of the RLP? It is known that the cap structure of eukaryotic mRNAs mediates melting of secondary structure in the 5′ UTR through the action of the cap-binding complex (eIF-4F). This facilitates binding of the 40S ribosomal subunit. This melting activity is dependent on two other initiation factors, eIF-4A and eIF-4B. It is also possibly true that in polio virus mRNA (cap-independent) translation, eIF-4F is not required while eIF-4A and eIF-4B are required. Based on these observations, the current hypothesis for poliovirus mRNA translation is: eIF-4A in conjunction with eIF-4B binds to an internal sequence of the mRNA (Fig. 9.19). However, there is no consensus sequence onto which ribosomes bind internally, and also the nature of the sequence is unclear.

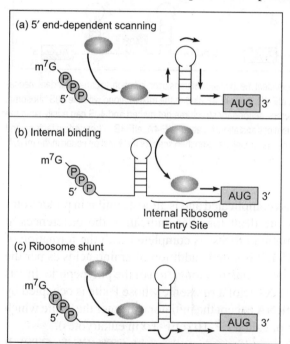

Figure 9.18 Various modes of ribosome scanning. Ribosome scans the mRNA in search of the initiation codon. (a) First mode is 5′ end– dependent scanning, wherein the small ribosomal subunit binds at the 5′ 7methyl guanosine cap and then scans the mRNA having secondary structure for initiation codon. (b) Second mode is internal binding, wherein the small ribosomal subunit directly binds to the mRNA at the internal ribosome entry site (IRES) and then it scans for the initiation codon. (c) Third mode is ribosome shunt, wherein the small ribosomal subunit binds at the 5′ 7methyl guanosine cap and jumps the secondary structure while searching for the initiation codon.

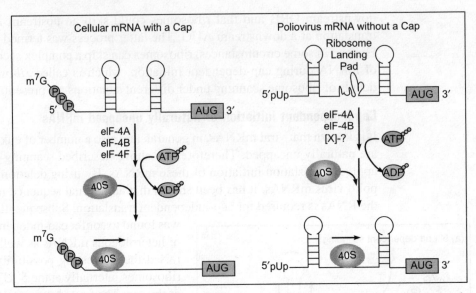

Figure 9.19 Cap-dependent and Cap–independent mechanisms of translation initiation. In cap-dependent mechanism of translation initiation, eIF-4A, eIF-4B, and eIF-4F bind to the mRNA along with the 40S ribosomal subunit, and due to these protein factors, secondary structures get melted and 40S can easily scan the mRNA. In cap-independent mechanism of translation initiation, eIF-4A, eIF-4B, other protein factors, and 40S ribosomal subunit bind at the ribosome landing pad and then scan for the initiation codon in the mRNA.

9.3.2 Elongation

This step of protein synthesis is less complicated and is almost similar in prokaryotes and eukaryotes. Therefore, they are dealt together, indicating the differences at appropriate places. Once the initiation process is completed with the formation of the complete ribosome at the AUG, it is simple addition of amino acids as per the codes in the mRNA in a cyclic manner causing elongation of the polypeptide. In this cycle, aminoacyl-tRNA enters the A site of a ribosome whose P site is occupied by peptidyl-tRNA. Any aminoacyl-tRNA except the initiator can enter the A site which is mediated by an elongation factor called EF-Tu (eEF-1α in eukaryotes).

EF-Tu is a G-protein, and it forms a binary complex with GTP (EF-Tu.GTP). It then binds to an aminoacyl-tRNA to form a ternary complex, aminoacyl-tRNA.EF-Tu.GTP. This complex binds only to the A site of the ribosome whose P site is already occupied by a peptidyl-tRNA. This precise activity ensures that the two RNAs are correctly positioned on the ribosome for peptide bond formation between two amino acids (Fig. 9.20). After aminoacyl-tRNA has been placed in the A site, GTP is hydrolyzed and the resulting inactive binary complex EF-Tu.GDP is released. The accuracy of protein synthesis is determined by the insertion of the correct aminoacyl-tRNA to the A site. Despite the specific interaction of the codon with the anticodon, there could be error, which is taken care by a 'decoding centre' in the small ribosomal subunit. This recognizes the correct codon-anticodon base pairs, and it can discriminate the wrongly added aminoacyl-tRNA. It appears that

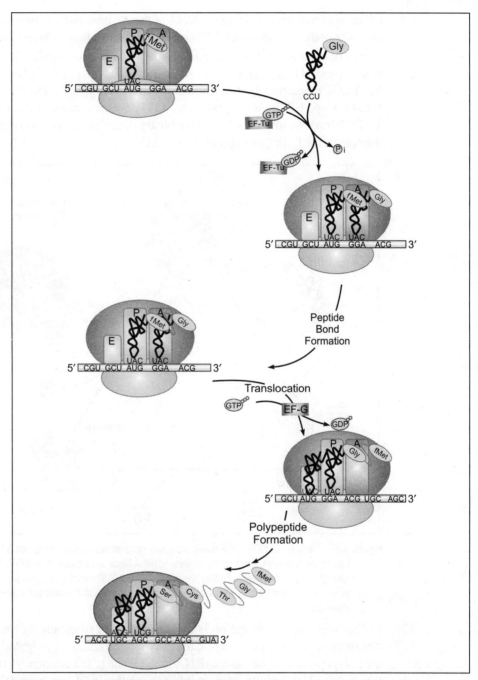

Figure 9.20 Elongation during translation. First, initiator tRNA binds at the P site in the ribosome. In the second step, the next aminoacyl tRNA binds at the A site. After this, there is peptide bond formation and further, the initiator tRNA and the glycylyl tRNA (shown here) are shifted to the E site and P site, respectively. Then, the next aminoacyl tRNA arrives at the A site and the process is repeated till the termination codon is reached. **(See Colour Plate 13)**

addition of a correct aminoacyl-tRNA at the A site causes a conformational change that induces the hydrolysis of GTP bound to EF-Tu, resulting in the release of EF-Tu.GDP.

EF-Tu is recycled by being activated by another factor called EF-Ts (eEF-1$\beta\gamma$ in eukaryotes). It takes place in two steps. In the first step, EF-Ts displaces GDP from EF-Tu.GDP complex by forming EF-Tu.EF-Ts complex, called the *T factor*. In the second step, EF-Ts is displaced by GTP, thereby regenerating the active EF-Tu.GTP binary complex. This active binary complex binds aminoacyl-tRNA, while the released EF-Ts can recycle (Fig. 9.21).

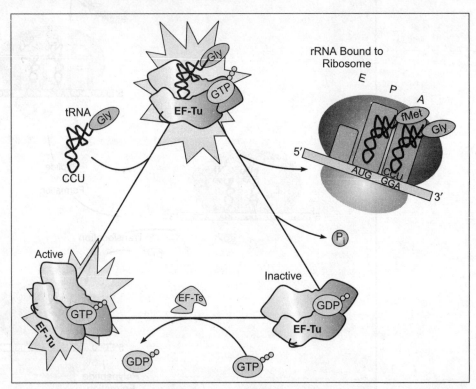

Figure 9.21 Recycling of EF-Tu.GTP during elongation of translation. During elongation, EF-Tu.GTP binds to the next aminoacyl tRNA, and if correct aminoacyl tRNA binds to the codon, then EF-Tu.GTP gets hydrolyzed and EF-Tu.GDP is released. Further, the inactive EF-Tu.GDP complex is converted to EF-Tu.GTP with the help of EF-Ts and thus, the EF-Tu.GTP is ready for binding to the next aminoacyl tRNA. **(See Colour Plate 15)**

The next step in elongation is the movement of ribosome by translocation. The ribosome remains in place, while the polypeptide chain is elongated by transferring the polypeptide attached to the tRNA in the P site to the aminoacyl-tRNA present in the A site. The activity responsible for synthesis of the peptide bond is called *peptidyl transferase*. This is a function of the large subunit of ribosome. Recent information on the structural studies of the ribosome as described earlier indicated that the peptidyl transferase activity is a function of the ribosomal RNA and not the proteins present in the large subunit of ribosomes.

Following peptide bond formation, a ribosome carries uncharged tRNA in the P site and peptidyl tRNA in the A site. The cycle of addition of amino acids to the growing polypeptide chain is completed by the translocation, in which the ribosome advances three nucleotides along the mRNA. In a concerted action, translocation simultaneously expels the uncharged tRNA from the P site and moves the peptidyl-tRNA into the P site. The ribosome then has an empty A site ready for entry of the aminoacyl-tRNA corresponding to the next codon. The discharged tRNA leaves the ribosome via another site called the *exit* (E) *site*. Now the ribosome is ready for insertion of the next amino acid in the growing polypeptide chain. For further details on the activities of these three sites, A,P and E, one may refer to the Moazed and Noller's model published in 1989.

Translocation requires GTP and another elongation factor EF-G. It binds to the ribosome to sponsor translocation, and then it is released when GTP gets hydrolyzed. It appears that this GTP hydrolysis is a function of the ribosome and not of the factor EF-G. Another potential factor involved in translocation is 4.5S RNA which is largely double stranded and is essential in *E. coli*. It is likely that this interacts with EF-G in some manner. The eukaryotic counterpart to EF-G is the protein eEF-2. This seems to function in a similar manner as described above.

mRNA is associated with the small subunit of ribosomes through about 30 bases being bound at any time, but only two molecules of tRNA are involved in peptide bond synthesis at any moment. Thus, polypeptide elongation involves reactions taking place at just two of the ten codons (roughly) covered by the ribosome. Each tRNA lies in a distinct site, and the two sites have distinct features. Much of the information regarding individual steps of bacterial protein synthesis has been obtained by using antibiotics that inhibit the process at particular stages. The details of the antibiotics and their mode of action are given in Table 9.4.

9.3.3 Termination

As described earlier, only 61 triplets are assigned to amino acids. The other three (UAG, UAA, and UGA) are termination codons that end protein synthesis. In every gene that has been sequenced, one of the termination codons lies immediately after the codon, representing the C-terminal amino acid. None of the termination codons is represented by a tRNA. They function in an entirely different manner from other codons, and are recognized directly by protein factors. Since the reaction does not depend on codon-anticodon recognition, there seems to be no particular reason why it should require a triplet sequence. Presumably, this reflects an aspect of the evolution of the genetic code.

In *E. coli*, there are two release factors, RF-1 and RF-2, which catalyze termination. While RF-1 recognizes UAA and UAG, RF-2 recognizes UGA and UAA. These factors act at the ribosomal A site, and require peptidyl-tRNA in the P site. These factors have 30% identity, and it is proposed that they have originated from a single factor and have evolved into two with codon-specific recognition. A

Table 9.4 Antibiotic inhibitors of protein synthesis

Chemical Class	Examples	Biological source	Spectrum	Mode of Action
Aminoglycosides	Streptomycin	*Streptomyces griseus*	Gram-positive and Gram-negative bacteria	Inhibits initiation and causes misreading
	Gentamycin	*Micromonospora species*	Gram-positive and Gram-negative bacteria esp. Pseudomonas	Binds to 30S ribosomal subunit
	Kanamycin	*Streptomyces kanamyceticus*	Gram-negative and partially against Gram-positive	Binds to 30S ribosomal subunit
	Hygromycin B	*Streptomyces hygroscopicus*	Bacteria, fungi, and higher eukaryotic cells	Induces misreading of aminoacyl tRNA by distorting the ribosomal A site.
	Kasugamycin		Prokaryotic cells	Inhibits 70S initiation complex formation
Lincomycins	Clindamycin	*Streptomyces lincolnensis*	Gram-positive and Gram-negative bacteria esp. anaerobic bacteroides	Binds to 50S ribosomal subunit
Macrolides	Erythromycin	*Streptomyces erythreus* Now called as *Saccharopolyspora erythraea*	Gram-positve bacteria, Gram-negative bacteria not enterics, *Neisseria, Legionella, Mycoplasma*	Binds to 23S rRNA molecule in the 50S ribosomal subunit, inhibiting the translocation of peptides
Tetracyclines	Tetracycline	*Streptomyces species*	Gram-positive and Gram-negative bacteria, Rickettsias	Inhibits action of the prokaryotic 30S ribosomal subunit, by binding to aminoacyl-tRNA
Semisynthetic Tetracycline	Doxycycline	....	Gram-positive and Gram-negative bacteria, Rickettsias, Borellia	Inhibits action of the prokaryotic 30S ribosomal subunit, by binding to aminoacyl-tRNA
Chloramphenicol	Chloramphenicol	*Streptomyces venezuelae*	Gram-positive and Gram-negative bacteria	Binds to 50S ribosomal subunit and interferes with peptidyl transferase activity
Everninomycins	Evernimicin Avilamycin		Gram-negative and Gram-positive bacteria	Interferes with the formation of 40S Initiation complex

Contd.

Table 9.4 *Contd.*

Chemical Class	Examples	Biological source	Spectrum	Mode of Action
Streptogramin	Quinupristin Dalfopristin		Prokaryotic organisms	Binds to 50S ribosomal subunit
Aminocyclitol	Spectinomycin	*Streptomyces spectabilis*	Prokaryotic organisms	Inhibits tRNA translocation
Cycloheximide	Cycloheximide	*Streptomyces griseus*	Eukaryotic organisms	Interferes with peptidyl transferase activity of the 60S ribosome, thus blocking translational elongation.
Aminonucleoside	Puromycin	*Streptomyces alboniger*	Prokaryotes and Eukaryotes	Causes premature chain termination

The antibiotics which inhibit translation, their source organisms, effector organisms, and the mode of action are detailed in this table.

third release factor, RF-3, which does not recognize any specific termination codon, acts along with RF-1 and RF-2.

In eukaryotes, two similar release factors, eRF-1 and eRF-2, are present. GTP is required for them to bind to the ribosomes, and their release is through GTP hydrolysis. eRF-1 alone recognizes all three termination codons. There is also eRF-3, which functions in a similar manner as the bacterial RF-3.

The termination relates to the release of the fully formed polypeptide from the last tRNA, ejection of the tRNA from the ribosome, and eventual dissociation of ribosome from mRNA (Fig. 9.22). The mechanism of the latter process is yet unclear. In prokaryotes, it is however postulated that this is achieved through an additional protein called *ribosome recycling factor* (RRF). RRF has a tRNA-like structure, and it appears to enter the P or A site and catalyze dissociation of the ribosome into its subunits. In eukaryotes, on the other hand, no RRF-like factor has yet been identified.

A single mRNA can be translated simultaneously by several ribosomes in both prokaryotes and eukaryotes. Therefore, the group of ribosomes bound to an mRNA at a particular time is called a *polysome*. In eukaryotes, generally there are 4-5 ribosomes in the polysomes (Fig. 9.23).

9.4 IN VITRO CELL-FREE TRANSLATION SYSTEMS

As mentioned before, various *in vitro* systems have been used to study and define the process of translation. Further, the comparison of *in vitro* and *in vivo* systems has also provided two important aspects of regulation and modifications of proteins.

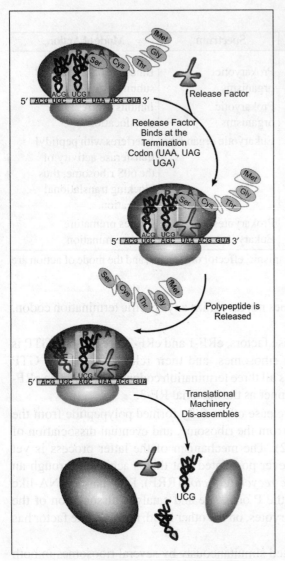

Figure 9.22 Termination of translation. When the termination codon is encountered by the ribosome, the release factor binds at the termination codon (UAA, UGA, and UAG). After the binding of release factor, translational machinery dis-assembles, as ribosome dissociates and tRNA is released. **(See Colour Plate 13)**

It is evidently demonstrated that some mRNAs although translatable *in vitro*, do not, or poorly translate *in vivo*. Such a phenomenon has been referred as *translation regulation* of gene expression. The above comparison has also indicated that there is *in vivo* processing of precursor protein into a functional one. The details of various *in vitro/in vivo* translation systems that are used are described below.

In the reconstituted cell-free systems, the main components are ribosomes, protein synthesis factors, and tRNAs. Among many such systems that are developed, the main ones are derived from: *E. coli*, yeast, wheat germ, and rabbit reticulocyte. The main drawback of such systems is their inefficiency; the mRNAs are translated at much lower rates, duration (90–120 min), and in number as compared to those in the *in vivo* conditions. However, more recently, Spirin and co-workers (1988) described a continuous flow cell-free translation system using rabbit reticulocyte lysates. This system uses a continuous flow of the feeding buffer including amino acids, ATP and GTP, through the reaction mixture and a continuous removal of the polypeptide product. As per these authors, by using such a system, mRNAs could be translated (in both eukaryotes as well as prokaryotes) for long times, at a high constant rate for tens of hours. As an example, using synthetic calcitonin mRNA, the corresponding polypeptide could be produced up to 300 copies/mRNA for 40 hrs.

Compared to the above systems, an *in vivo* or rather an *in ovo* system, using intact Xenopus oocyte, has been extensively used for translation of exogenous mRNA introduced through microinjection. Such a system does not discriminate the exogenous mRNA, and thus, it can translate the same like any other endogenous mRNA through repeated rounds of translation. This system remains active almost till 24 to 48 hrs. The detailed comparisons of all the above are presented in Fig. 9.24.

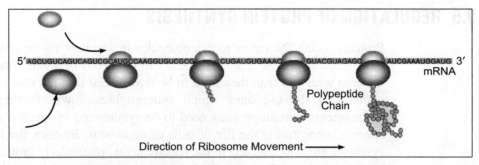

Figure 9.23 Polysomes and protein synthesis. During protein synthesis, many ribosomes bind to the mRNA consecutively for translation. This structure, i.e. ribosomes bound to mRNA, is called as polysomes.

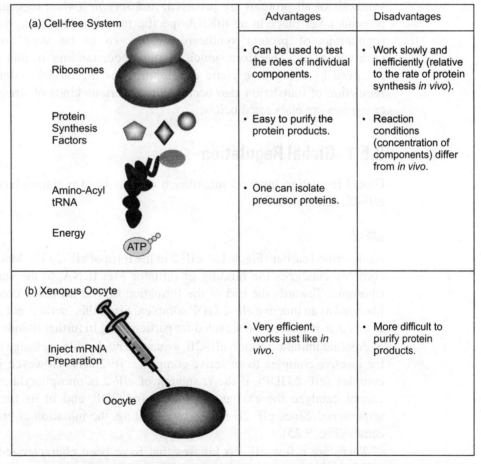

	Advantages	Disadvantages
(a) Cell-free System Ribosomes Protein Synthesis Factors Amino-Acyl tRNA Energy	• Can be used to test the roles of individual components. • Easy to purify the protein products. • One can isolate precursor proteins.	• Work slowly and inefficiently (relative to the rate of protein synthesis *in vivo*). • Reaction conditions (concentration of components) differ from *in vivo*.
(b) Xenopus Oocyte Inject mRNA Preparation Oocyte	• Very efficient, works just like *in vivo*.	• More difficult to purify protein products.

Figure 9.24 *In vitro* and *in ovo* protein synthetic systems. *In vitro* protein synthetic systems use purified ribosomes, protein synthesis factors, aminoacyl tRNA, and energy. *In ovo* systems utilize *Xenopus* oocyte. The advantages and disadvantages of both the systems are mentioned in the figure.

9.5 REGULATION OF PROTEIN SYNTHESIS

Proteins, being the major macromolecules in a cell or an organism, need to be maintained in precise amounts to cater to the needs of a cell. Since the proteins undergo wear and tear, they need to be replenished through new synthesis, at the expense of the old ones which undergo breakdown. Further, for specific requirements, certain proteins need to be synthesized in specific amounts and at specific times during the life of cells or organisms. Besides, the rates of protein synthesis are also adjusted to the requirements, particularly growth phase versus stationary phase. To meet all these requirements, protein synthesis in a cell/organism is highly regulated.

There are two broad categories of regulation: *global* (when regulation of synthesis of all proteins are involved) and *specific* (when regulation of specific proteins takes place in an mRNA-specific manner). Interestingly, both types of regulations of protein synthesis are known to be very common during embryogenesis. Therefore, much of our understanding in this area has been achieved by using embryonic development as the model system. The global regulation of translation also occurs during various kinds of stresses. These two categories are elaborated below.

9.5.1 Global Regulation

Global regulation involves modulation of activity of initiation factors, eIF-2 and eIF-4E, in particular.

eIF-2

As described earlier (Fig. 9.16), eIF-2 in the form of eIF-2.GTP.Met tRNA$_i$ ternary complex catalyzes the binding of initiator Met tRNA$_i$ to the small subunit of ribosome. Towards the end of the formation of 80S initiation complex, eIF-2 is liberated as an inactive eIF-2.GDP complex. Since this factor is not abundant in the cytoplasm, it needs to be recycled for participating in further rounds of initiation.

Another initiation factor, eIF-2B, a guanine nucleotide exchange factor, converts the inactive complex to an active complex, eIF-2.GTP. However, in this inactive complex (eIF-2.GDP), if the α subunit of eIF-2 is phosphorylated, then eIF-2B cannot catalyze the exchange of GTP for GDP, and in its turn eIF-2B gets sequestered. Since eIF-2B is also rate limiting, the initiation of protein synthesis ceases (Fig. 9.25).

There are a few eIF-2α kinases that have been characterized and they are activated differently and are involved in specific cell types or conditions. They are:

(a) the heme-regulated eIF-2α kinase (abundant in erythroid cells), which gets activated during heme-deficiency, stress, etc.

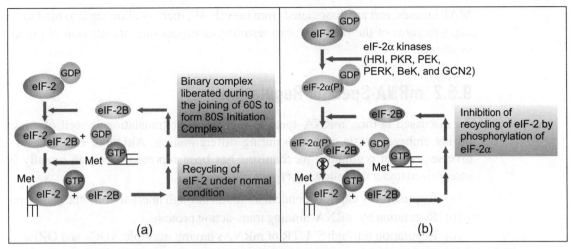

Figure 9.25 Role of eIF-2α kinases in initiation of protein synthesis in eukaryotes. **(a)** Recycling of eIF-2 by exchange of GDP for GTP from the binary complex catalyzed by eIF-2B. **(b)** Inhibition of GDP-GTP exchange occurred because of phophorylation of eIF-2α at Ser51 catalyzed by eIF-2α kinases during different cytoplasmic stresses. Phosphorylated eIF-2 tightly binds to eIF-2B and therefore sequesters it. As eIF-2B is a rate limiting factor, unavailability of free eIF-2B inhibits recycling of eIF-2 by inhibition of GDP-GTP exchange and hence inhibits global translation initiation.

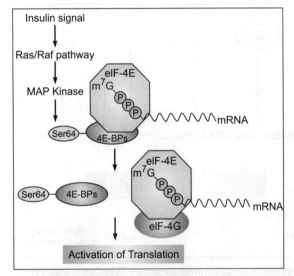

Figure 9.26 Stimulation of translation initiation by more availability of eIF-4E. eIF-4E, the cap binding protein, is crucial for cap-dependent translation of mRNA. eIF-4E-BPs bind to eIF-4E and inhibit its cap binding activity. MAP kinase phosphorylates eIF-4E-BPs at serine 64. Thus, phosphorylated eIF-4E-BPs cannot bind to eIF-4E and translation begins.

(b) the double-stranded RNA-dependent protein kinase, PKR, activated by viral RNA,

(c) the endoplasmic reticulum-specific kinase, PERK, regulated by ERK stress, and

(d) the yeast eIF-2α kinase (GCN2) activated by amino acid starvation.

Thus, during certain conditions, such as heme-deficiency, environmental stress, viral infection, and amino acid starvation, etc., the global protein synthesis is regulated by eIF-2α kinases through modification of the eIF-2.

eIF-4E

eIF-4E, the cap binding protein which is crucial for cap-dependent translation of mRNA, actually controls interaction of mRNA with the ribosomal complex (Fig. 9.16). The binding of eIF-4E to the cap structure is regulated by two protein factors called the eIF-4E-BPs (4E-BP1 and 4E-BP2). These protein factors can bind to eIF-4E and inhibit its cap binding (Fig. 9.26). However, during growth factor stimulation, these factors get phosphorylated by

MAP kinases, and are dissociated from the eIF-4E, thereby allowing it to bind to the cap structures of the mRNAs, thus resulting in higher rate of initiation of protein synthesis.

9.5.2 mRNA-Specific Regulation

As discussed before, mRNA-specific regulation of translation is well observed during embryogenesis, and also during differentiation. Although existence of a diverse array of regulatory mechanisms has been reported, they can broadly be categorized under the following types:

(a) Regulation by cap-independent initiation through internal ribosome entry sites,

(b) Regulation by mRNA-binding trans-acting proteins,

(c) Regulation through 5′ UTR of mRNAs having multiple AUGs and ORFs,

(d) Regulation by controlling the size of poly A tail of mRNA, and

(e) Regulation by complementary non-coding RNAs.

A schematic presentation (Fig. 9.27) summarizes these strategies of translation regulation. The details are described below.

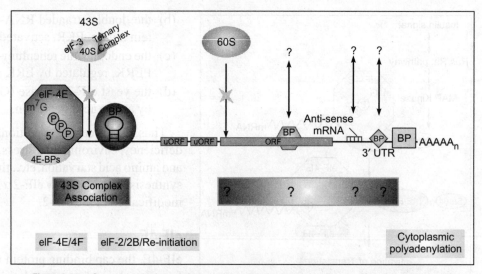

Figure 9.27 Summarized scheme of translation regulation in eukaryotic cells. **(See Colour Plate 15)**

Regulation by cap-independent initiation through internal ribosome entry sites As discussed earlier (Fig. 9.19), members of picornavirus family and other viruses produce non-capped mRNAs with long (600–1200 nucleotides) 5′ UTR containing multiple non-initiating AUGs. The translation efficiency of these mRNAs is dependent on the specific sequences within the 5′ UTR called *internal ribosome entry sites* (IRES). It turns out that a number of eukaryotic mRNAs of proteins that are important for stress response, embryonic development, cell cycle, etc. have IRES. These mRNAs are not affected for translation under adverse conditions when other mRNAs are unable to undergo translation. Further, the efficiency of translation of such mRNAs are much higher than that of capped mRNAs.

Regulation by mRNA-binding transacting proteins It is known that ribosome binding is most sensitive to secondary structure of mRNA near the cap. However, stable stem-loop structures anywhere in the 5′ UTR can block AUG scanning by ribosome (Kozak, 1989). Therefore, transacting protein factors that stabilize the secondary structures in the 5′ UTR can be potent inhibitors of translation. The best studied example is the regulation of mammalian ferritin mRNA translation (Fig. 9.28). This is regulated by an RNA-binding repressor protein which is regulated by iron concentration in the cell and is thus called the *iron regulatory protein* or *iron regulatory factor* (IRP/IRF). IRP interacts with a stretch of nucleotides that can fold into a stem-loop structure at the 5′ UTR of ferritin mRNA called the *iron responsive element* (IRE) and thereby inhibits its translation. In high iron concentration, iron binds to IRP and prevents its interaction with IRE which results in translation of mRNA.

In a reverse scenario, there may be positive regulation by binding of protein factors to the 5′- or 3′ UTR of mRNAs leading to stimulation of translation. An

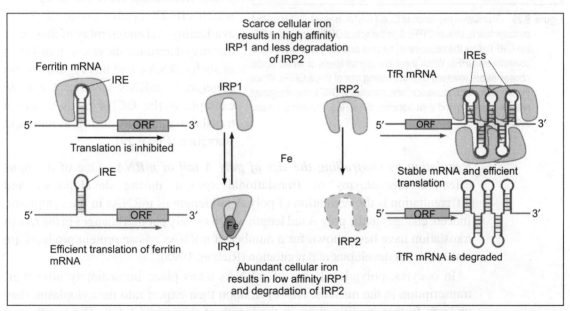

Figure 9.28 Translation regulation of ferritin and transferrin receptor mRNAs. 5′ UTR of ferritin mRNA has IRE (iron responsive element). The IRP (iron regulatory protein) binds to IRE and inhibits the translation of the mRNA. When iron is present in excess, iron binds to IRP, thus IRP cannot bind to IRE and translation occurs. 3′ UTR of transferrin receptor mRNA also contains IREs and IRPs binding to it stabilize the mRNA and thus efficient translation occurs. But when iron is present in excess, IRPs do not bind to IREs and mRNA gets degraded.

excellent example of such regulation is the translation of transferrin receptor mRNA, which has a very elaborate secondary structure in the 3′ UTR region containing IREs. The same protein factor which negatively regulates ferritin mRNA translation, regulates transferrin receptor mRNA translation positively by interaction with IREs located in the 3′ UTR (Fig. 9.28). Such an interaction prevents degradation of the mRNA and stabilizes it.

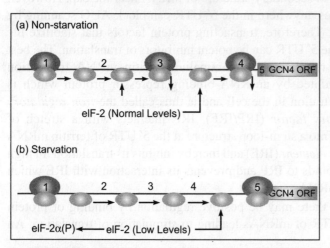

Figure 9.29 Translation regulation of GCN4 mRNA in yeast. GCN4 mRNA contains four upstream ORFs 5′ to the actual ORF. Sequences 3′ to the ORF1 allow the ribosome to resume scanning and to initiate at downstream ORFs. When there are normal levels of amino acids, ribosomes are released after translating one of these ORFs. When there is amino acid starvation, after translating ORF1, the ribosomes resume scanning and then translate the ORF5, the actual coding sequence.

Regulation through 5′ UTR of mRNAs having multiple AUGs and ORFs Eukaryotic ribosomes have a low capacity to reinitiate translation at downstream cistrons; upstream ORFs reduce the translation of downstream major ORF. Thus, mRNAs which contain multiple non-initiating AUGs 5′ to the actual initiating AUG resulting in a number of additional 5′ upstream ORFs are highly regulated for translation. In such mRNAs, reinitiation into the correct ORF would necessitate recycling of eIF-2 in which eIF-2B is also involved. Thus, availability and an interplay of these factors may determine the rate of translation of such mRNAs, and this regulation can therefore be mRNA-specific. A notable example is the GCN4 mRNA, which translates manifold during amino acid starvation (Fig. 9.29).

Regulation by controlling the size of poly A tail of mRNA One of the most effective mechanisms of translational control during development and differentiation is the regulation of poly A tail length of mRNAs in the cytoplasm. Indeed, changes in the poly A tail length which closely parallel changes in the rate of translation have been shown for a number of mRNAs whose protein products are invaluable for developmental regulation (Richter, 1995).

In oocytes, polyadenylation of mRNAs takes place immediately after their transcription in the nucleus. However, upon their export into the cytoplasm, they undergo further modifications in the length of their poly A tail. The length may increase due to further addition of adenine residues, or it may decrease due to deadenylation (Fig. 9.30). It is likely that both these processes that are mediated by

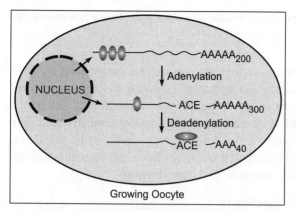

Figure 9.30 Translation regulation by changing the length of poly A tail. In oocytes, polyadenylation takes place after the transcription of the mRNA. The length of the poly A tail is altered by adenylation or deadenylation by cytoplasmic enzymes, thereby regulating its translation.

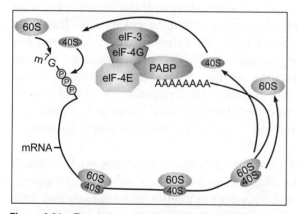

Figure 9.31 Translation regulation by eIF-4G and PABP through circularization of mRNA. Multiple ribosomes can simultaneously translate the mRNA. The circular form generated due to interaction of eIF-4G and PABP is stabilized by proteins which are bound to the 5'- and 3' end of the mRNA. The ribosome which dissociates from the 3' end of mRNA can easily find the neighbouring ribosomal subunits for re-assembling at the 5' end.

cytoplasmic enzymes occur continuously, and a balance between them determines the steady state length of the poly A tail of mRNA which is favourable for its translation.

It has been shown that there is a direct correlation between the length of the poly A tail and the rate of initiation of translation of capped mRNAs. This indicates that poly A tail enhances the formation of cap binding complex. Further, it is known that poly A binding protein (PABP) is necessary for efficient translation. It appears that PABP and eIF-4G interact in bringing together the 5' and 3' ends of mRNA physically, which enhances the rate of translation of mRNA. This increased translation efficiency of some mRNAs may be due to shunting the ribosomes from the 3' end to the 5' end for reinitiation. A number of additional proteins may be involved in this complex (Fig. 9.31). It has also been shown by electron microscopic analysis that a number of polysomes are circular.

Regulation by complementary non-coding RNAs Another unique mechanism of translation regulation of certain mRNAs through their interaction with complementary RNA (anti-sense) has been reported, once again from the developing embryos. In *Caenorhabditis elegans*, translation of *lin 14* mRNA is regulated through some sequences at its 3' UTR. It has been shown by genetic analysis that a negative regulation of *lin 14* is mediated by the products of another gene *lin 4*. Two transcripts of *lin 4* (22 and 61 nucleotides long) which do not produce proteins, bind to complementary sequences at the 3' UTR of *lin 14* mRNA. Such an interaction leads to inhibition of translation of *lin 14* mRNA (Fig. 9.27).

SUMMARY

Protein synthesis

Translation of mRNAs into proteins represents the second major step of gene expression. This step of gene expression occupies a pivotal place during many important processes in cells/organisms, such as gametogenesis, fertilization, early embryonic development, differentiation, cancers, and environmental stresses. This process that consumes a major part of the cell's energy needs to be regulated precisely in quantity and quality. Thus, both protein synthesis and its regulation are very important for the cell's growth, survival, and reproduction.

Components of protein synthesis machinery

The three major components required for protein synthesis are: (a) mRNA, (b) tRNA, and (c) ribosomes. Besides, there are a number of protein factors, as cohorts, and energy sources in the form of ATP and GTP that are essential for protein synthesis.

mRNA being transcribed from DNA (gene) carries information for the synthesis of polypeptides, and therefore, it acts as the template. mRNAs in prokaryotes are polycistronic, while they are mostly monocistronic in eukaryotes. Unlike in eukaryotes in which translation takes place in the cytoplasm, in prokaryotes, it is coupled with transcription. Eukaryotic mRNAs which contain introns get matured through splicing and modifications at the 5'-end (capping) and 3'-end (polyadenylation) before being transported to the cytoplasm. A typical mRNA contains a coding sequence that gives rise to the amino acids of the polypeptide, and 5' and 3' untranslated regions that are required for stability and regulation of translation.

The coding sequence of mRNA contains sets of trinucleotides, each of which codes for a particular amino acid. These trinucleotide sets are called the genetic codes. There are 20 amino acids and 61 codons in a cell, which indicates that there is redundancy of genetic code. Genetic codes are universal although there are exceptions in the mitochondria and chloroplasts and lower organisms.

tRNAs act as the adaptors, which match the amino acids in the cytoplasm with the genetic codes in the mRNA during translation. The characteristic adaptor function of tRNA is due to its characteristic folded structure. The specific recognition of the amino acids by tRNAs is contributed by the specific aminoacyl-tRNA synthetases.

The platform for protein synthesis is provided by the ribosomes, which are high molecular weight nucleoprotein complexes. The ribosomes attach to the mRNAs and migrate along from 5' to 3' direction while catalyzing the formation of polypeptides. Therefore, these are like mobile factories. Ribosomes are composed of a large and a small subunit. Each subunit is composed of 2–3 ribosomal RNAs and 30–50 proteins. Recently, it has been established that the ribosomal RNAs have the required catalytic activity for protein synthesis; ribosomes are thus called as ribozymes.

Mechanism of protein synthesis

It is the concerted actions of the major components and the associated cohorts that result in translation of mRNA into proteins. For our convenience of understanding, protein synthesis can be divided into three steps, namely initiation, elongation, and termination.

Although there are differences in the initiation process for prokaryotes and eukaryotes, the general pattern remains the same. Initiation refers to the formation of a complete 80S ribosome (in eukaryotes) at the initiation codon AUG of mRNA onto which the initiator tRNA is aligned with its anticodon. This process takes place through the following steps: (a) formation of eIF-2.GTP.Met tRNA$_i$ ternary complex, and its binding to the small subunit of ribosomes, (b) binding of such pre-initiation/ribosomal complex to the cap structure of mRNA followed by scanning for the initiation codon, and (c) joining of the large subunit of ribosome to form the initiation complex. Each of these steps is catalyzed by a number of protein factors called the initiation factors. In prokaryotes, where there are no caps in the mRNA, ribosome binds to the Shine-Dalgarno sequence present at the −7 position in the mRNA.

Elongation refers to the elongation of the polypeptide through successive addition of amino acids by the tRNAs and the formation of peptide bonds between amino acids. This process is helped by three elongation factors. The ribosome moves after each addition of amino acids towards the 3′ direction until it encounters the termination codon.

Termination is caused due to the lack of any tRNA for the termination codons, and it is aided by protein factors called the release factors. At the end, the polypeptide and the mRNA are released into the cytosol, the ribosome dissociates, and its subunits are recycled for further translation.

In vitro cell-free translation systems

Most of our understanding on the mechanism of protein synthesis has been acquired by studying various *in vitro* translation systems. The advantages of such cell-free systems are many, and they can be manipulated based on our requirement. The routinely used *in vitro* translation systems are cell-free extracts or lysates of *E. coli*, yeast, wheat germ, and rabbit reticulocytes. The latter two are frequently used for eukaryotic protein synthesis. In addition to these systems, an *in ovo* system made of *Xenopus* oocytes is also used. This system has a higher translation rate and any exogenous mRNA can be translated over a long period of time, unlike the *in vitro* systems.

Regulation of protein synthesis

Proteins, being the most abundant and important macromolecules, need to be maintained in a precise amount to cater to the needs of a cell. Due to wear and tear, they undergo degradation, and thus proteins need to be synthesized in an appropriate amount and at specific times to maintain a steady-state level of proteins. Further, under certain circumstances, specific proteins are selectively synthesized, while the rest are shut down. Thus, protein synthesis is a highly regulated process.

There could be two types of regulations: global- and mRNA-specific regulations. The former is carried out mostly by various factors, and components, i.e. their availability or non-availability in active forms. The latter is mostly dependent on the cis-sequences in the mRNA and trans-acting factors that interact with the cis-sequences. There are many strategies of translation regulation, which can be broadly classified as the following: (a) regulation by the modifications of initiation factors, (b) regulation by changing the length of poly(A) of mRNA, (c) regulation by masking proteins, and (d) regulation by specific sequences and secondary structures in 5′ and 3′ UTRs in the mRNA.

■■■ SHORT ANSWER QUESTIONS ■■■

Test questions

1. How do mRNAs differ structurally and functionally in eukaryotes and prokaryotes? Describe briefly with sketches.
2. Describe the structural organization of ribosomes in detail. What are the two main activities of ribosomes during the elongation step of protein synthesis?
3. What is the recent revision of our concept regarding the catalytic activity of ribosomes? What kind of experiments led to this revision?
4. Describe briefly the experiments that led to the cracking of the genetic code. Explain the redundancy of the genetic code.
5. Explain how tRNA acts as an adaptor molecule. Describe the specific correlation of tRNA structure with its adaptor function.
6. Define charging of tRNA. Explain the two steps through which charging takes place.
7. What is wobble concept? How does it explain the ease of interaction between mRNA and tRNA?
8. Describe translation initiation in eukaryotes with a detailed diagram.
9. What is Kozak's scanning hypothesis? Explain how it fails to explain IRES-mediated initiation of translation.
10. What are the major strategies that are known for translation regulation? Explain three regulatory strategies with suitable diagrams and examples.
11. Explain the mechanism of recycling of EF-Tu during bacterial protein synthesis.

Study questions

1. Give three examples in which translation deregulation results in diseases. Explain the mechanism of deregulation causing those diseases.

2. Name the different eIF2α kinases and the specific conditions in which they get activated. How do they regulate translation during stress? What would happen if such mechanisms were absent in cells?

3. How do uORFs modulate translation? Describe with an example and a detailed diagram.

4. A scientist is successful in making an eIF4E mutant cell, a cell in which eIF4E is constitutively over-expresssed, and another in which both eIF-4E and eIF-4E-BP are over-expresssed. What would be the fate of translation in such cells? Explain your answer with reasons.

5. Draw and label a typical eukaryotic mRNA. What are the functions of 3'-UTR? Explain in detail how various motifs present in the mRNA are involved in translation regulation during embryogenesis and deveoplment.

FURTHER READING

Components of protein synthesis machinery

- Ban, N., Nissen, P., Hansen, J., Moore, P.B., and Steitz, T.A. (2000). The complete atomic structure of the large ribosomal subunit at 2.4 Å resolution. *Science* **289**, 905–920.

- Clark, B.F.C. (2001). The crystallization and structural determination of tRNA. *Trends Biochem. Sci.* **26**, 511–514.

- Hale, S.P., Auld, D.S., Schmidt, E., and Schimmel, P. (1997). Discrete determinants in transfer RNA for editing and aminoacylation. *Science* **276**, 1250–1252.

- Ibba, M. and Soll, D. (2000). Aminoacyl-tRNA synthetases. *Annu. Rev. Biochem.* **69**, 617–650.

- Moore, P.B. and Steitz, T.A. (2003). The structural basis of large subunit function. *Annu. Rev. Biochem.* **72**, 813–850.

- Percudani, R. (2001). Restricted wobble rules for eukaryotic genomes. *Trends Genet.* **17**, 133–135.

- Wimberly, B.T., Broderson, D.E., Clemons, V.M., Morgan-Warren, R.J., Carter, A.P., Vonrhein, C., Hartsch, T., and Ramakrishnan, V. (2000). Structure of the 30S ribosomal subunit. *Nature* **407**, 327–339.

- Yusupov, M.M., Yusupova, G.Z., Baucom, A., Lieberman, K., Earnest, T.N., Cate, J.H., and Noller, H.F. (2001). Crystal structure of the ribosome at 5.5 Å resolution. *Science* **292**, 883–896.

Mechanism of protein synthesis

- Anderson, G.R., Nissen, P. and Nyborg, J. (2003). Elongation factors in protein biosynthesis. *Trends Biochem. Sci.* **28**, 434–441.
- Frank, J. and Agarwal, R.K. (2000). A rachet-like inter-subunit reorganization of the ribosome during translocation. *Nature* **406**, 318–322.
- Green, R. and Noller, H. F. (1997). Ribosomes and translation. *Annu. Rev. Biochem.* **66**, 679–716.
- Ibba, M. and Soll, D. (1999). Quality control mechanisms during translation. *Science* **286**, 1893–1897.
- Kapp, L.D. and Lorsch, J.R. (2004). The molecular mechanics of eukaryotic translation. *Annu. Rev. Biochem.* **73**, 657–704.
- Kozak, M. (1989). Structural features in eukaryotic mRNAs that modulate the initiation of translation. *J. Cell Biol.* **108**, 229–241.
- Moazed, D. and Noller, H.F. (1989). Intermediate states in the movement of transfer RNA in the ribosome. *Nature* **342**, 142–148.
- Nakamura, Y. and Ito, K. (2003). Making sense of mimic in translation termination. *Trends Biochem. Sci.* **28**, 99–105.
- Nissen, P., Hansen, J., Ban, N., Moore, P.B. and Steitz, T.A. (2000). The structural basis of ribosome activity in peptide bond synthesis. *Science* **289**, 920–930.
- Pelletier, J. and Sonnenberg, N. (1988). Internal initiation of translation of eukaryotic mRNA directed by a sequence derived from poliovirus RNA. *Nature* **334**, 320–325.
- Polacek, N., Gaynor, M., Yassin, A., and Mankin, A.S. (2001). Ribosomal peptidyl transferase can withstand mutations at the putative catalytic nucleotide. *Nature* **411**, 498–501.
- Ramakrishnan, V. (2002). Ribosome structure and the mechanism of translation. *Cell* **108**, 557–572.
- Rodnina, M.V. and Wintermeyer, W. (2001). Ribosome fidelity: tRNA discrimination, proofreading and induced fit. *Trends Biochem. Sci.* **26**, 124–130.
- Spirin, A. S., Baranov, V.I., Ryabova, L.A., Ovodov, S.Y., and Alakhov, Y.B. (1988). A continuous cell-free translation system capable of producing polypeptides in high yield. *Science* **242**, 1162–1164.
- Steitz, T.A. and Moore, P.B. (2003). RNA, the first macromolecular catalyst: the ribosome is a ribozyme. *Trends Biochem. Sci.* **28**, 411–418.

Regulation of protein synthesis

- Farabaugh, P.J. (1996). Programmed translational frameshifting. *Annu.Rev. Genet.* **30**, 507–528.
- Geabauer, F. and Hentze, M.W. (2004). Molecular mechanisms of translational control. *Nat. Rev. Mol. Cell. Biology* **5**, 827–835.

- Herr, A.J., Atkains, J.F., and Gesteland, R.F. (2000). Coupling of open reading frames by translational bypassing. *Annu. Rev. Biochem.* **69**, 343–372.
- Hershey, J.W.B. (1991). Translational control in mammalian cells. *Annu. Rev. Biochem.* **60**, 717–55.
- Holcik, M. and Sonenberg, N. (2005). Translational control in stress and apoptosis. *Nat. Rev. Mol. Cell. Biology* **6**, 318–327.
- Kozak, M. (2005). Regulation of translation via mRNA structure in prokaryotes and eukaryotes. *Gene* **361**, 13–37.
- Richter, J.D. (1995). In: Translational control (Mathews, M., Hershey, J., and Sonnenberg, N. eds.). Cold Spring Harbor Laboratory Press, CSH, New York.

10

Protein Folding and Modifications

OVERVIEW

- Functionality of a protein depends mostly on its life after synthesis.

- Protein life cycle begins with protein folding, continues with processing and/or modifications, and ends with degradation.

- We are beginning to understand that often these processes are linked and very important for proper functioning of a cell.

Protein synthesis, though represents the final step of gene expression, does not necessarily result in functional polypeptides which are actually useful for the cell. There are certain co- and post-translational processes, namely protein folding, protein subunit assembly, cleavage of precursor proteins, and various covalent modifications, which are very important in the formation of functional polypeptides. We will explain these processes in this chapter.

10.1 PROTEIN FOLDING

Protein folding refers to the formation of three-dimensional conformation of proteins due to interaction of the side chains of amino acids present in the protein. Thus, protein folding is principally controlled by the primary structure of proteins. Contrary to the initial thought, this process does not take place spontaneously in the proteins. There are a group of proteins in the cell that help in protein folding, and they are called *molecular chaperones*. These chaperones act as catalysts and they facilitate the process of folding of polypeptides right from the time they are synthesized on the ribosomes. They bind to and stabilize unfolded or partially folded polypeptides, and thereby lead to the formation of correctly folded proteins.

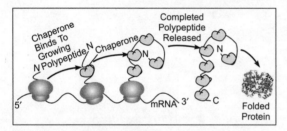

Figure 10.1 Chaperones and protein folding during translation. During protein synthesis, chaperones start binding to the polypeptide from the N-terminus, thus stabilizing it in an unfolded state. After the completion of the polypeptide synthesis, it is released from the ribosome and is then folded correctly.

While any protein is synthesized, chaperones bind to the completed N-terminus of the peptide and help in folding of the parts of the peptide from 50 to 300 amino acid residues, and thereby protect this part while the rest of the peptide is being synthesized, and eventually the entire polypeptide is correctly folded (Fig. 10.1). These molecules also stabilize unfolded polypeptide chains while they

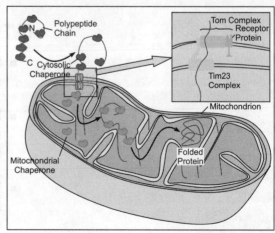

Figure 10.2 Chaperones during protein transport into mitochondria. For the proteins which have to be transported to the mitochondria, the proteins in the unfolded state bound to the chaperones come to the mitochondria. TOM complex in outer mitochondrial membrane and TIM complex in inner mitochondrial membrane provide help in the transport of the polypeptide. Cytosolic chaperones are left outside the mitochondria. After transport, the mitochondrial chaperones bind again to the polypeptide till the transport is complete. Then the polypeptide is folded into a correct form. **(See Colour Plate 16)**

are transported from the site of synthesis in the cytoplasm to various subcellular organelles, such as mitochondria (Fig. 10.2), chloroplasts, etc. Further, chaperones also help in the assembly of proteins that are composed of multiple polypeptide chains.

Chaperones belong to the class of *heat shock proteins* (Hsps), which were discovered in the context of heat shock response. These proteins are now known to be expressed under normal conditions as well, and therefore, they have roles beyond the management of heat shock or other stresses. The more natural role of these proteins is as molecular chaperones. Among various Hsp families, Hsp60 and Hsp70 families are involved in protein folding pathways of both prokaryotes and eukaryotes. Details of various members of these families are given in Table 10.1.

Table 10.1 Chaperone proteins in prokaryotes and eukaryotes.

Protein Family	Prokaryotes/Organelles	Eukaryotes
Hsp70	DnaK BiP (Endoplasmic Reticulum) SSC1 (Mitochondria) ctHsp70 (Chloroplasts)	Hsc73 (cytosol)
Hsp60	GroEL Hsp60 (Mitochondria) Cpn60 (Chloroplasts)	TriC (cytosol)
Hsp90	HtpG Grp94 (Endoplasmic Reticulum)	Hsp90 (cytosol)

Figure 10.3 Structure of a chaperonin. GroEL is a member of the Hsp60 family. **(See Colour Plate 16)**

The members of Hsp70 family stabilize unfolded polypeptide chains during translation as well as during transport of polypeptides to various subcellular compartments. They bind to the short hydrophobic segments of the nascent unfolded polypeptides, keep the polypeptide in unfolded conformation, and also prevent aggregation. In contrast, the members of Hsp60 family, which are also known as *chaperonins*, help folding of proteins into their native conformations. A chaperonin has a 'double doughnut' structure composed of 14 subunits (60 kDa each) being stacked in the form of two rings (Fig. 10.3).

There are processes, such as folding of nascent proteins in *E. coli* and transport of proteins in mitochondria, where members of both the families of Hsp70 and Hsp60 are involved sequentially (Fig. 10.4). This phenomenon appears to be more

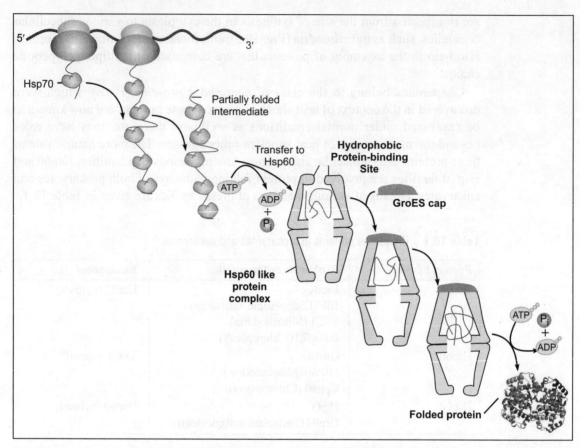

Figure 10.4 Sequential involvement of Hsp70 and Hsp60 chaperones in protein folding. During protein synthesis, Hsp70 binds to the synthesizing polypeptide to keep it in an unfolded state. After protein synthesis, the polypeptide is transferred to Hsp60 using ATP as an energy source for correct folding of the polypeptide. **(See Colour Plate 17)**

general pathway of protein folding in prokaryotes, eukaryotes, and the organelles as well. However, there exists an alternative pathway, particularly in the cytosol and endoplasmic reticulum, where both Hsp70 and Hsp90 family members are involved. For example, proteins involved in cell signaling including steroid hormone receptors and a number of protein kinases use this alternative pathway for protein folding.

Besides the chaperones, there are two important enzymes that are involved in protein folding: *protein disulfide isomerase* and *peptidyl prolyl isomerase*. For secreted and some membrane proteins, formation of disulfide bonds (S-S linkage) between cysteine residues which takes place in the endoplasmic reticulum is very important for stabilizing the folded structure of proteins. The cytosolic proteins, on the other hand, in general, are free of S-S linkages due to the presence of reducing agents. It is in this context, the enzyme protein disulfide isomerase (PDI) catalyzes the breakage and reformation of these bonds (Fig. 10.5). For proteins with multiple

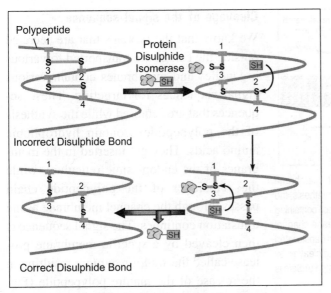

Polypeptide

Protein Disulphide Isomerase

Incorrect Disulphide Bond

Correct Disulphide Bond

Figure 10.5 Protein disulphide isomerase and disulphide bond formation. Protein disulphide isomerase catalyzes the formation and breakage of disulphide bonds. Incorrectly formed disulphide bonds are broken and correct disulphide bonds are formed by exchanging the cysteine residues used in bond formation.

number of cysteine residues, PDI promotes rapid exchange between paired disulfides, thus allowing the protein to attain the correct pattern of protein folding involving the formation of disulfide bonds with compatible pattern.

The other enzyme, peptidyl prolyl isomerase, catalyzes the isomerization of peptide bonds involving proline residues (Fig. 10.6). Peptide bonds between amino acids in general are in the trans form. However, proline residue is an exception: peptide bonds preceding proline residues can be of either cis or trans conformation. Thus, isomerization of the cis and trans configurations of prolyl-peptide bonds, which could otherwise be a rate-limiting step in protein folding, is catalyzed by this enzyme.

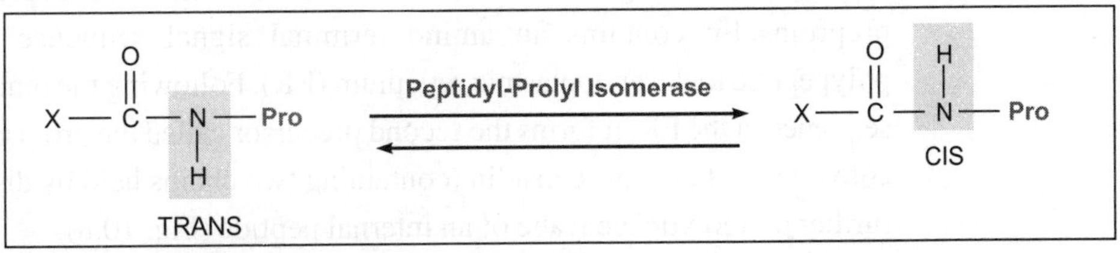

Figure 10.6 Peptidyl prolyl isomerase and isomerisation of peptide bonds. Peptidyl prolyl isomerase catalyzes the peptide bond isomerisation that involves proline between *cis* and *trans* configurations.

10.2 PROTEIN PROCESSING

Processing of precursor proteins to a mature active protein by proteolytic cleavage is important for many proteins. There are three major types of proteolytic cleavages that are common in cells, namely cleavage of the initiator methionine, cleavage of the signal sequence, and cleavage of larger precursor proteins.

Cleavage of the initiator methionine

This type of cleavage from the N-terminus of proteins is common for most of the proteins. It takes place immediately after the N-terminus of the growing polypeptide chain emerges from the ribosome. Following the removal of methionine, in many proteins, addition of acetyl groups or fatty acid chains to the N-terminus takes place.

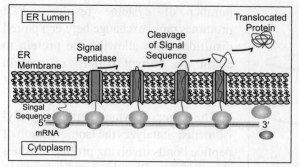

Figure 10.7 Signal sequences and mechanism of membrane translocation of nascent proteins. During protein synthesis, the N-terminus of the polypeptide, i.e. the signal sequence containing hydrophobic amino acids, inserts into the membrane channel from the ribosome. The remaining polypeptide is then translocated through the channel. Then the signal sequence is cleaved by signal peptidase and the properly folded translocated protein is released.

Cleavage of larger precursor proteins

Cleavage of the signal sequence

We know that the proteins that are secreted outside the cell or are transported to various organelles and membranes contain various signal sequences for targeting. These sequences that are acquired while the synthesis of the polypeptides contain hydrophobic amino acids. They get inserted in the membranes of the endoplasmic reticulum, while the remainder of the polypeptide chain passes through the channel membrane as the translation continues. The signal sequence is then cleaved by a specific membrane protease called the *signal peptidase* resulting in the release of the mature polypeptide (Fig. 10.7).

The best example under this category is insulin. Insulin precursor protein undergoes cleavages twice to give rise to the mature insulin. The nascent precursor called the preproinsulin contains an amino terminal signal sequence that targets the polypeptide to the endoplasmic reticulum (ER). Following the removal of the signal sequence in the ER, it forms the second precursor called the proinsulin which is then converted to the mature insulin (containing two chains held by disulfide bonds) by further proteolytic cleavage of an internal peptide (Fig. 10.8).

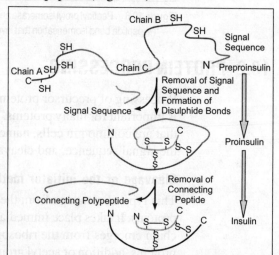

Figure 10.8 Conversion of precursor preproinsulin to functional insulin by proteolytic processing. The preproinsulin contains the chains A, B, C, and the signal sequence. After cleavage of the signal sequence and the disulphide bond formation, proinsulin is formed. Matured insulin results after the removal of internal connecting polypeptide (chain C) from proinsulin.

10.3 PROTEIN MODIFICATIONS

A large number of proteins, in eukaryotic cells in particular, are known to undergo various types of modifications prior to becoming functional. A few major modifications are described below. These modifications actually work as switches for initiating various metabolic processes.

10.3.1 Glycosylation

Protein modification by the addition of carbohydrates is known as glycosylation. These proteins called the glycoproteins are usually secreted or localized to the cell surface. However, there are glycoproteins in the cytosol and nucleus as well. The carbohydrate moieties of these proteins are important for protein folding in the ER, and protein targeting to their precise compartments in the cell. Further, they are highly important for cell-cell recognition.

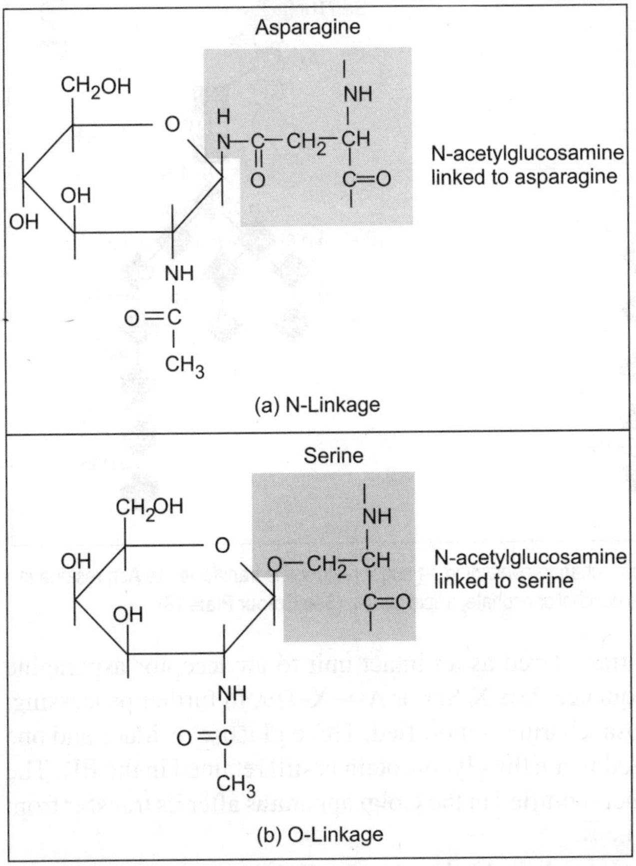

Based on the site of attachment of the carbohydrate side chain, glycoproteins are either N-linked or O-linked (Fig. 10.9). In the former, the carbohydrate is attached to the nitrogen atom in the side chain of asparagine. While in the latter, the carbohydrate is attached to the oxygen atom in the side chain of serine or threonine. The sugars that are directly attached to these positions are usually N-acetylglucosamine or N-acetylgalactosamine, respectively.

As discussed above, the glycoproteins are destined for secretion or incorporation into the plasma membrane. They are thus transferred to the ER, where the glycosylation is initiated.

In the first step, there is the transfer of a common oligosaccharide consisting of 14 sugar residues (2N-acetylglucosamine, 3 glucose, and 9 mannose) to an asparagine residue of the growing polypeptide chain (Fig. 10.10). The oligosaccharide is assembled within the ER on a lipid carrier (dolichol phosphate).

Figure 10.9 Attachment of carbohydrate side chains to glycoproteins through specific linkages. The carbohydrate chains are attached to the side chain of asparagine to form N-linked glycoproteins. In O-linked glycoproteins, the carbohydrates are attached to the side chain of either serine or threonine.

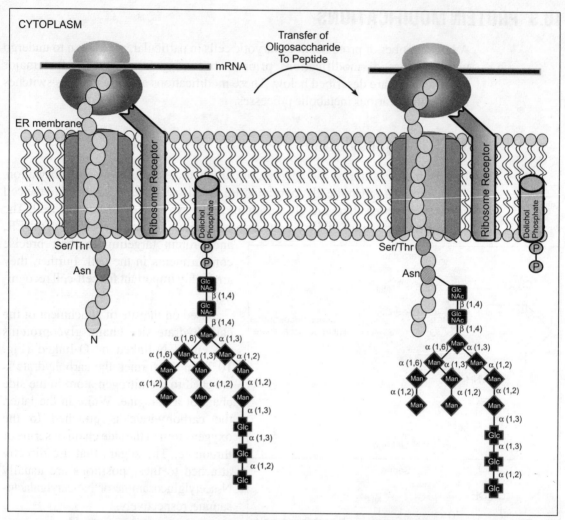

Figure 10.10 Synthesis of N-linked glycoproteins. Oligosaccharide containing 14 sugar residues is transferred to Asn residue in a polypeptide within the sequence Asn-X-Ser/Thr from dolichol phosphate, a lipid carrier. **(See Colour Plate 18)**

In the second step, it is transferred as an intact unit to an acceptor asparagine (Asn) residue within the sequence Asn-X-Ser or Asn-X-Thr. In further processing, the common N-linked oligosaccharide is modified. Three glucose residues and one mannose residue are removed while the glycoprotein is still retained in the ER. The oligosaccharide is then further modified in the Golgi apparatus after its transfer from the ER to Golgi (Fig. 10.11).

O-linked oligosaccharides are also added within the Golgi apparatus. In contrast to the N-linked oligosaccharides, these are formed by the addition of one sugar at a time, and usually consist of only a few residues (Fig. 10.12).

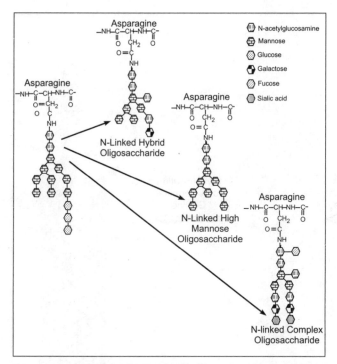

Figure 10.11 Examples of N-linked oligosaccharides. After the addition of oligosaccharide containing 14 sugar residues, further modifications of oligosaccharide occurs in many ways. In N-linked high mannose, oligosaccharide glucose and some mannose residues are removed. In N-linked complex, oligosaccharide glucose and more mannose residues are removed and other sugars are added. Hybrid oligosaccharides are in between these two.

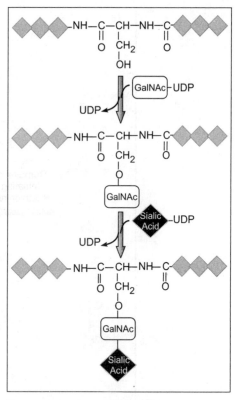

Figure 10.12 Examples of O-linked oligosaccharides. In this class, sugars are added to a polypeptide one after another in the form of sugar–UDP complex.

10.3.2 Attachment of Lipids

Modifications of some eukaryotic proteins by attachment of lipids target and anchor these proteins to the plasma membrane. There are three types of lipid additions for the proteins associated with the cytoplasmic face of the plasma membrane: *N-myristoylation*, *prenylation*, and *palmitoylation*. Besides these, another type of modification, the addition of glycolipids, is important for anchoring some cell surface proteins to the extracellular surface of the plasma membrane.

N-myristoylation

In this process a myristic acid, a 14 carbon fatty acid is attached to an N-terminal glycine residue of some proteins during translation (Fig. 10.13). The glycine which is generally the second amino acid is available for fatty acid addition only after removal of the initiator methionine by proteolysis. Many such N-myristoylated

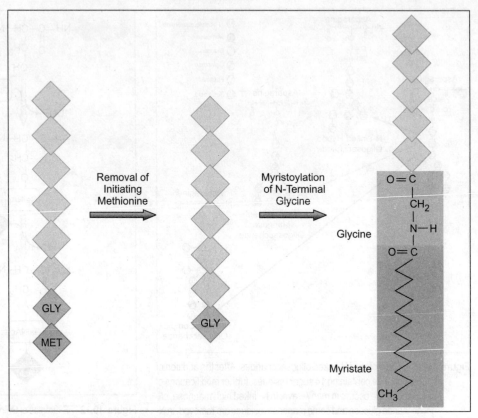

Figure 10.13 Addition of a fatty acid by N-myristoylation. First step is the removal of initiating methionine and in the second step, myristate is added to the N-terminal glycine, leading to N-myristoylation.

proteins are associated with the inner face of the plasma membrane, and the fatty acid plays the main role in this association.

Prenylation

In this process, specific types of lipids, prenyl groups are attached to the sulphur atoms in the side chains of cysteine residues located near the C-terminus of the polypeptide chain (Fig. 10.14). As seen in this figure, prenylation of proteins is a multi-step procedure. In the first step, the prenyl groups [farnesyl (15 carbons) or geranyl-geranyl (20 carbons)] are added to a cysteine located 3 amino acids from the carboxy terminus of the protein. In the second step, the amino acids following the cysteine residue are removed. In the third step, a methyl group is added to the carboxyl group of the C-terminal cysteine residue.

Among many such modified proteins that are associated with the plasma membrane and are involved in the regulation of cell growth and differentiation, *Ras oncogene* proteins that are linked with many human cancers, are the best

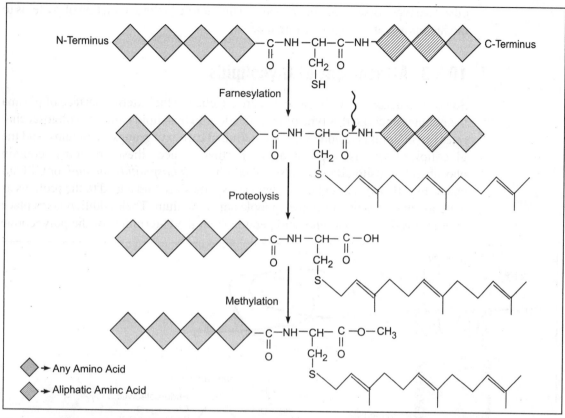

Figure 10.14 Prenylation of a C-terminal cysteine residue. In the first step, farnesylation occurs at the sulphydryl group of cysteine in the polypeptide within the sequence Cys-Aliphatic-Aliphatic amino acid. Then proteolytic cleavage occurs after cysteine residues towards C-terminus. In the third step, methylation occurs at the carboxyl group of cysteine.

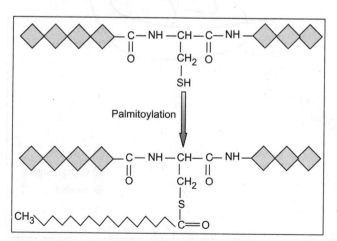

Figure 10.15 Palmitoylation. In this process, palmitate is added at the sulphydryl group of cysteine.

examples. Considering the rarity of this type of modifications of proteins, it is possible to target Ras protein function by using inhibitors of the enzyme farnesyl transferase. In fact, such research has progressed to the extent of human clinical trials using this approach.

Palmitoylation

It is a type of modification in which *palmitic acid* (a 16 carbon fatty acid) is added to the sulphur atoms of the side chains of internal cysteine residues (Fig. 10.15). This modification similar to the other two types discussed above, plays a similar role in the association of proteins to the membrane at the cytoplasmic face.

10.3.3 Attachment of Glycolipids

Some proteins particularly those that are attached to the external surface of plasma membranes undergo this type of modification. In this, lipids linked to oligosaccharides (glycolipids) are added to the C-terminal carboxyl groups of proteins, and the glycolipids act as the anchors. Further, since these proteins contain phosphatidylinositol, they are also called *glycosylphosphatidylinositol*, or GPI anchors (Fig. 10.16). These GPI anchors are synthesized and added to the proteins as a pre-assembled unit within the endoplasmic reticulum. Their addition takes place after removal of a 20 amino acid peptide from the C-terminus of the polypeptide

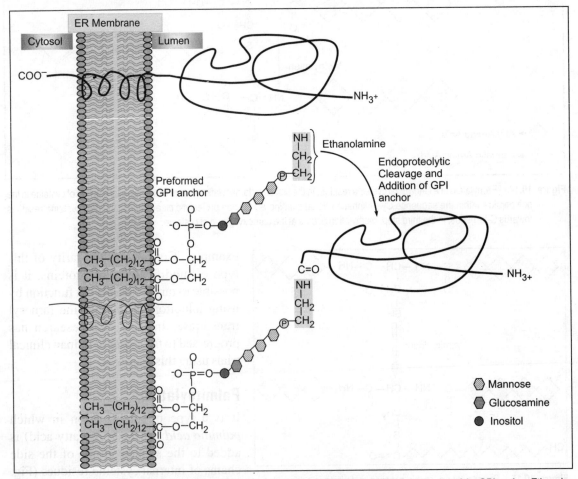

Figure 10.16 Structure of a GPI anchor. The protein is attached from the C-terminus to the ethanolamine of the GPI anchor. Ethanolamine is in turn attached to oligosaccharides and then to the inositol of phosphatidylinositol.

chain. The modified protein is then transported to the cell surface, where the fatty acid chains of the GPI anchor mediate its attachment to the plasma membrane.

10.3.4 Protein Phosphorylation

Protein phosphorylation is an example of several types of covalent protein modifications that occur routinely as required by the cells. The other types of commonly occurring protein modifications of this type are, methylation, acetylation,

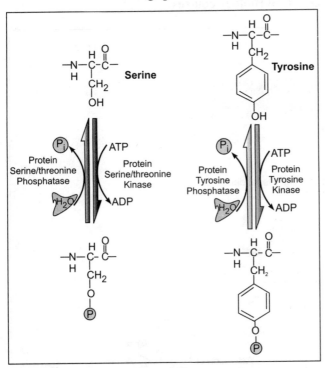

adenylation, etc. These modifications are reversible, and both the forward and reverse states are highly regulated in a signal-dependent manner. While the forward reaction, i.e. phosphorylation is brought about by protein kinases, the reverse reaction, i.e. dephosphorylation is catalyzed by protein phosphatases (Fig. 10.17). Since the protein kinases catalyze phosphorylation to specific substrate proteins, there are a large number of protein kinases in the form of families present in a cell. In fact, about 2% of the total genes code for protein kinases in eukaryotes. On the contrary, there are a very limited number of protein phosphatases that suffice the needs.

Protein kinases catalyze transfer of phosphate groups from ATP to the hydroxyl groups of the side chains of 3 amino acids, namely serine, threonine, and tyrosine. Those kinases which phosphorylate serine or threonine amino acid residues are called *Ser/Thr protein kinases*, while those phosphorylating tyrosine residues are called *Tyr protein*

Figure 10.17 Protein phosphorylation and dephosphorylation by protein kinases and protein phosphatases, respectively. Protein kinase catalyzes the phosphorylation of serine/threonine/tyrosine using ATP, while protein phosphatases catalyze the removal of phosphate group from the amino acids by hydrolysis.

kinases. Regulation of these molecules takes place through various signals in the cell, and therefore, are highly important for various vital cellular processes. The detailed studies on protein kinases were initially done by two scientists, Ed Fischer and Ed Krebs, who received Nobel prize for their pathbreaking work on glycogen metabolism in muscle cells.

Epinephrine acts as the regulatory signal for the breakdown of glycogen to glucose-1-phosphate that provides the required energy for increased muscular activity. However, the activation of this kinase is regulated by another kinase, cAMP-dependent kinase, which in turn is regulated by a signal, epinephrine hormone (Fig. 10.18). Secretion of this hormone and exposure of muscle cells by it, triggers the cascade of reactions leading to increased muscular activity. Similar to this, there exist a number of such regulatory cascades in a cell that result in execution of various cellular processes, upon stimulation by specific signals received by a cell from the exterior. These cascades that involve extensive protein-protein interactions are known as *signal transduction pathways*. Readers can get further details on these topics in Cell Biology courses.

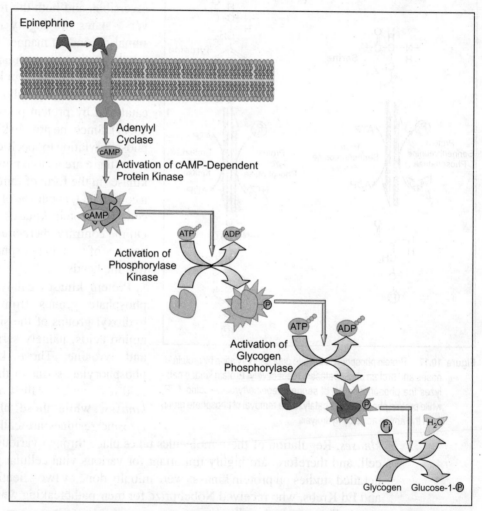

Figure 10.18 Glycogen breakdown regulated by protein phosphorylation. After the binding of epinephrine to the cell surface receptors, cAMP is produced. This activates the cAMP-dependent protein kinase. This in turn activates phosphorylase kinase by phosphorylation, which subsequently activates glycogen phosphorylase. This glycogen phosphorylase then breaks down glycogen into glucose-1-phosphate.

10.4 PROTEIN DEGRADATION

Protein degradation is as important as protein synthesis for a cell. It is particularly essential under the following situations:

(a) Proteins in general have specific half-lives. When they undergo wear and tear, they need to be degraded and replenished with new proteins.

(b) Many proteins are synthesized transiently for specific time-bound purpose (e.g., transcription factors, protein hormones, growth factors, etc.). They therefore need to be degraded in a regulated manner.

(c) Improperly folded, truncated, or otherwise faulty proteins when synthesized, need to be degraded.

For these purposes, there are two pathways of protein degradation, namely *lysosomal pathway* and *ubiquitin-proteasome pathway*.

10.4.1 Lysosomal Pathway

This is a non-specific and a more general pathway of proteolysis. Proteins are taken up by the lysosomes which are membrane-bound organelles. These contain a diverse array of digestive enzymes including various proteases. In addition to lysis of cellular proteins, lysosomes also degrade cell organelles and extracellular proteins that are endocytosed. The major pathway of uptake of cellular proteins is called *autophagy*, in which small areas of cytoplasm or organelles are enclosed in small vesicles derived from the endoplasmic reticulum. These vesicles called the autophagosomes fuse with the lysosomes forming phagolysosomes (Fig. 10.19).

In this structure, the proteins are degraded by the lysosomal proteases in a non-specific and slow manner. Autophagy is usually activated under nutrient starvation. In such conditions, the deficiency of components like amino acids can be made available through degradation of non-essential organelles and proteins. Further, during development and particularly during metamorphosis in insects, autophagy plays a crucial role in tissue remodeling and degradation of cellular components.

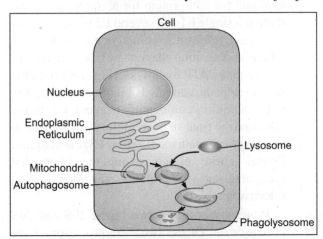

Figure 10.19 The lysosomal pathway of protein degradation. Lysosomes in the cell contain various digestive enzymes including proteases, nucleases, lipases, carbohydrases, and phosphatases. Autophagosomes are fused with lysosomes to form phagolysosomes, where the proteins get degraded.

10.4.2 Ubiquitin-proteasome Pathway

This pathway eliminates specific proteins for a specific purpose. For example, many cellular processes, such as cell cycle, gene expression, etc. that are regulated by specific proteins are controlled by their synthesis and degradation as well. As the name suggests, there are two components which are major players of this pathway: *ubiquitin* and *proteasomes*. Ubiquitin, a small polypeptide (75 amino acid long) tags the potential proteins that need to be degraded by the proteasomes, the large, multisubunit multicatalytic proteases. Ubiquitin which is ubiquitous in eukaryotic cells binds to the amino terminus of lysine residues (side chain amino acids) in proteins. Multiple ubiquitins bind to the proteins subsequently prior to their degradation by proteasomes.

Ubiquitination of proteins is a multi-step process, and it involves three enzymes called E1, E2, and E3. Of these, E3 is highly protein-specific, and therefore present in large numbers in a cell; while the other two enzymes are common for any protein for ubiquitination, and there is a single E1 and several E2 in a cell. In the first step, ubiquitin is activated by E1, the ubiquitin-activating enzyme, by being attached to it. It requires ATP hydrolysis (Fig. 10.20). In the second step, the activated ubiquitin is transferred to the ubiquitin-conjugating enzyme, E2. In the third and the final step, ubiquitin is transferred to the specific protein mediated by E3, the ubiquitin ligase. These enzymes are the ones that recognize target proteins for degradation selectively.

Proteasomes are of two types: 20S and 26S proteasomes. Of the two, 20S proteasomes are the core proteasomes, and 26S proteasomes are formed by addition of 19S regulatory caps on both top and bottom of the core proteasomes. The core proteasomes have α and β rings, that assemble in a doughnut-shaped complex. The catalytic activity lies in the inside of the core

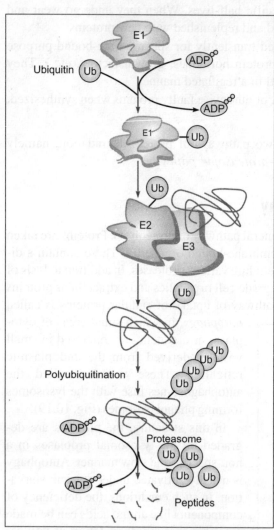

Figure 10.20 The Ubiquitin-Proteasome pathway of protein degradation. Proteins, which are to be degraded, are marked by covalent attachment of several molecules of ubiquitin. Ubiquitin is first activated by enzyme E1. The ubiquitin is then transferred to enzyme E2 (conjugating enzyme). Then with the help of enzyme E3 (ubiquitin ligase), ubiquitins are transferred to the protein, after which the ubiquitinated protein is degraded by the proteasomes. **(See Colour Plate 16)**

proteasomes. These structures are highly conserved from archaebacteria to man. For protein degradation, 26S proteasomes require ATP, while the 20S proteasomes are ATP-independent.

As stated earlier, the selective degradation of proteins control various cellular processes. For example, during cell cycle regulation, for entry into mitosis, activity of a factor called *mitosis promoting factor* (MPF) is essential; while for exit from mitosis, MPF activity is to be destroyed. MPF is a complex of cyclin and a protein kinase called cyclin-dependent kinase (Cdk). Cyclin which is synthesized and degraded in synchrony with mitosis actually controls the MPF activity. Particularly for exit from mitosis, inactivation of MPF is caused due to the degradation of cyclin by proteasomes (Fig. 10.21). It is also known that the anaphase promoting complex (APC) is the E3 enzyme which ubiquitinates cyclin for its degradation by proteasomes.

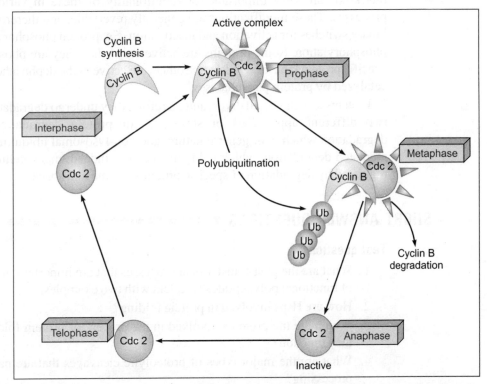

Figure 10.21 Cyclin degradation by Ubiquitin-Proteasome pathway during cell cycle. The progression of eukaryotic cell into the cell cycle is partially governed by the synthesis and degradation of the cyclin B, which is a regulatory subunit of Cdc2 kinase. Synthesis of cyclin B in the interphase leads to the formation of cyclin B-Cdc2 active complex in the prophase. During metaphase, cyclin B gets ubiquitinated and gets degraded by proteasomes. Thus, from anaphase till interphase, there is a formation of inactive Cdc2 kinase. **(See Colour Plate 17)**

SUMMARY

Proteins undergo changes of various types including folding into 3-dimensional conformation, processing from precursor to actual form, modifications of various kinds in order to become functional. These processes are either co-translational or post-translational. There are a variety of catalytic enzymes that work towards such modifications. For example, the regulatory molecules that catalyze protein folding are called chaperones. Besides these, there are two important enzymes that participate in protein folding, and they are: protein disulfide isomerase and peptidyl prolyl isomerase.

Processing of precursor proteins to the active forms are catalyzed by three major types of proteolytic cleavages: cleavage of initiator methionine, cleavage of signal sequence, and cleavage of larger precursor proteins. Protein modifications of various kinds are very important for functionality of these in various biological processes. These modifications are generally reversible, and therefore, they act as binary switches for activation and inactivation, e.g. protein phosphorylation and dephosphorylation. Some proteins are active only when they are phosphorylated by specific protein kinases, while some others are active in the dephosphorylated forms catalyzed by protein phosphatases.

Proteins also have half-lives, and therefore, they undergo degradation. There are two different approaches or strategies for protein degradation: lysosomal degradation which is of general nature, and non-lysosomal ubiquitin-proteasome-mediated degradation, which is highly specific. Indeed, many cellular events are regulated by degradation of specific proteins in time and space.

SHORT ANSWER QUESTIONS

Test questions

1. What are the post-translational processes that are important in the formation of functional polypeptides? Explain with two examples.
2. How are Hsps involved in protein folding?
3. Which are the proteins involved in the process of protein folding? How do they function?
4. What are the major types of proteolytic cleavages that are parts of protein processing?
5. Which are the different types of protein modifications that occur in eukaryotic cells?
6. What is N-linked and O-linked glycosylations and how do they occur? Draw the structure of their linkages.
7. What kind of modifications occur in the eukaryotic proteins for anchoring the plasma membrane?

8. What kind of modifications occur in the proteins that are attached to the external surface of the plasma membrane?

9. Which are the different families of protein kinases?

10. Name and explain two cellular processes in which specific protein degradation is necessary.

11. Describe the ubiquitin-proteasome pathway of protein degradation.

Study questions

1. How does binding of chaperones to proteins help in its correct folding?

2. What will be the fate of the cell if,

   (a) a eukaryotic cell is mutated in protein folding machinery?

   (b) there is an over-expression of chaperones in the cell?

3. Nature has adopted an extra step in the proteolytic processing of the proteins for its functionality. First, a polypeptide is synthesized, and then, it is cleaved and degraded for the protein to be functional. Why is this extra step involved?

4. What is so special in the signal peptide (its sequence), that it determines the location of the protein in the cell?

5. What is the factor that determines which kind of oligosaccharide sequence is to be attached to a protein?

6. What can happen if degradation of protein does not occur in a cell?

▰▰▰ FURTHER READING ▰▰▰

Protein folding

- Anfinsen, C.B. (1973). Principles that govern the folding of protein chains. *Science* **181**, 223–230.

- Daggett, V. and Fersht, A.R. (2003). Is there a unifying mechanism for protein folding? *Trends Biochem. Sci.* **28**, 18–25.

- Dobson, C.M. (2003). Protein folding and misfolding. *Nature* **426**, 884–890.

- Frydman, J. (2001). Folding of newly translated proteins *in vivo*: the role of molecular chaperones. *Annu. Rev. Biochem.* **70**, 603–649.

- Sigler, P.B., Xu, Z., Rye, H.S., Burston, S.G., Fenton, W.A., and Horwich, A.L. (1998). Structure and function in GroEL-mediated protein folding. *Annu. Rev. Biochem.* **67**, 581–608.

- Thirumalai, D. and Lorimer, G.H. (2001). Chaperonin-mediated protein folding. *Annu. Rev. Biophys. Biomol. Struct.* **30**, 245–269.

- Xu, Z., Horwich, A.L., and Sigler, P.B. (1997). The crystal structure of the asymmetric GroEL-GroES-(ADP)7 chaperonin complex. *Nature* **388**, 741–750.

Protein processing and modifications

- Brooks, D. A. (1997). Protein processing: a role in the pathophysiology of genetic disease. *FEBS Letters* **409**, 115–120.
- Drickamer, K. and Taylor, M.E. (1998). Evolving views of protein glycosylation. *Trends Biochem. Sci.* **23**, 321–324.
- Hart, G. W. (1997). Dynamic O-linked glycosylation of nuclear and cytoskeletal proteins. *Annu. Rev. Biochem.* **66**, 315–335.
- Helenius, A. and Aebi, M. (2004). Roles of N-linked glycans in the endoplasmic reticulum. *Annu. Rev. Biochem.* **73**, 1019–49.
- Parodi, A. J. (2000). Protein glucosylation and its role in protein folding. *Annu. Rev. Biochem.* **69**, 69–93.
- Patwardhan, P. and Miller, W. T. (2007). Processive phosphorylation: Mechanism and biological importance. *Cellular Signalling.* **19**, 2218–2226.
- Paulus, H. (2000). Protein splicing and related forms of protein autoprocessing. *Annu. Rev. Biochem.* **69**, 447–496.
- Raina, S. and Missiakas, D. (1997). Making and breaking disulfide bonds. *Annu. Rev. Microbiol.* **51**, 179–202.
- Seidah, N. G. and Chretien, M. (1997). Eukaryotic protein processing: endoproteolysis of precursor proteins. *Current Opinion in Biotechnology.* **8**, 602–607.

Protein degradation

- Bonifacino, J. S. (1998). Ubiquitin and the control of protein fate in the secretory and endocytic pathways. *Annu. Rev. Cell Dev. Biol.* **14**, 19–57.
- Hay, R. T. (2007). SUMO-specific proteases: a twist in the tail. *Trends Cell Biol.* **17**. doi:10.1016/j.tcb.2007.08.002.
- Hochstrasser, M. (1996). Ubiquitin-dependent protein degradation. *Annu. Rev. Genet.* **30**, 405–39.
- Schwartz, A. L. (1999). The ubiquitin-proteasome pathway and pathogenesis of human diseases. *Annu. Rev. Med.* **50,** 57–74.
- Varshavsky, A. (1997). The ubiquitin system. *Trends Biochem. Sci.* **22**, 383–387.
- Voges, D., Zwicki, P., and Baumeister, W. (1999). The 26S proteasome: a molecular machine designed for controlled proteolysis. *Annu. Rev. Biochem.* **68**, 1015–1068.

Genomics and Proteomics

11

OVERVIEW

- Genomics and proteomics are the recent extensions of molecular biology that help us understand biology at a more global level.

- The advanced tools in these areas are instrumental in unravelling structure-function relationships of genome and its evolution, and the cellular dynamics.

- Data generated through genomics and proteomics research are analysed by using modern tools in bioinformatics. Results obtained from such analyses have a direct bearing on human health and diseases.

11.1 GENOMICS

Genomics is the study of structure and function of a genome of an organism. The first genome to be sequenced was that of a bacteriophage called ØX174. This DNA was completely sequenced by Sanger in 1977. From the sequence of DNA, the amino acid sequence of proteins could be deciphered using information on genetic code. Analysis of this simple genome revealed the fact that there is an overlap in the phage genes which means that the coding region for one gene lies within another gene.

With the techniques of automated sequencing much larger genomes such as that of *Hemophilus influenzae* and *Mycoplasma genitalium* could then be sequenced in 1995. In 1996, the first eukaryotic genome of *Sacchromyces cerevisiae* was entirely sequenced. Later in 1996, the first genome of an archaebacteria *Methanococcus jannaschii* was sequenced and in 1997, 4.6 million base pair *E. coli* genome was reported. In 1998, the sequence of first animal genome from *Caenorhabditis elegans* was reported. The first plant genome of *Arabidopsis thaliana* was completed in 2000. Around the same time, a rough draft of the human genome was announced. By 2001, the working draft of the human genome was published and since then, many other genomes have been sequenced.

11.1.1 The Human Genome Project

The ambitious project of mapping and ultimately sequencing the entire human genome began in 1990. This was an enormous effort and required a lot of money. The human genome is around 3 million base pairs and therefore sequencing this genome was a difficult task. The original plan for sequencing human genome was systematic and conservative. To begin with, the genetic and physical map of the genome was prepared. This would contain the markers, which would then allow DNA sequences to be put together in a proper order. The bulk of the sequencing could be done only after the mapping was completed, and the clones representing all points on the map were then systematically restored. This massive job was expected to be finished by 2005.

However, around 1998, Craig Venter proposed to sequence the genome by shotgun sequencing approach. Instead of relying on a map and the clones used to build it, Venter proposed that human genome would be chopped, cloned, and the clones would be sequenced at random, and the sequences would then be joined together using powerful computer analysis. Finally Craig Venter and Francis Collins, Director of the publicly financed Human Genome Project together announced the completion of a rough draft in 2001. Thus, one approach to sequence the large genome was mapping and then sequencing clone by clone, whereas the second approach involved shotgun cloning and sequencing of the clones (Fig 11.1).

For both these approaches, special cloning vectors were developed. The two most commonly used vectors for this project were *yeast artificial chromosomes* (YACs) and *bacterial artificial chromosomes* (BACs). Both these vectors were very useful in mapping the human genome because they could accommodate hundreds of thousands of bps each. BACs are even preferred over YACs as they are more stable, easy to isolate from the host cells, more efficient as compared to YACs, and more stable both *in vitro* and *in vivo*.

Clone-by-clone Strategy

In this systematic approach, the whole genome is mapped by finding markers regularly spaced along each chromosome and the entire clone corresponding to these markers are collected. They are then sequenced and the sequences are put in their proper places in the genome. Different methods are used for mapping the large genomes, such as restriction fragment length polymorphism (RFLP), making use of sequence tagged sites (STS) and expressed sequence tags (EST), etc.

RFLP means that a genetic locus has different forms and therefore cutting the DNA from two individuals with a restriction enzyme may yield fragments of different length. Analyzing the whole genome for RFLP is difficult. Typically a small portion of

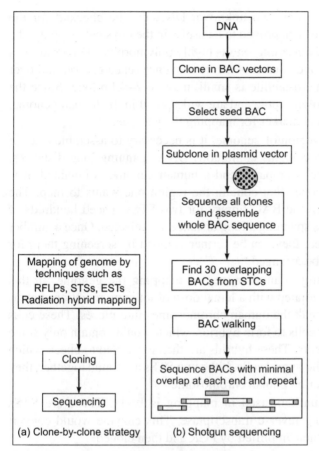

Figure 11.1 Different strategies used for human genome sequencing project **(a)** In this strategy, the genome is first mapped, cloned and then, the clones are sequenced. **(b)** In shotgun sequencing, the whole genome is first cloned and the clones are randomly sequenced. These sequences are then assembled by computational analysis.

a genome is often checked for RFLPs. This is a labourious process; however, it can provide a starting point and has been proved to be a useful method for finding the genes responsible for several genetic diseases. The greater the degree of polymorphism of RFLP, more useful it becomes for mapping the genome.

RFLPs called variable number tandem repeats (VNTRs) are very useful. These are derived from minisatellites, stretches of DNA that contain a short core sequence repeated over and over again. The number of repeats of the core sequence in a VNTR differs among individuals and therefore easy to map. Another type of useful marker is STSs. STSs are short sequences and can be detected by PCR. One must first know enough about the DNA sequence in the region being mapped to design short primers, which will hybridize a few hundred bps apart and cause amplification of a predictable length of DNA in between.

Using these two primers, any two unknown DNA can be checked for the presence of STSs. STSs have been proved to be useful in the physical mapping of a gene. A class of STSs called *microsatellites* is highly polymorphic. Microsatellites are similar to minisatellites and consist of a core sequence repeated over and over many times. The core in microsatellite is small, may be 2–4 bp long. Since the microsatellites are highly polymorphic and are widespread in the human genome, they have proved to be a good marker for mapping the genome.

For sequencing a given region of genome, it is necessary to assemble a set of clones called *contig*, which contains contiguous DNA spanning long distances. Thus, once the BAC library is prepared and a human genome is obtained, it is necessary to identify the clones that contain the region one wants to map. The simple approach is to screen the BAC library for two STSs spaced hundreds of kilobases apart so that BACs spanning a long distance are selected. Once a number of positive BACs is obtained, they can be further mapped by screening them for several STSs and then can be arranged.

Radiation hybrid mapping is another way of mapping the genome. In this method, human cells are irradiated with a lethal dose of ionizing radiation such as X-rays and γ rays, which break the human chromosomes into pieces. These cells are then fused with hamster cells to form hybrids, which would contain only some human chromosome fragments. These hybrids are then examined to check which STSs are found together in the hybrid cells. If the two STSs are found together, they are likely to be close together on human chromosome.

Another class of STS markers used in mapping is ESTs. These are STSs generated from mRNA using reverse transcriptase. This enzyme would convert mRNA to cDNA, which is then amplified by PCR and then cloned.

Shotgun Sequencing

Shotgun sequencing proposed by Venter bypasses the mapping stage and begins with the sequencing stage. BAC clones containing the inserts of 150 kilobases are sequenced on both ends using an automated sequencer. All available BAC clones are sequenced in this manner and these sequences serve as identity tags called *sequence tag collector* (STC) for each BAC clone.

The next stage is to fingerprint each clone by digesting with restriction enzymes. This gives information about the insert size. After this, the entire sequence of a BAC is obtained, referred to as *seed BAC*. For this, a BAC is further subdivided into smaller clones. This whole BAC sequence allows identification of many more BACs that overlap with the seed, i.e. they are the ones with STCs that occur somewhere in the seed BAC.

In the next step, one selects other BACS with minimal overlap with the original ones and proceeds to sequencing. This process is referred to as *BAC walking*. Venter and colleagues modified the procedure by sequencing BACs at random until they had about 35 billion nucleotide of sequence. This covers the human genome ten times over, giving a high degree of coverage and accuracy. Then they fed all the sequences in a computer with a powerful program that found areas of overlap between clones and fit their sequence together, building the whole sequence of a genome.

The rough draft of human genome meant that only 90% of it was complete and had some amount of error. However, it is agreed that the final draft should have a error rate of not more than 0.5% and should have as few gaps as possible. Even in the final draft available today, some gaps remain to be filled because some region of DNA, for mostly unexplained reasons, resists cloning. The human genome draft available now is considered to be functionally approved with all the sequences possible with the cloning and sequencing tools currently available, even though some gaps remain.

11.1.2 An Overview of the Human Genome

The human genome is the total genetic information in a human cell. It comprises of a complex nuclear genome with about 30,000 genes and a simple mitochondrial genome with 37 genes. The nuclear genome provides a great bulk of essential genetic information, most of which specifies polypeptide synthesis. Mitochondrial genome specifies a very small portion of specific mitochondrial function. The bulk of mitochondrial polypeptides are encoded by nuclear genes, synthesized on cytoplasmic ribosomes, and then imported into mitochondria.

The majority of nuclear DNA is used to make RNAs for synthesizing polypeptides, but a significant amount of human genes specify non-coding RNAs. A very large component of the human genome is made up of non-coding DNA.

The cells of an organism show little variation in their DNA content; nevertheless, an appreciable difference in both mitochondrial and nuclear DNA content is seen in different cell types. Certain cells, e.g. terminally differentiated skin cells lack any mitochondria, whereas the mitochondrial DNA copy number in other somatic cells varies from 1,000 to 10,000. Between the gamete cells, sperm cells have a few hundred copies of mitochondrial DNA, whereas oocytes have 1,00,000 copies of mitochondrial DNA. Each mitochondrial genome is a circular double-stranded DNA molecule of 16.6 kilobases.

The nuclear genome is distributed between 24 different types of linear double-stranded DNA molecules. These 24 chromosomes can be differentiated by chromosome banding and are classified based on their size and centromere position. The total nuclear genome size is 3,200 Mb of which 3,000 Mb is euchromatic portion and about 200 Mb is constitutive heterochromatin and is transcriptionally inactive. The average size of the human chromosome is about 140 Mb, though there is a considerable size difference between the chromosomes. The genome sequence suggests an average of 41% GC content for the euchromatin component. The total number of genes is considered in the range of 30,000–35,000.

Human genes are not evenly distributed on the chromosome. The gene density varies substantially between chromosomal regions and also between the whole chromosomes. Some regions are gene-rich, while some are gene-poor. A great majority of human genes encodes polypeptides, but a significant minority specifies RNA molecules as their end products, described as RNA genes. There are approximately 700–800 rRNA genes organized in tandemly repeated clusters, and there are around 497 nuclear genes encoding cytoplasmic tRNA molecules. The tRNA genes are dispersed throughout the genome.

In addition to rRNA and tRNA, small nuclear RNA (snRNA) and small nucleolar RNA (snoRNA) are also encoded and are involved in regulation of gene expression. Additionally, some microRNAs (miRNA) which are very small, approximately 22 nucleotide long RNA molecules, are also encoded and are involved in regulating gene expression. These are also found dispersed in the human genome. Most of the human genes contain introns, though there is a small minority of human genes that lack introns and are small in size. The genes which possess introns have an inverse correlation between gene size and the fraction of coding DNA. The average exon size in human genes is around 200 bp, although some very long exons are known. There is a huge variation in intron lengths, and large genes tend to have very large introns.

Genes often have repetitive DNA components within non-coding introns and flanking sequence, but in addition, they are also found to a different extent in coding DNA. In human genome, functionally similar genes are sometimes found to be clustered, but are frequently dispersed over different chromosomes. A large fraction of human genes are members of gene families where individual genes are closely related but not identical in sequence. Many such genes are clustered and have arisen by tandem gene duplication, e.g. members of α-globin and β-globin clusters.

Overlapping genes are not found in the human genome; however, very closely neighbouring genes are found with their 5′ ends separated by a few hundred nucleotides and transcribed from opposite strands. Such an organization is called bidirectional organization and is found in DNA repair genes. The small nucleolar genes are unusual, as majority of them are located within other genes which encode a ribosome-associated protein or a nucleolar protein. Highly repeated non-coding human DNA often occurs in arrays of tandem repeats of sequence which may be 1–10 nucleotides long.

According to the array size, three major subclasses are defined, such as satellite DNA, minisatellite DNA, and microsatellite DNA. Satellite DNA makes up most of the heterochromatic region of the genome and is found in the vicinity of centromeres. Minisatellite DNA comprises a collection of moderately sized arrays of tandemly repeated DNA sequence which are dispersed over a considerable portion of the nuclear genome. They are normally not transcribed. A major family of minisatellite DNA is found at the termini of chromosome and is referred to as *telomeric DNA*. Length varies between 3–20 kb consisting of tandem hexanucleotide repeat units. Microsatellite DNAs are small arrays of tandem repeats of simple sequence, usually less than 10 bp. They are interspersed throughout the genome and account for 2% of the genome. Mostly, microsatellite DNA is found in intergenic DNA or within the introns of genes. It is also present within the coding sequence of the gene and are the sites for mutation hotspots.

Almost the entire interspersed repeated non-coding DNA in the human genome is derived from transposable region which can migrate to other parts of the genome. This accounts for 45% of the genome. Four major classes of transposons are seen. They are: long interspersed nuclear elements (LINEs), short interspersed nuclear elements (SINEs), retrovirus-like elements containing long terminal repeats, and DNA transposons.

Thus, the total human genome (Fig. 11.2) consists of 3200 Mb nuclear genome and 16.6 kb mitochondrial genome. Mitochondrial genome is small, circular, and codes for around 37 genes. Nuclear genome of size 3,200 Mb contains 1200 Mb corresponding to genes and gene-related sequences, whereas 2000 Mb is intergenic DNA. In genes and genes related apart, only 48 Mb comprises of exons, while the remaining 1152 Mb comprises of pseudogenes, gene fragments, introns, and untranslated regions. In 2000 Mb intergenic region, 1400 Mb is occupied by genome

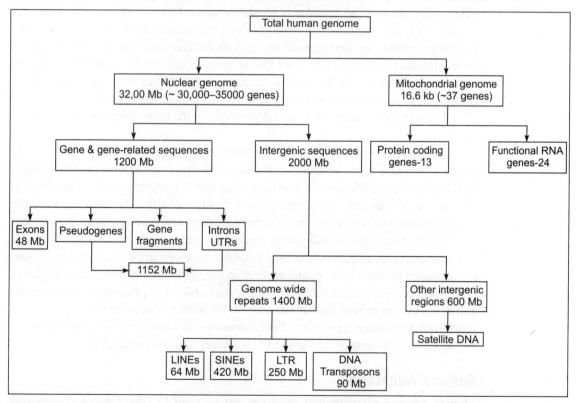

Figure 11.2 Composition of human genome. The total human genome consists of 3200 Mb nuclear genome and 16.6 kb mitochondrial genome.

with repeat sequences including LINEs, SINEs, LTRs, and DNA transposons, whereas the remaining 600 Mb is occupied by satellite DNA. Thus, around 1.5% of the genome is only coding, whereas 45% of the genome is occupied by genome wide repeat sequences. It was possible to understand the composition of the human genome because of the massive sequencing data obtained through the human genome project.

11.1.3 Applications of Genomics

The process of finding out the sequence of a genome is referred to as *structural genomics* which is a massive task. Once the sequence of a genome is available, the major task is to identify all the genes that represent the functional units of the genome. The exact number of human genes is still not known, but most estimates agree on the presence of approximately 30,000–35,000 genes. Having understood the number of genes, the next task is to determine the precise function of each of the genes, a process referred to as *functional annotation*.

The function of a gene depends on the product it codes for. Though most of the genes code for proteins, many more produce non-coding RNA molecules. Even when every gene in the genome is identified and a function is assigned to it, it is difficult to figure out how the gene products coordinate the biological activities. Therefore, efforts are being made to study and understand the function of all the gene products simultaneously. This global analysis of gene function is the basis of *functional genomics*. In functional genomics, one studies the expression of the genome to produce *transcriptome* and *proteome*. Transcriptome is the collection of all mRNAs in a particular cell, whereas proteome is the total collection of proteins in a cell.

Both transcriptome and proteome are more complex than the genome. A single gene can produce many different mRNAs and the proteins synthesized from these mRNAs can be modified in different ways. Unlike the genome which is identical in all cells, transcriptome and proteome are highly variable. By studying how the transcriptome and the proteome change in healthy and diseased state, it becomes possible to understand the individual function of genes in greater details.

Functional genomics has benefited from the development of various experimental strategies to investigate gene function and development of new technologies to understand protein-protein interactions. Similarly, bioinformatics support is essential and development of new algorithms and establishment of new databases have revolutionized functional genomics. Bioinformatics is necessary for comparison of sequences and structures, modeling of structures and interactions, etc.

Genome Annotation

Though various experimental methods can be used to detect genes in genomic DNA, initial annotation is carried out using computer algorithms which can process the sequence very rapidly. These algorithms identify sequences with homology to known genes by searching through databases. Homologous genes have similar sequences, as they are derived from a common evolutionary ancestor, and two sequences related evolutionarily have similar functions. So, the simplest way to assign a function to a gene is to look for the related sequences which have already been annotated.

Comparisons are generally made at protein level, as amino acid sequences are more conserved than nucleotide sequences. If the sequences of two genes are very similar, they might represent homologous genes carrying out identical functions in different species. Such genes are known as *orthologues*. A lower degree of similarity might indicate that the genes are homologous but have diverged in functional terms and are known as *paralogues*.

The principle of functional annotation based on homology searching is that the conserved structure always reflects conserved functions. However, this may not always be true and therefore, confirming the function by experimental data is necessary. Closely related genes from different species can be used to assign quite precise functions to uncharacterized genes. Orthologues are not identical because separate mutations accumulate in each evolutionary lineage after the speciation event. Therefore, the degree of similarity of orthologues provides a useful measure of evolutionary time and is used to build phylogenetic trees.

Comparative genomics tries to study the similarities and differences between genome to obtain structural, functional, and evolutionary information. It is believed that genetic similarities between species extend much further than the gene level. Closely related species have similar genes and genomes, and the degree of similarity would depend on the evolutionary divergence of the species. Similarities are often seen both at the level of sequences and at the gross level of genome organization such that the related species often demonstrate conserved *syntany*, i.e. conserved gene order.

Comparative genomics also help in identifying gene regulatory elements. Orthologous genes in related species can be compared and looked for the most conserved sequence motifs outside the coding region of the gene. Comparative genomics have also been exploited to identify and characterize human disease genes. For example, the insulin signaling pathway is fully conserved between humans and nematodes, so mutant worms impaired for insulin signaling are useful models for type II diabetes, and they can be used to screen many potential drugs to identify compounds that can revert insulin resistance to normal.

Functional annotations based on sequence and structural comparisons are informative; however, experimental evidences are required to determine the gene function at a cellular and the whole organism level, and to investigate how the activities of individual gene products are coordinated.

Global Gene Expression

To find out the pattern of expression of a gene in a given cell at a given time, the simple technique of *dot blot analysis* can be used. In this method, membrane filters

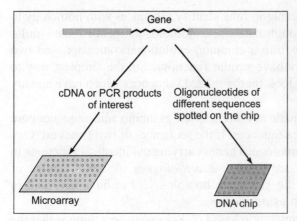

Figure 11.3 Microarray and DNA chips. In microarray, each position contains a cDNA or PCR product for a gene of interest, whereas DNA chip contains at each position a mixture of oligonucleotides where the sequences match different segments of genes of interest. **(See Colour Plate 18)**

are spotted with single-stranded DNA from the gene in question and are hybridized to labeled RNAs made from the cell in question. However, if one wants to look at the global pattern of expression of all the genes in the tissue, a very large dot blot will have to be used which is very difficult to make. To simplify this problem and to be able to analyze the expression of whole genomes, DNA microarrays and microchips are being developed (Fig 11.3).

In DNA microarray, different DNAs are spotted on one glass chip using robotic system. Each spot may contain 0.25–1.0 nl of DNA and the spots are very small of around 100 micrometer in diameter and are separated from each other by 100 µm. After spotting, the DNAs are air dried and are covalently attached by UV radiation to a thin layer on top of glass. Another strategy has been to synthesize tiny oligonucleotides on the surface of the chip simultaneously. This chip is referred to as *DNA microchip* or *oligonucleotide array*. As many as 3,00,000 oligonucleotides can be attached to a chip of ½ inches square. The oligonucleotides spotted are at least 16 bases long and are unique for different genes. DNA microchips are now commercially available for different tissues from different animals which can be bought and used for analysis of gene expression.

The oligonucleotide on a microchip can be hybridized to labeled RNA isolated from a cell to see which genes in the cell were being transcribed. For example, to see the effect of serum on RNAs made by a human cell, an RNA was isolated from cells grown in presence or absence of serum. This RNA was then reverse transcribed in presence of nucleotides tagged with a fluorescent dye so that the cDNA prepared is labeled with it. cDNAs from serum-starved and serum-stimulated cells were labeled with different fluorescin tags (green fluorescin nucleotide were used to label cDNA from serum-deprived cells and red fluorescin to label cDNA from serum-stimulated cells). Both cDNAs were then hybridized to microarray containing cDNAs corresponding to several different human genes and the resulting fluorescence due to hybridization was detected. Red spots in hybridization correspond to genes that are turned on by serum stimulation, while green spots represent transcriptionally active genes in serum-deprived cells (Fig. 11.4). The yellow spots result from the hybrid of both probes to the same spot and therefore represent the genes that are active in absence as well as in the presence of serum.

Once the sequence of all the genes in a human cell was made available, it was possible to design a comprehensive DNA microchip to assay transcription from every gene. Thus, the power of functional genomics increased with the knowledge

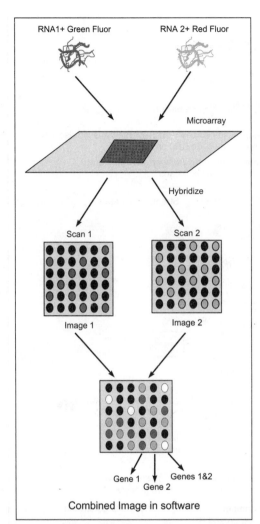

RNA1+ Green Fluor

RNA 2+ Red Fluor

Microarray

Hybridize

Scan 1

Scan 2

Image 1

Image 2

Gene 1

Genes 1&2

Gene 2

Combined Image in software

Figure 11.4 Analysing gene expression by microarray. Comparative expression is carried out by differentially labeling two mRNA or cDNA samples with different fluorophores. These are hybridized to arrays spotted on the glass slides and scanned to detect both the fluorophores independently. Coloured dots observed in image 1 and image 2 correspond to gene 1 and gene 2 expressed at increased levels. Combined image confirms the expression of both the genes 1 and 2 expressed simultaneously. **(See Colour Plate 20)**

of sequence of organism's genome. DNA arrays have been widely used to characterize the given transcriptome under different conditions. The comparison of healthy and diseased tissue represented by tissue biopsies or animal models allows the genes to be identified that are specifically expressed in the diseased state. This data can be meaningfully used to develop diagnostic markers and help to identify potential targets for therapeutic interventions.

Sometimes, single markers are not useful in the diagnosis of all diseases, particularly for those which are closely related such as different types of cancer. In such cases, arrays can be used to obtain transcriptional profiles of several genes that offer greater discrimination. Using this approach, it was possible to discriminate between acute myeloid leukemia and acute lymphoid leukemia. Although the diseases are similar, they respond to different therapies and therefore, accurate diagnosis is essential for successful treatment.

Another method to analyze the range of genes expressed in a given cell is called as *serial analysis of gene expression* (SAGE) (Fig. 11.5). To perform a SAGE analysis, polyA$^+$ RNA is extracted from the concerned cell and converted to cDNA using biotinylated oligo(dT) primer. The cDNA is then cleaved with restriction enzyme which cuts frequently as it recognizes four nucleotide sequence sites. This is referred to as *anchoring enzyme*. Biotinylated 3′ ends of cDNAs are bound to streptavidin beads. The fragments in one pool are ligated to linker A *en masse* and the other to linker B. Both linkers contain a recognition site for restriction endonuclease. Each linker has an overhang that matches the overhang generated by the anchoring enzyme and immediately adjacent to this 5 bp recognition site for enzymes like FOK1. This enzyme has an unusual property of recognizing a sequence but cleaving the DNA outside this sequence at a particular number of bases downstream. Cleavage with this enzyme called as *tagging enzyme* (TE) generates a short sequence defined as a *SAGE tag* attached to the linker.

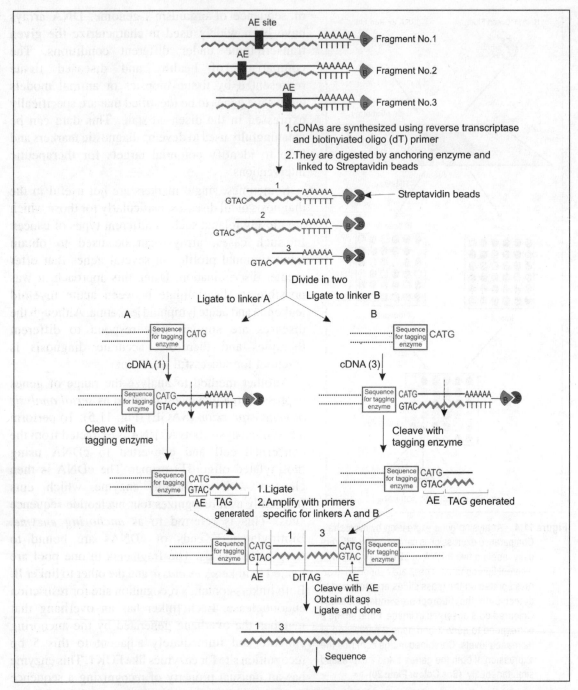

Figure 11.5 Serial analysis of gene expression. cDNA prepared form mRNA is converted to a short sequence tag. Individual tags are then joined together into a single long DNA clone. The clone is then sequenced which provides information on the differential sequence tags which can identify the presence of corresponding mRNA sequences.

In the next stage, the two pools of linker tags are mixed and ligated to produce *ditags* (two SAGE tags joined together flanked on each side by one of the linkers). The PCR is then used to amplify these molecules. The ditags are released from amplified products by cleavage with anchoring enzyme. The purified ditags are ligated together and cloned. These are then sequenced which allows the tags to be read in a serial manner. Relative transcript levels are decided from the frequency with which the tag occurs. The SAGE method of expression profiling has been applied to many different systems, such as characterizing transcriptional profiles associated with diseases, and characterizing the genes associated with developmental arrest and longevity in *Caenorhabditis elegans*, etc.

Pharmacogenomics

The perfect therapeutic drug is one that effectively treats the disease and is free from unwanted side effects. Many new classes of drugs are already in use. However, even the most effective drug provides optimal therapy to a subset of those treated. Some individuals with a particular disease may receive little or no benefit from a drug, while others may suffer from drug-related adverse effects. Thus, there is individual variation in response to a drug because of which there are high failure rates during a clinical trial stage. Pharmacogenomics is the study of the association between genomic, genetic, and proteomic data on one hand, and the drug response pattern on the other. The objective of the study is to explain variability in drug response seen among the individuals and to predict the likely response in individuals receiving a particular medicine. There are two fundamental causes of individual variations in response to drugs. The first is the variation in the structure of target molecule. For example, if the drug acts by blocking a particular receptor and if the receptors in the individuals are not identical, then they will show differential response to the drug. The second reason is differences in pharmacokinetics that is the way the drug is absorbed, distributed, metabolized, and excreted by the body. If the drug is not absorbed and metabolized quickly, it will not be effective. On the other hand, if it is poorly metabolized, it could get accumulated and cause adverse effects.

The variations in adsorption, distribution, metabolism, and excretion can be due to several reasons. One of them could be polymorphism in drug transport and drug metabolism. Polymorphism in drug transport is due to the variation in multi-drug resistant genes, whereas polymorphism in drug transport is due to the variation in the genes coding for cytochrome P450 group of enzymes involved in metabolizing the drug (MDR1). The product of this gene is an ATP-dependent membrane efflux pump whose function is to export the substance from the cell, thereby preventing accumulation of toxic substances. Similarly, genes coding for cytochrome P450 group of enzymes are highly polymorphic.

Over 70 allelic variants have been described, and phenotypically, four different types can be recognized which are extensive metabolizers, intermediate metabolizers, poor metabolizers and ultra rapid metabolizers. Each group of these individuals will require different doses of a drug to be administered. If the genotype and phenotype of an individual is not known, then there is a significant risk of over-dosing and toxicity in case of poor metabolisers and under-dosing in case of ultra rapid metabolizers. Therefore, the knowledge of polymorphism of cytochrome P450 gene locus is important to decide the administration of a drug. Knowing a genotype of an individual in advance allows better clinical decisions to be made. Thus, the genetic analysis of an affected individual can suggest the drug that could be used to treat the disease, and pharmacogenomic analysis will determine the drug to be used.

11.2 PROTEOMICS

Proteomics, an area that is being developed rapidly in the post-genomic era, envisages to identify and catalogue all the proteins synthesized in a cell, called the proteome. In essence, proteomics covers functional annotation of genes based on protein structure, protein expression, protein modifications, protein-protein interactions, localization and functions on a global scale in a cell/tissue/whole organism. Thus, proteomics is equally, if not more, relevant than genomics.

In a cell, the global gene expression profiling (transcriptomics) fails to tally, more often than not, with the proteome profile. This phenomenon in particular necessitates the development of proteomics. It is known that gene expression does not culminate in the production of its transcripts. It is completed only when a functional polypeptide or RNA contribute to the functionality of a cell/organism. Thus, for proteins, protein modifications, protein-protein interactions are far more important than the mRNA product with the genetic code for the protein.

Unlike genomes, proteomes of various types of cells vary widely, and even the proteome of a single cell type differs under different conditions. It is therefore very important to have proteome data in the context of normal cell functions and various pathological diseases. In order to achieve this knowledge, proteomics may likely to play a monumental role. Thus, proteomics is one of the thrust areas in the post-genomic era. For convenience, proteomics is often subdivided into the following categories: (a) protein expression proteomics, (b) structural proteomics, and (c) functional proteomics.

11.2.1 Objectives of Proteomics

The primary objectives of proteomics are cataloguing and functional annotation of proteome in a cell/tissue/organ/organism-specific manner. The source of proteome could be normal, abnormal/diseased samples. The more specific objectives are as follows:

(a) Determine the fraction of the proteome and the specific proteins that are expressed in a sample at a specific time.

(b) Determine the ratios between the unmodified and post-synthetically modified forms of these proteins, and their implications.

(c) Determine the interactions of these proteins with other proteins in well-identified multi-protein complexes, and also their individual/collective functions.

(d) Determine the proteome profiles including expression and modifications (protein fingerprint) of any sample cell/tissue at different physiological conditions, such as development, differentiation, stress, and also in diseased conditions.

(e) Identify and develop drug targets based on proteome profiling.

11.2.2 Methodologies for Proteomics

Studies on proteins in a cell/organism are not new. Various methods, both analytical and preparative, for isolation, purification, and characterization of individual proteins have been in use since long. However, in the post-genomic era, the approach for looking at a cell in terms of protein-mediated activities has changed. The approach has been to know the entire proteome, protein-protein interactions, their localization and functions as global entity than individual protein functions (proteomics approach). Thus, development of high throughput protein profiling methods is the primary necessity in proteomics. Nonetheless, the existing methods still play important roles.

The progress in proteomic research has however been slow due to inherent difficulty in developing suitable techniques. As compared to transcription profiling by DNA microarray methods, protein profiling is difficult. Unlike the straightforward approach of transcription profiling based on nucleic acid hybridization principle, the antibody-based approach of protein profiling is highly unsuccessful. This is particularly so due to high specificity of the antigen-antibody recognition, and de-recognition of even the slightly varied protein by the monoclonal antibodies. These together befit the use of a standard set of reaction conditions employed for one antibody, for the other antibodies to recognize their cognate antigens. These inherent difficulties have led to the search of a suitable alternative approach. The most widely used technology for proteome analysis involves separation of proteins through a high resolution two-dimensional gel electrophoresis followed by their identification and characterization through high-throughput mass spectrometry.

Two-dimensional (2D) Gel Electrophoresis

For global protein profiling, 2D gel electrophoresis is the method of choice. This method which is a combination of isoelectric focusing in the first dimension (separation of proteins based on charges in a gel with a pH gradient) and SDS PAGE (separation based on molecular weights) in the second dimension was developed in the 1970s. The one that has been widely used was developed by O'Farrell in 1975, which could resolve around 1000 proteins from *E. coli* by 2D gel electrophoresis. The details of this method have been described in Chapter 4. The steps involved in this method are presented here (Fig. 11.6).

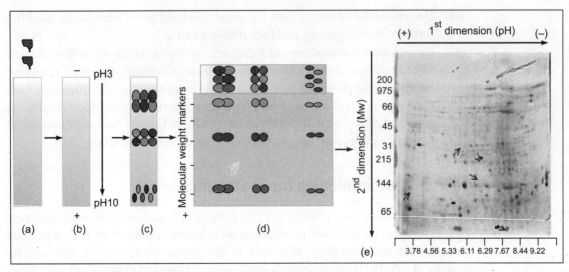

Figure 11.6 Schematic representation (a–d) of the procedure followed during 2-D polyacrylamide gel electrophoresis. **(a)** Sample loaded on the IEF strip. **(b)** pH gradient established during the run. **(c)** Proteins get focused on the gel based on their isoelectric points. **(d)** IEF strip equilibrated with sample buffer placed on top of a SDS polyacrylamide gel for the second dimension and electrophoresed. **(e)** An actual 2D gel after silver staining. **(See Colour Plate 19)**

More recently, there has been further improvement of the procedure including the use of immobilized pH gradient (for the first dimension). Such improved procedures allow resolution of up to 10,000 proteins in a large format gel. At the end of the procedure, the proteins are visualized by protein stains in the form of thousands of spots representing various individual proteins/polypeptides of the proteome of a cell/tissue. Traditionally, Coomassie brilliant blue and silver stains are in use for detection of proteins. In recent times, many sensitive detection methods have been developed, so as to detect all the proteins including the low abundant ones. These include the use of radioactivity, fluorescent dyes, modified silver stain, and antibody. Among these, non-covalent fluorescent dyes known as SYPRO dyes are promising for high throughput analysis.

Among a few drawbacks of this technique, reproducibility has been a major problem. It has been possible to resolve this problem, partly by using freely available gel matching software programmes, such as MELANIE II (http://www.expasy.ch/ch2d/melanie) and Java applet CAROL (http://gelmatching.inf.fuberlin.de/Carol.html). Further, a list of current 2D electrophoresis databases is maintained on the WORLD-2D PAGE (http://www.expasy.ch/ch2d/2d-index.html) with links to further resources.

The next major tasks are to identify, quantify, and characterize these protein spots. Attempts have been made to identify each cell proteome by: (a) creating 2D fingerprints, and (b) cutting each spot followed by amino acid sequencing by an improved method of mass spectrometry (MS). Two-dimensional fingerprints are useful for identifying new proteins in a cell under various conditions including changes in the environment, differentiation, and diseases. In the latter situation, specific proteins expressed during diseases could be drug targets.

Mass Spectrometry

For identification and characterization of proteins separated by 2D gel electrophoresis, mass spectrometry is a powerful method. Recently, with the improvement of this technique and a parallel development of suitable algorithms for searching protein databases based on molecular mass, there has been a breakthrough in proteomics research. There are two major components that are achieved through mass spectrometry:

(a) Proteins can be analyzed and sorted based on their mass to charge ratio.

(b) Proteins can be fragmented into peptides, thereby aiding in their identification.

There are four key components of a mass spectrometer (Fig. 11.7): (a) an ionizer, (b) a mass analyzer, (c) a detector, and (d) a computerized data system. The ionizer converts the sample into gas phase ions. This occurs in the presence of a high

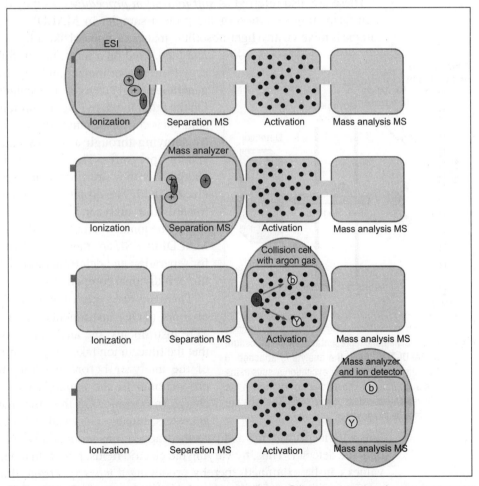

Figure 11.7 Tandem mass spectrometry for protein identification. Different components of a mass spectrometer are labelled in a sequential manner as highlighted in the figure.

electric field, and the charged ions are then sent to the second component, the mass analyzer. The mass analyzer which is always in a high vacuum separates the ions according to their mass:charge (m/z) ratio. These separated ions are then directed to the third component, the detector, which measures the relative quantities of each of the ions in the sample. The computerized data system acquires, stores, and processes the resulting data including automatic collection of additional specific data regarding the sample analyzed. This kind of automatic system is used for the tandem MS (MS/MS) peptide sequencing methods.

There are two methods that are most frequently used for generating ions of proteins and protein fragments:

(a) matrix-assisted laser desorption/ionization (MALDI), and

(b) electrospray ionization (ESI).

These are also referred as *soft ionization procedures*, as they do not cause any significant fragmentation of the protein sample. In MALDI, the protein/peptide sample is mixed with a light-absorbing matrix compound like dihydroxybenzoic acid, and then dried on a metal target. Subsequently, energy from a laser ionizes and vaporizes the sample, generating singly charged molecular ions (Fig. 11.8). On the other hand, in ESI, the protein/peptide in the solution is converted into a fine mist of small droplets by spraying through a narrow capillary at atmospheric pressure. The droplets become highly charged, as they are formed in presence of a high electric field. The droplets evaporate while moving towards the analyzer, forming charged ions (Fig. 11.9). The ionized samples produced, through either MALDI or ESI, are then passed on to the analyzer for separation and determination of charge:mass ratios of individual components.

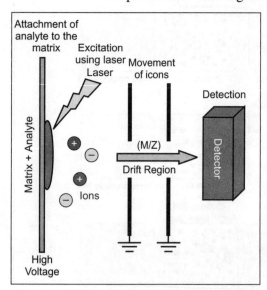

Figure 11.8 Schematic representation of the procedure of MALDI-TOF. First, the analyte is attached to the matrix and then, laser excitation results in the formation of ions. These ions travel through the drift region according to m/z ratio. The ions are detected by the detector.

There are two types of mass analyzers: (a) time-of-flight (TOF) instruments, and (b) ion traps. In TOF instruments, the analysis is based on the fact that the time an ion takes to pass through the length of the analyzer before reaching the detector is proportional to the square root of the mass-to-charge (m/z) ratio. On the other hand, in ion trap analyzers, tunable electrical fields are used to trap ions with a specific m/z and pass the trapped ions sequentially out of the analyzer onto the detector. Thus, by varying the electric fields, ions with a wide range of m/z values can be examined, thereby producing a *mass spectrum*, drawn as a graph with m/z plotted along the X-axis and relative abundance along the Y-axis (Fig. 11.10A).

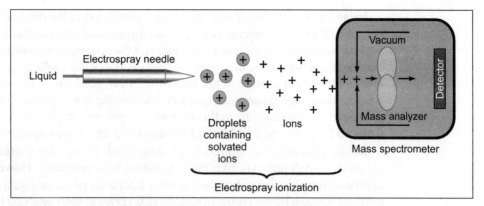

Figure 11.9 Diagrammatic representation of the electrospray ionization. The protein is sprayed and converted into mist, and these small droplets get evaporated and ions are formed. These ions are passed through the analyzer and then detected.

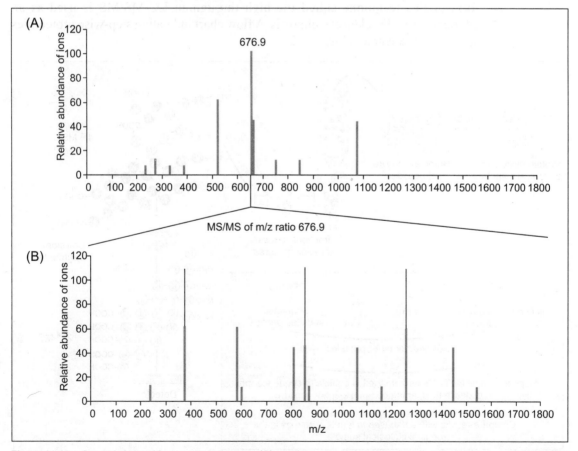

Figure 11.10 A protein is used for mass spectrometry **(A)** and then one component is selected, fragmented, and MS (MS/MS) is done again to determine its sequence **(B)**.

In tandem or MS/MS spectrometry, any parent ion in the original mass spectrum (Fig. 11.10A) can be selected by its mass, fragmented into smaller ions by collision with an inert gas, and thereafter, the m/z and the relative abundance of the resulting fragment ions are measured (Fig. 11.10B). This entire procedure is carried out within the same machine at the rate of 0.1s per parent ion. The second round of fragmentation and its analysis helps in determining the sequences of short peptides of around 25 amino acids. In this manner, the primary sequence of a protein can be deduced by using the MS/MS and the available data base sequence information.

In the conventional approach, as described above, the protein samples are subjected to MS analysis after their digestion with proteases. However, in another approach, the complex mixture of proteins after their protease digestion are partially purified through liquid chromatography (LC) prior to their analysis through tandem MS/MS. Essentially, in this technique called the LC-MS/MS, the eluted samples can directly be transferred to the mass spectrometer, thereby allowing continuous analysis of a complex mixture of proteins (Fig. 11.11). More recently, an advanced form of this technique called the high throughput LC-MS/MS is used as an alternative to 2-D gel electrophoresis. A flow chart indicating step-wise proteomics analysis is presented in Fig. 11.12.

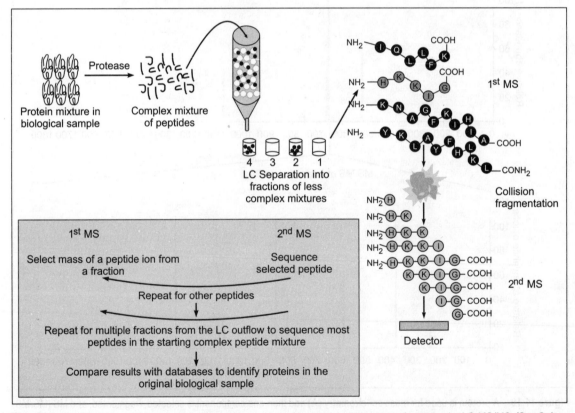

Figure 11.11 Analysis and identification of proteins in a sample containing a mixture of proteins by LC-MS/MS. **(See Colour Plate 19)**

Colour Plate 13

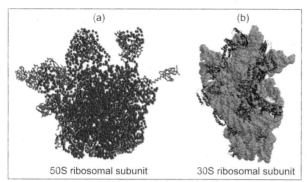

50S ribosomal subunit   30S ribosomal subunit

3-dimensional high resolution X-ray crystal structures of prokaryotic ribosomal subunits. (a) 50S ribosomal subunit. (b) 30S ribosomal subunit (Chapter 9, p. 268)

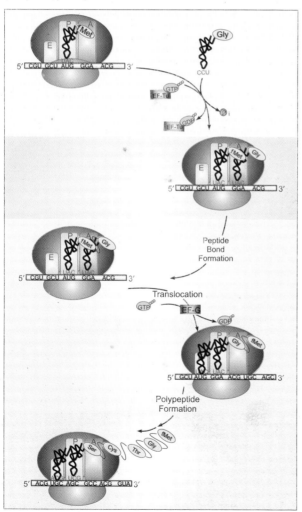

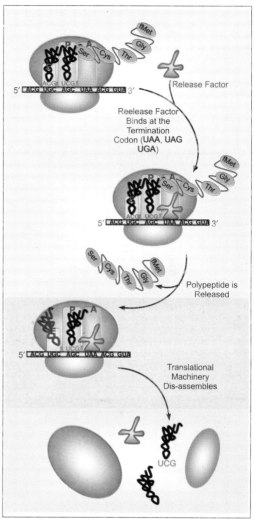

Termination of translation. When the termination codon is encountered by the ribosome, the release factor binds at the termination codon (UAA, UGA, and UAG). After the binding of release factor, translational machinery dis-assembles, as ribosome dissociates and tRNA is released. (Chapter 9, p. 284).

Elongation during translation. First, initiator tRNA binds at the P site in the ribosome. In the second step, the next aminoacyl tRNA binds at the A site. After this, there is peptide bond formation and further, the initiator tRNA and the glycylyl tRNA (shown here) are shifted to the E site and P site, respectively. Then, the next aminoacyl tRNA arrives at the A site and the process is repeated till the termination codon is reached. (Chapter 9, p. 279).

Colour Plate 14

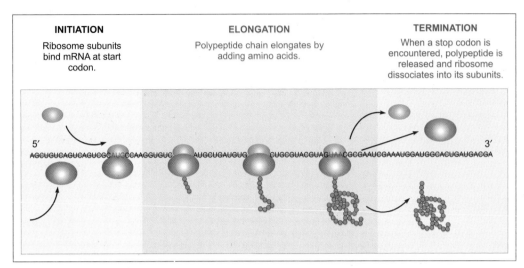

INITIATION	ELONGATION	TERMINATION
Ribosome subunits bind mRNA at start codon.	Polypeptide chain elongates by adding amino acids.	When a stop codon is encountered, polypeptide is released and ribosome dissociates into its subunits.

Overview of translation. The first step is initiation in which ribosomal subunits bind to the mRNA and translation starts at the initiation codon (AUG). The second step is elongation, in which the polypeptide is elongated by the addition of amino acids. In the third step, termination occurs at the termination codon (UAA, UAG, and UGA) by binding of release factor, and then the polypeptide is released and ribosomes dissociate. (Chapter 9, p. 272)

Initiation of translation in prokaryotes. In the first step, three initiation factors, IF-1, IF-2, and IF-3 bind to the 30S ribosomal subunit. In the second step, mRNA and the initiator tRNA (fMet-tRNA$_f$) bind, and thereafter, the large subunit of ribosome joins to form the initiation complex. The factors are released subsequently. (Chapter 9, p. 274).

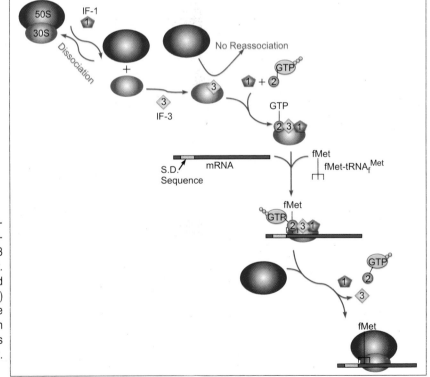

Colour Plate 15

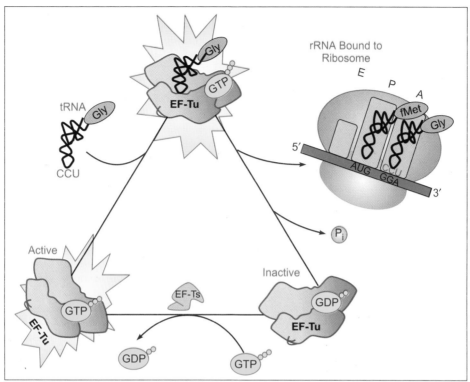

Recycling of EF-Tu.GTP during elongation of translation. During elongation, EF-Tu.GTP binds to the next aminoacyl tRNA, and if correct aminoacyl tRNA binds to the codon, then EF-Tu.GTP gets hydrolyzed and EF-Tu.GDP is released. Further, the inactive EF-Tu.GDP complex is converted to EF-Tu.GTP with the help of EF-Ts and thus, the EF-Tu.GTP is ready for binding to the next aminoacyl tRNA. (Chapter 9, p. 280).

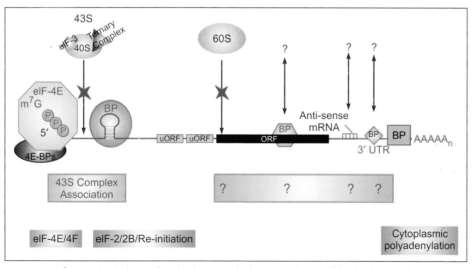

Summarized scheme of translation regulation in eukaryotic cells. (Chapter 9, p. 288).

Colour Plate 16

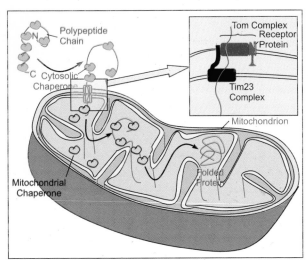

Chaperones during protein transport into mitochondria. For the proteins which have to be transported to the mitochondria, the proteins in the unfolded state bound to the chaperones come to the mitochondria. TOM complex in outer mitochondrial membrane and TIM complex in inner mitochondrial membrane provide help in the transport of the polypeptide. Cytosolic chaperones are left outside the mitochondria. After transport, the mitochondrial chaperones bind again to the polypeptide till the transport is complete. Then the polypeptide is folded into a correct form. (Chapter 10, p. 300).

Structure of a chaperonin. GroEL is a member of the Hsp60 family. (Chapter 10, p. 301).

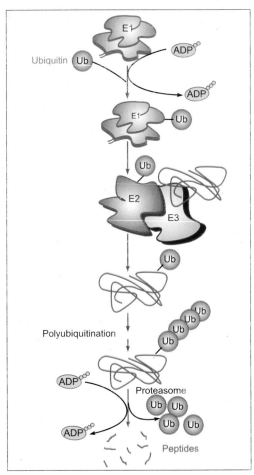

The Ubiquitin-Proteasome pathway of protein degradation. Proteins, which are to be degraded, are marked by covalent attachment of several molecules of ubiquitin. Ubiquitin is first activated by enzyme E1. The ubiquitin is then transferred to enzyme E2 (conjugating enzyme). Then with the help of enzyme E3 (ubiquitin ligase), ubiquitins are transferred to the protein, after which the ubiquitinated protein is degraded by the proteasomes. (Chapter 10, p. 314).

Colour Plate 17

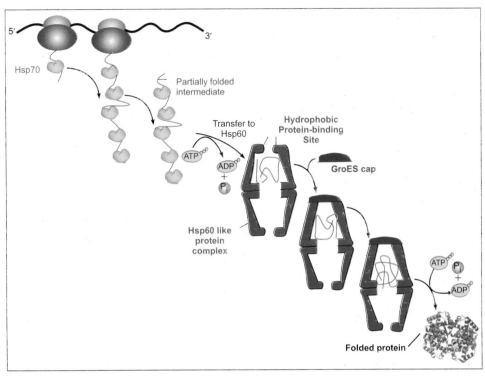

Sequential involvement of Hsp70 and Hsp60 chaperones in protein folding. During protein synthesis, Hsp70 binds to the synthesizing polypeptide to keep it in an unfolded state. After protein synthesis, the polypeptide is transferred to Hsp60 using ATP as an energy source for correct folding of the polypeptide. (Chapter 10, p. 302)

Cyclin degradation by Ubiquitin-Proteasome pathway during cell cycle. The progression of eukaryotic cell into the cell cycle is partially governed by the synthesis and degradation of the cyclin B, which is a regulatory subunit of Cdc2 kinase. Synthesis of cyclin B in the interphase leads to the formation of cyclin B-Cdc2 active complex in the prophase. During metaphase, cyclin B gets ubiquitinated and gets degraded by proteasomes. Thus, from anaphase till interphase, there is a formation of inactive Cdc2 kinase. (Chapter 10, p. 315).

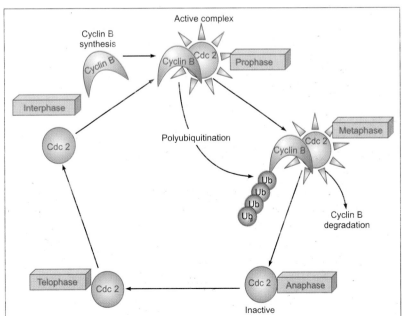

Colour Plate 18

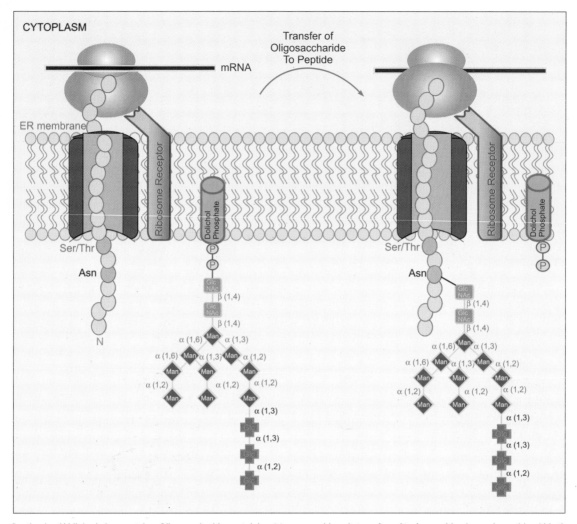

Synthesis of N-linked glycoproteins. Oligosaccharide containing 14 sugar residues is transferred to Asn residue in a polypeptide within the sequence Asn-X-Ser/Thr from dolichol phosphate, a lipid carrier. (Chapter 10, p. 306).

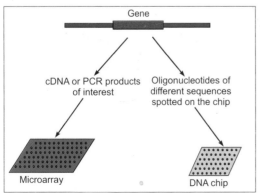

Microarray and DNA chips. In microarray, each position contains a cDNA or PCR product for a gene of interest: whereas DNA chip contains at each position a mixture of oligonucleotides where the sequences match different segments of genes of interest. (Chapter 11, p. 328).

Colour Plate 19

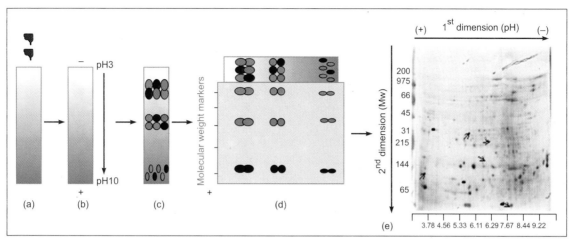

Schematic representation (a-d) of the procedure followed during 2-D polyacrylamide gel electrophoresis. **(a)** Sample loaded on the IEF strip. **(b)** pH gradient established during the run. **(c)** Proteins get focused on the gel based on their isoelectric points. **(d)** IEF strip equilibrated with sample buffer placed on top of a SDS polyacrylamide gel for the second dimension and electrophoresed. **(e)** An actual 2D gel after silver staining. (Chapter 11, p. 334).

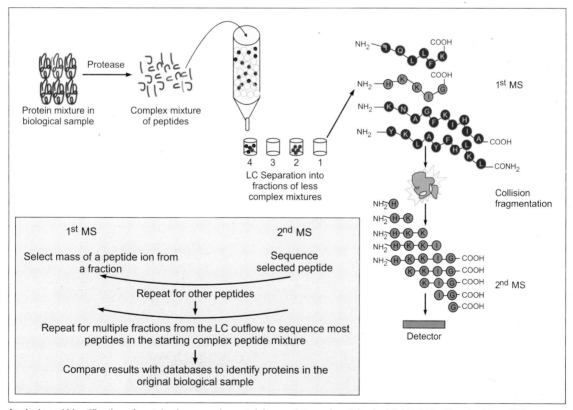

Analysis and identification of proteins in a sample containing a mixture of proteins by LC-MS/MS. (Chapter 11, p. 338).

Colour Plate 20

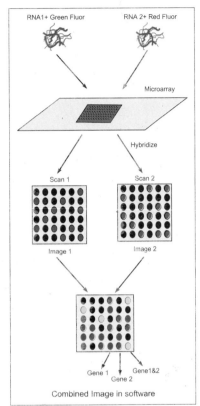

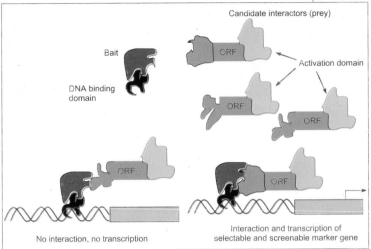

The principle of a basic yeast two-hybrid system. Bait is the protein, for which one wants to find interacting protein partners. It is fused with a DNA binding domain. Other proteins are fused with an activation domain. If the two proteins are interacting, then the reporter gene will be transcribed. (Chapter 11, p. 341).

Analysing gene expression by microarray. Comparative expression is carried out by differentially labeling two mRNA or cDNA samples with different fluorophores. These are hybridized to arrays spotted on the glass slides and scanned to detect both the fluorophores independently. Coloured dots observed in image 1 and image 2 confirm the expression of both the genes 1 and 2 expressed simultaneously. (Chapter 11, p. 329).

Principle of GST pulldown assay. Glutathione conjugated sepharose beads bind to protein X expressed as a GST fusion protein. Subsequently, the sample lysate is passed through the column and the protein which interact with protein X binds to it and remain in the column. SDS PAGE is used to determine the interactor or protein X. (Chapter 11, p. 342).

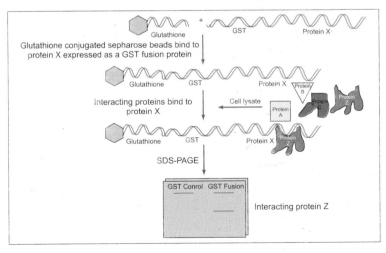

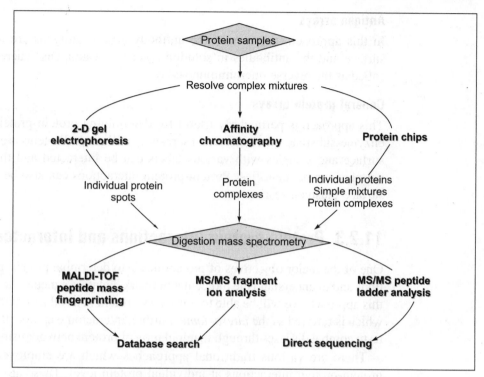

Figure 11.12   A flow chart of procedures for protein annotation in proteomics.

In addition to these methods as described above, there are a few recent proteomics procedures in their developmental phase. Broadly, they are called as *protein arrays* and *protein chips*. Similar to DNA arrays, protein arrays are miniature devices in which proteins or other molecules that recognize proteins are arrayed on the surface (Wilson and Nock, 2001; Templin *et al.*, 2002). These molecules are arrayed by contact printing, inkjet printing, and photolithography. As described above, due to the inherent dynamic characteristics of proteins, development of protein arrays as compared to DNA microarrays has been very difficult. Therefore, a number of diverse approaches have been attempted, and a few of them are briefly described below.

Antibody arrays

Antibody arrays (microimmunoassays) have a great promise as a tool for bulk analysis of proteins. These are relatively inexpensive. In this approach, antibodies are arrayed on the surface. The antibodies react to the proteins from the cell/tissue samples and thus, capture the respective antigens from the sample solution. In this manner, protein profiles of the samples can be identified. Proteins can also be quantified when the sample proteins are fluorescently labelled. This approach is somewhat similar to immunoassay and thus, is referred as *solid state microimmunoassay*. Antibody arrays have been used for identification of biomarkers in the blood serum/plasma for various diseases including cancers.

Antigen arrays

In this approach, contrary to the antibody arrays, antigens are arrayed on the surface and the antibodies in solution react to the antigens. Therefore, it is also called as the reverse microimmunoassay.

General protein arrays

This approach is particularly useful for determining protein-protein interactions, enzyme-substrate interactions and screening of ligands. Proteins are arrayed on the surface and samples with various labels can be interacted and thus, interacting partners can be identified. Protein-protein interactions can also be determined by surface plasmon resonance.

11.2.3 Protein-protein Interactions and Interactome

One of the major objectives of proteomics is to determine protein-protein interactions and to understand the molecular network dictating a metabolic process. Using this approach, we will be able to link every protein in a cell to a functional network which is referred as the *interactome*. Further, interactome approach will also allow us to combat diseases through manipulation of protein network using drugs.

There are various traditional approaches which are employed for studying protein-protein interactions at individual protein level. These are grouped under genetic methods (e.g., various types of mutations), biochemical methods (e.g., co-immunoprecipitations, affinity chromatography, cross-linking), and physical methods (X-ray crystallography and nuclear magnetic resonance spectroscopy). Some of them are described in details in Chapter 4. These methods, however, are unsuitable for high throughput analysis as well as interactome analysis. For high throughput purpose, the following approaches are being developed: (a) Library-based interaction mapping, (b) High throughput protein analysis and annotation from GST-pulldown approach, (c) Bioinformatics tools and databases of interacting proteins to generate interaction data.

Library-based Interaction Mapping

This approach helps in parallel screening of hundreds of proteins and linking them to their genes/cDNAs. Thereafter, the corresponding clones of these interacting proteins from the library can be isolated, sequenced and compared with the DNA sequence databases. There are two types of libraries: *in vitro* library and *in vivo* library.

In vitro library type

In *in vitro* library type, classically, an expression library is screened using antibodies. This approach has been extended for high throughput assays by using *phage display techniques*. In this, a fusion protein is expressed in such a way that a foreign peptide is displayed on the bacteriophage surface. Such libraries expressing many proteins can be probed with protein-specific antibodies, thereby identifying various interacting proteins in a microtitre plate. Such proteins can be sequenced for characterizing the interacting proteins. Using this approach, libraries of high complexity to the extent of 10^{12} can be generated and many rounds of screening can be performed. This procedure can be carried out in an array format for high throughput analysis.

In vivo library type

In *in vivo* library type, the interaction assay appears more natural. One example is the yeast two-hybrid (Y2H) system. The principle of this system is as follows. Protein X is expressed as a hybrid with the DNA binding domain of a transcription factor to generate a *bait*. In parallel, a library of *pray* is generated, in which each clone is expressed as a fusion protein with the transactivation domain of the transcription factor. The third component of the system is a reporter gene whose activation is mediated specifically by the two-hybrid transcription factor. Mating between haploid yeast cells carrying the bait construct and those carrying the library of prey results in diploid cells carrying both the components. In such cells where the bait interacts with the prey, the transcription factor is constituted, and it results in activation of transcription of the given reporter gene. This way, the cells can be identified and the DNA sequence of the interactor identified (Fig. 11.13).

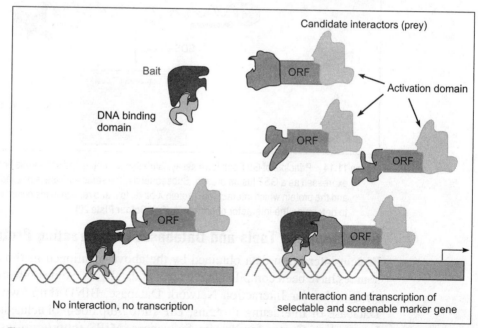

Figure 11.13 The principle of a basic yeast two-hybrid system. Bait is the protein, for which one wants to find interacting protein partners. It is fused with a DNA binding domain. Other proteins are fused with an activation domain. If the two proteins are interacting, then the reporter gene will be transcribed. **(See Colour Plate 20)**

High Throughput Protein Analysis and Annotation

GST-pulldown assay, although initially developed for single proteins, has now been extended for genomic scale applications. In this method, a bacterial enzyme glutathione-S-transferase (GST) is used as a fusion tag for expression of recombinant proteins in bacteria. Such GST-tagged proteins are then purified by affinity chromatography using a matrix coupled with the substrate, glutathione. Using the same principle, a specific bait protein is expressed as a GST-fusion protein, it is passed through a glutathione affinity matrix, and subsequently copurified with other interactive protein(s) from a cell lysate (Fig. 11.14). These new proteins can be subsequently analyzed by MALDI-TOF MS. By this approach, from organisms like yeast, a number of new genes have been functionally annotated and networks established (Martzen *et al.*, 1999).

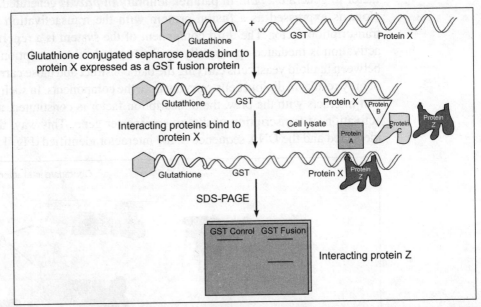

Figure 11.14 Principle of GST pulldown assay. Glutathione conjugated sepharose beads bind to protein X expressed as a GST fusion protein. Subsequently, the sample lysate is passed through the column and the protein which interact with protein X binds to it and remain in the column. SDS PAGE is used to determine the interactor of protein X. **(See Colour Plate 20)**

Bioinformatics Tools and Databases of Interacting Proteins

Protein interaction data obtained by the above-mentioned methods from various sources have been compiled in the form of databases. Some of them are as follows: (i) Bimolecular Interaction Network Database, BIND (http://www.bind.ca), (ii) Database of Interacting Proteins, DIP (http://dip.doe-mbi.ucla.edu), (iii) Munich Information Centre for Protein Sequences, MIPS (http://mips.gsf.de/proj/yeast/CYGD/db/index.html). To extract information from the databases, a number of informatics tools have been developed. However, they are still at the developing stage.

Further, another problem has been to represent the protein interaction data in a comprehensible manner. In order to do that, the interaction network has been simplified by restricting to clusters of a few related processes, such as cell cycle (To and Vohradsky, 2008), mitosis, DNA synthesis or to various subcellular structures or components (Fig. 11.15). Although there is a long way to go before obtaining a fully satisfied protein interaction network, establishment of some basic foundation network is necessary. These can further be extended with the increasingly added information available in future. As an example, an interactive proteome is given in Fig. 11.16.

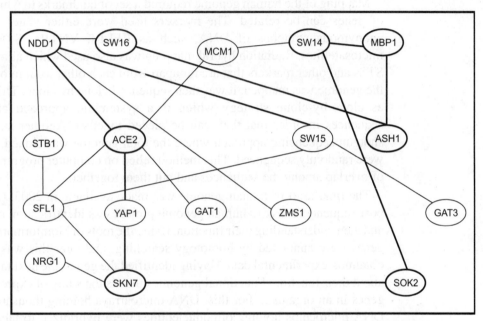

Figure 11.15 Cell cycle interaction networks with sub-networks (bold) used as a training set.

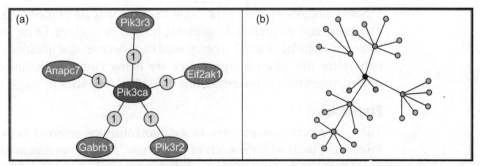

Figure 11.16 Network of protein interactions. **(a)** Protein interactions for protein pik3ca using MINT Molecular INTeraction database. **(b)** Protein interaction network of a protein in yeast using DIP.

SUMMARY

Genomics

The technique of automated DNA sequencing allowed scientists to obtain the DNA base sequence of organisms, ranging from simple phages to bacteria to yeast and some animals and plants. Finally, the human genome was also sequenced by Craig Venter and Francis Collins. Two high capacity vectors were extensively used in human genome project. Most of the mapping was done with YAC vectors, whereas most of the sequencing work was done using BAC vectors.

Mapping of the human genome required a set of landmarks to which the position of genes can be related. The markers used were either genes themselves or anonymous stretches of DNA, such as RFLPs, VNTRs, STSs, ESTs, and microsatellites. Radiation hybrid mapping was also used which allows mapping of STSs and other markers that are far apart from each other to fit in one BAC. Once the genome was mapped, it was then sequenced clone by clone. This is referred to as clone-by-clone strategy which is a systematic approach and places the sequences in order that they can be pieced together. Another strategy was the shotgun sequencing approach where the total genome was cloned and the clones were randomly sequenced. This method relied on computer programs to find areas of overlap among the sequences and put them together.

The final draft of human genome was made available in 2003. One of the first post-sequencing task to human genome project was identification of all the genes and then understanding their function. Using the tools of bioinformatics, initially the genes were annotated by homology searching, and later this was confirmed by obtaining experimental data. Having identified the genes, the next task was to learn about their function. Functional genomics thus is the study of expression of all the genes in an organism. For this, DNA microarrays holding thousands of DNA or DNA microchips holding oligonucleotides were hybridized to label RNAs from different cells. The intensity of hybridization to each spot reveals the extent of expression of the corresponding gene.

Serial analysis of gene expression (SAGE) was performed to determine genes expressed and the extent of their expression in a given tissue in a given time. Genomic techniques are also being used to answer several questions in biology, and new terms like pharmacogenomics are being coined. Pharmacogenomics has proved to be useful in understanding individual responses to drugs.

Proteomics

The total protein complement in a cell/organism is referred to as the proteome. Proteomics deals with the study of proteomes. This allows functional annotation of genes based on protein structure, protein expression, protein modifications, protein-protein interactions, localization and functions on a global scale in a cell/tissue/whole organism.

For such an extensive and global analysis of proteins, specific methodologies have been developed that allow high throughput analysis. Some of the methods are: 2-D gel electrophoresis, various types of mass spectrometry, protein/antibody arrays, yeast two hybrid assays, GST-pulldown assay, etc.

In parallel, an extensive advancement in bioinformatics tools has helped in projecting and organizing a large amount of data obtained through proteomics approach. The role of proteomics appears to be monumental, not only in understanding biology of organisms but also in various applications in health and disease, in the post-genomic era.

SHORT ANSWER QUESTIONS

Genomics

1. What is the basic difference between the two strategies used for sequencing the human genome?
2. The human genome contains about 35,000 genes as predicted from the available sequence. Why was the initial prediction of number of genes so high?
3. How is the presence of a gene identified in a given sequence?
4. How does the mitochondrial genome differ from the nuclear genome?
5. Differentiate between EST and STS used to map the genome.
6. What is the difference between DNA microarray and DNA microchip?
7. What is the advantage of using radiation hybrid mapping?
8. Differentiate between structural genomics and functional genomics.

Proteomics

1. Define proteomics and enlist its various applications.
2. Proteomics is more important than genomics in the context of understanding biology and also in drug discovery. Why is it so?
3. What are the additional features of LC-MS/MS as compared to MS/MS alone?
4. Name two databases that are useful for analyzing 2D gel electrophoreris data.
5. Explain how yeast two-hybrid system can be extended for high throughput analysis.

FURTHER READING

Genomics

■ Blattner, F.R, Plunkett, G., Bloch, C.A., *et al.* (1997). The complete genome of *Escherichia coli* K12. *Science* **277**, 1453-62.

- IHGSC (International Human genome sequencing consortium) (2001). Initial sequencing and analysis of the human genome. *Nature* **409,** 860–921.
- Iyer, V.R., Eiscn, M.B., Ross, D.T., *et al*. (1999). The transcriptional program in the response of human fibroblast serum. *Science* **283**, 83–87.
- Judson, R., Stephens, J.C., and Windemuth, A. (2000). The predictive power of haplotypes in clinical response. *Pharmacogenomics*. **1**,15–26.
- Leung, Y.F. and Cavalieri, D. (2003). Fundamentals of cDNA microarray data analysis. *Trends Genet*. **90**, 649–59.
- Venculescu, V.E., Vogelstein, B., and Kinzler K.W. (2000). Analysing uncharted trans-criptome with SAGE. *Trends Genet*. **16**, 423–25.
- Venter, J.C., Adams, M.D., Myers, E.W., *et al*. (2001). The sequence of Human genome. *Science* **291,** 1304–51.
- Zhang, C. and Kim, S.H. (2003). Overview of structural genomics: From structure to function. *Curr. Opinion Chem. Biol*. **7**, 28–33.

Proteomics

- Anderson, J.S. and Mann, M. (2000). Functional genomics by mass spectrometry. *FEBS Lett*. **480**, 25–31.
- Boulton, S.J., Vincent, S., and Vidal, M. (2001). Use of protein interaction maps to formulate biological questions. *Curr. Opinion Chem. Biol*. **5**, 57–62.
- Campbell, A.M. and Heyer, L.J. (2007). Discovering Genomics, Proteomics, & Bioinformatics (second edn.). Person Education, Inc., India.
- Chalmers, M.J. and Gaskel, S.J. (2000). Advances in mass spectrometry for proteome analysis. *Curr. Opinion Biotech*. **11**, 384–390.
- Foster, L.J. *et al*. (2006). A mammalian organelle map by protein correlation profiling. *Cell* **125**, 187–199.
- Fu, Q. and Van Eyk, J.E. (2006). Proteomics and heart disease: identifying biomarkers for clinical utility. *Expert Rev. Proteomics* **3**, 237–249.
- Haab, B.B., Dunham, M.J., and Brown, P.O. (2001). Protein microarrays for highly parallel detection and quantitation of specific proteins and antibodies in complex solutions. *Genome Biol*. **2**, 0004.1–0004.13.
- Kolkar, E., Higdon, R., and Hogan, J.M. (2006). Protein identification and expression analysis using mass spectrometry. *Trends Microbiol*. **14**, 229–235.
- Krogan, N.J., *et al*. (2006). Global landscape of protein complexes in the yeast *Saccharomyces cerevisiae*. *Nature* **440**, 637–643.
- Lee, K.H. (2001). Proteomics: a technology-driven and technology-limited discovery science. *Trends Biotech*. **19**, 217–222.
- Lodish, H., Berk, A., Kaiser, C.A., Kreiger, M., Scott, M.P., Bretscher, A., Ploegh, H., and Matsudaira, P. (2008). Molecular Cell Biology (Sixth edn.). W.H. Freeman, New York.

- MacBeath, G. and Schreiber, S.L. (2000). Printing proteins as microarrays for high-throughput function determination. *Science* **289**, 1760–63.
- Mayer, M.L. and Hieter, P. (2000). Protein networks–Built by association. *Nature Biotech.* **18**, 1242–43.
- Ong, S.E. and Mann, M. (2005). Mass spectrometry-based proteomics turns quantitative. *Nat. Chem. Biol.* **1**, 252–62.
- Pelletier, J. and Sidhu, S. (2001). Mapping protein-protein interactions with combinatorial biology methods. *Curr. Opinion Biotech.* **12**, 340–47.
- Primrose, S.B. and Twyman, R.M. (2006). Principles of Gene Manipulation and Genomics (7th Edn.). Blackwell Publishing, UK.
- Rifai, N., Gillette, M.A., and Carr, S.A. (2006). Protein biomarker discovery and validation: the long and uncertain path to clinical utility. *Nat. Biotech.* **24**, 971–83.
- Schweitzer, B. and Kingsmore, S.F. (2002). Measuring proteins on microarrays. *Curr. Opinion Biotech.* **13**, 14–19.
- Templin, M.F., Stoll, D., Schrenk, M., Traub, P.C., Vohringer, C.F., and Joos, T.O. (2002). Protein microarray technology. *Trends Biotech.* **20**, 160–66.
- To, C.C. and Vohradsky, J. (2008). Supervised inference of gene-regulatory networks. *BMC Bioinformatics* **9**:2. doi:10.1186/1471-2105-9-2.
- Uetz, P. (2001). Two-hybrid arrays. *Curr. Opinion Chem. Biol.* **6**, 57–62.
- Wilson, D.S. and Nock, S. (2001). Functional protein microarrays. *Curr. Opinion Chem. Biol.* **6**, 81–85.
- Zhou, H., Roy, S., Schulman, H., and Natan, M.J. (2001). Solution and chip arrays in protein profiling. In: *A trends Guide to Proteomics II*, supplement to *Trends Biotech.* **19**, 534–39.
- www.expasy.ch/ch2d/2d-index.html : Worldwide 2DE homepage.
- www.biacore.com: Biacore, a company that manufacture chips for the detection of protein interactions.
- www.bind.ca and http://dip.-doe-mbi.ucla.edu : BIND and DIP – two databases listing protein interactions with proteins and other molecules.

Glossary

3′-terminus The end of a polynucleotide which carries the hydroxyl group attached to the 3′ position of the sugar.

5′-terminus The end of a polynucleotide which carries the phosphate group attached to the 5′ position of the sugar.

A site The binding site for the aminoacyl-tRNA on the ribosome.

Acrylamide Used to make polyacrylamide gels for separation of macromolecules by electrophoresis. Polyacrylamide gels are produced by polymerization of acrylamide into linear chains and crosslinking the acrylamide chains with bis-acrylamide.

Activator A gene product (usually protein) that positively regulates transcription. Activators may either increase binding of RNA polymerase to the promoter (closed complex formation) or stimulate RNA polymerase to begin transcription (open complex formation).

Adenosine triphosphate (ATP) A compound with high energy phosphate bonds that provide the energy for many cellular processes.

Affinity chromatography The separation of soluble macromolecules by use of a stationary phase that is designed to interact specifically with, and thus retard the elution of, the desired material.

Agarose An uncharged polysaccharide purified from agar used as an electrophoresis and chromatography support.

Alleles Alternative forms of a gene.

Alpha-helix A secondary structure in proteins; the right-handed helical folding of a polypeptide such that amide nitrogens share their hydrogen atoms with the carbonyl oxygens of the fourth amide bonds towards the C-terminal end of the polymer.

Alpha-complementation The ability of a short N-terminal fragment of beta-galactosidase to interact with the C-terminal portion of beta-galactosidase to form a functional enzyme.

Ampicillin An antibiotic that inhibits crosslinking of peptidoglycan chains in the cell wall of eubacteria.

Angstrom A unit of measurement that was widely used until recently to describe molecular dimensions, but the unit nanometer (nm) is now more commonly used. One angstrom equals 10 nm.

Annealing Pairing of a single-stranded nucleic acid with the complementary strand to form a duplex.

Antibiotic A substance that interferes with a particular step of cellular metabolism, causing either bactericidal or bacteriostatic inhibition; sometimes restricted to those having a natural biological origin.

Antibody A protein that specifically interacts with an antigen.

Anticodon The three bases in a tRNA that are complementary to those in the codon of the mRNA (a triplet) that specifies an amino acid.

Antigen A substance that interacts with an antibody.

Antisense RNA An endogenous or semi-synthetic oligoribonucleotide complementary to mRNA and capable of base-pairing and annealing with mRNA to prevent translation

AP endonuclease An endonuclease which recognizes an AP site and cuts the defective strand on the 5′ side of the missing base.

Apoptosis Cell death due to an intracellular developmental program or induced by other cells.

Aporepressor A protein that requires binding of a small molecule co-repressor to negatively regulate gene expression.

Apurinic or apyrimidinic (AP) site   A molecule of single-stranded or double-stranded DNA missing a purine or pyrimidine base.

Attenuation   A mechanism of regulating the level of transcription by interfering with mRNA elongation in prokaryotes.

Attenuator   A region of leader mRNA that can form alternative secondary structures that determine whether transcription is terminated or proceeds into downstream genes.

Autoradiography   A technique for visualization of radioactivity in histological preparations (DNA, RNA, or protein), paper chromatograms or slab gels from electrophoresis by overlaying the surface with X-ray film and allowing the radiation to form an image on the film.

Bacterial artificial chromosome (BAC)   A cloning vector that can accept up to 350 kb fragments for cloning and sequencing of fragments of the human genome.

Bacteriophage   A virus that infects bacteria. Many phages have proved useful in the study of molecular biology and as vectors for the transfer of genetic information between cells.

Base pair   A complementary purine and pyrimidine that are hydrogen-bonded to form double-stranded DNA or RNA.

B-DNA   A right-handed helix; the dominant conformational variant of DNA in solution, in which the base pairs are stacked nearly perpendicular to the axis of the helix.

Beta-turn (reverse turn)   A beta-turn is a region of the protein involving four consecutive residues where the polypeptide chain folds back on itself by nearly 180 degrees.

Bidirectional replication   Two replication forks proceed in opposite directions from the same origin of replication.

Bioinformatics   The management and analysis of data using advanced computing techniques. Bioinformatics is an important field of genomics research, because of the complexity of searching and comparing the large number of sequences generated.

Biotechnology   The use of living organisms (often microbes) in industrial processes.

Blunt (flush) end   The end of a DNA molecule at which both strands terminate at the same nucleotide position with no single-stranded extension. It is often the product of cleavage by a restriction endonuclease.

bZIP protein   A type of Y-shaped homodimeric DNA-binding protein in which the stem is composed of two intertwined α-helices held together as a leucine zipper

C value   The calculated value of the total DNA content of a cell per haploid number of chromosomes.

CAAT box   A regulatory sequence upstream from some eukaryotic structural genes.

Caenorhabitis elegans   A nematode; a small worm favoured by developmental and molecular biologists because of its ability to grow under laboratory conditions, its short generation time and its transparency.

Capsid   The protein layer that encloses the nucleic acid of a phage or virus and protects it from the environment.

Cap   The 7-methylguanine nucleoside attached to the 5′-end of a mRNA by a 5′-5′-triphosphate bond.

Carbohydrate   One of a class of biological materials comprising sugars, polymers of sugars, and compounds related to them. The name derives from the basic sugar structure, $(CH_2O)n$.

Carbon source   A nutrient which provides the carbon required for cellular biosynthesis.

Carboxy-terminal domain (CTD)   The domain of a protein which includes the carboxy-terminal amino residue.

Carcinogen   A physical or chemical agent that causes cancer.

cDNA cloning   A technique involving reverse transcription of purified mRNA into the corresponding complementary DNA prior to insertion into a vector.

cDNA library   A collection of cells, usually *E. coli*, transformed by DNA vectors each of which contains a different cDNA insert synthesized from a collection of mRNA species.

cDNA   Complementary DNA; DNA that is synthesized by reverse transcriptase, from an mRNA template, and therefore has no introns.

Cell cycle The necessary sequence of growth and synthetic stages through which a cell passes, from its origin by mitosis of a parent cell until its own division into daughter cells.

Cell extract A preparation consisting of a large number of broken cells and their released contents.

Cell-free translation system A cell extract containing all the components required for protein synthesis (i.e. ribosomal subunits, tRNAs, amino acids, enzymes and cofactors) capable of translating any exogenously added mRNA molecules.

Central dogma In biology, the proposition that the permanent repository of genetic information is DNA which can be replicated, and that the information is expressed unidirectionally by transcription into RNA and thence by translation into protein.

Centromere The region of a eukaryotic chromosome responsible for attachment of the sister chromatids together and also to the mitotic or meiotic spindle leading to controlled partitioning of chromosomes during nuclear division.

Chain-termination codon A codon signalling the termination of polypeptide synthesis. The three standard stop codons are UAG (amber), UAA (ochre), and UGA (opal). Also called *nonsense* or *stop codon*.

Chaperone A protein that facilitates the folding of other proteins or assembly of multi-protein complexes.

Chaperonin A chaperone machine of a prokaryote mitochondrion or plasmid; a class of molecular chaperones, a complex of proteins which functions to properly fold denatured or improperly folded globular proteins.

Chemiluminescence The light emitted by an exergonic chemical reaction; contrasted with *photoluminescence* (fluorescence or phosphorescence), which is the light emitted as a fluorophore falls to a lower energy state after having been excited by irradiation.

Chimeric plasmid A plasmid used in gene cloning and constructed from two or more different plasmids.

Chloroplast An organelle of a green plant cell in which light harvesting and ATP synthesis occur.

Chromatid One of the two identical strands of a newly replicated chromosome.

Chromatin The complex of DNA and associated proteins, most notably histones, that occurs in the nuclei of eukaryotic cells.

Chromogenic substrate A substrate that changes color when modified by a specific enzyme.

Chromosome A self-replicating DNA molecule that carries essential genetic information for growth and replication of a cell or virus.

Circular dichroism (CD) A technique for determination of the asymmetry of a molecule. Unlike in a conventional polarimeter, where light is restricted to oscillation in a plane, in CD the direction of oscillation turns clockwise around the direction of the beam.

cis-**acting** Descriptive of the controlling effect of a regulatory gene on a structural gene that is adjacent to it; later broadened to distinguish intramolecular (*cis*) from intermolecular (*trans*) actions.

cis In stereochemistry, descriptive of a substituent on the same side of a structure that prevents equalization of positions by restricted rotation.

Cistron An obsolete term for a gene. Often used to refer to the number of genes encoded by a single transcript; monocistronic refers to a transcript that encodes one gene, and polycistronic refers to a transcript that encodes multiple genes.

Clonal A population of cells derived from a single cell and thus expected to be genetically identical.

Clone In microbiology, a colony of cells all descended from a single ancestral cell and therefore genetically homogeneous; in molecular biology, an exact replica of a DNA fragment.

Cloning vector A DNA molecule that is capable of replication in a suitable host cell, that has suitable site(s) for the insertion of DNA fragments by recombinant DNA techniques, and that has genetic markers that allow selection for the vector in a host cell.

Cloning The production of multiple, genetically identical molecules of DNA, cells, or organisms.

Cloverleaf The formalized pattern assumed by a tRNA molecule viewed in two dimensions that shows the regions of internal complementarity that allow the polynucleotide to fold back upon itself into base-paired double helices.

Coding region A sequence of DNA that encodes a polypeptide.

Codon The three consecutive nucleotides (triplets) in DNA or RNA that encode a particular amino acid or signal the termination of polypeptide synthesis.

Colony hybridization A technique for screening bacterial colonies for those that contain a desired polynucleotide sequence.

Colony A visible group of cells arising from a single cell plated on solid medium.

Compatibility The ability of two different types of plasmids to coexist in the same cell.

Competence The transient physiological state required for a bacterial cell to take up transforming DNA.

Complementary Two polynucleotide chains that can base-pair to form a double-stranded molecule.

Complementation The ability of a gene to produce a functional gene product which compensates for the mutant phenotype caused by a mutation in another gene.

Complex carbohydrate An oligomer or higher polymer of more than one kind of sugar moiety, or a glycoside formed with a non-sugar compound.

Composite transposon A transposable element flanked by two copies of an IS element.

Conformation For a compound or macromolecule that has at least limited freedom of rotation about its chemical bonds, one alternative arrangement in space of its constituent atoms and groups.

Conjugative plasmid A bacterial plasmid that encodes functions required for conjugation.

Conjugative transposon A transposon that encodes functions required for conjugation.

Consensus sequence An idealized nucleotide sequence that represents a sequence that serves some particular function (e.g. a promoter) at multiple places in a genome. Each position of the consensus sequence represents the nucleotide most often found at that position in the real sequences. The precise sequence will vary from site to site, but they all are similar to the consensus sequence.

Conservative transposition A transposition event where the transposable element is lost from its original location and inserted at a new location.

Conserved sequence A nucleotide or protein sequence which is shared between different regions or organisms, typically because the sequence fulfills an important function.

Constitutive gene expression A gene or operon which is expressed at all times independent of the environmental conditions.

Contig A set of DNA fragments that overlap to yield a continuous sequence without gaps.

Coordinate gene expression Transcription of a group of genes at the same time due to a common regulatory mechanism.

Copy number The number of molecules of a particular plasmid present in a bacterium.

Co-repressor A small molecule that binds to an aporepressor protein, producing a conformational change that causes it to function as a transcriptional repressor.

Cos site The sequence that is cut to produce the cohesive, single-stranded extensions located at the ends of the linear DNA molecules of certain phages (e.g. lambda).

Cosmid A cloning vector consisting of the phage lambda cos site inserted into a plasmid. Such vectors can be packaged into lambda phage or maintained as plasmids. Cosmids are often used to clone large DNA fragments.

Covalent modification As applied to enzymes, the regulation of activity by modifications that may be reversible (e.g. phosphorylation or adenylation) or irreversible (e.g. limited proteolysis).

Covalently closed-circular (CCC) A completely double-stranded circular DNA molecule, with no nicks or discontinuities, usually with a supercoiled conformation.

CRP Catabolite receptor protein, also called CAP or catabolite activator protein. The interaction of CRP with cAMP modulates many aspects of catabolite repression in enteric bacteria.

Cryptic A function that is silent. For example, a cryptic gene may have an intact coding sequence but lack an active promoter. Cryptic functions can be turned on by an appropriate mutation.

C-terminus The end of a polypeptide chain that has a free carboxylic acid (-COOH) group.

Cyclic AMP (cAMP) An adenosine monophosphate molecule with the phosphate covalently attached to both the 3' and 5' carbons of the ribose.

Defective phage A phage that lacks one or more functions that are required to become an infectious virus but can replicate indefinitely within the host as a prophage.

Degeneracy   Redundancy of the genetic code, in that each amino acid is specified by more than one codon.

Deletion mapping   The use of deletion mutations to determine the position of overlapping point mutations for fine structure mapping of a genetic locus.

Deletion   The loss of one or more bases or base pairs from a molecule of DNA.

Denaturation of DNA or RNA   Separation of the two strands of a double-stranded DNA or RNA molecule by disrupting the hydrogen bonds that join the complementary bases from the two opposite strands, usually by heat or alkali.

Denaturation of proteins   The destruction of the ordered folding of a protein molecule, usually by high temperature or ionic detergents such as sodium dodecyl sulfate (SDS).

Denaturing gel electrophoresis   It is a type of electrophoresis in which the native structure of macromolecules that are run within the gel is not maintained. For example, gels used in SDS-PAGE will unfold and denature the native structure of a protein.

Density-gradient centrifugation   Separation of molecules and particles on the basis of buoyant density, by high-speed centrifugation in a density gradient formed by the concentration gradient of a solute. It is also known as isopycnic centrifugation or zonal centrifugation.

Differential centrifugation   The fractionation of subcellular components according to their sedimentation behaviour; separation into nuclei, mitochondria, lysosomes, microsomes (endoplasmic reticulum), ribosomes, cytosol, etc. by removal of sedimenting material after cycles of processing at progressively increasing centrifugal force.

Diffraction pattern   The array of reflections obtained by crystallography. Each reflection indicates an intensity and, by its location, the angle with respect to the incident beam.

Diploid   An organism which contains a pair of each chromosome within its somatic cells.

Disulphide bridge   An inter- or intra-polypeptide cross-link formed by oxidation of the thiol groups of two cysteine residues to a single cystine residue.

DNA (deoxyribonucleic acid)   A macromolecule usually made up two chains of repeating deoxyribose sugar linked by phosphodiester bonds between the 3-hydroxyl group of one and the 5-phosphate group of the next. The two chains are held antiparalell to each other by weak hydrogen bonds.

DNA chip   An arrayed series of thousands of microscopic spots of DNA oligonucleotides, each containing a specific DNA sequence.

DNA fingerprinting (DNA profiling)   A method to generate a pattern of DNA restriction fragments that is unique to an individual, especially for forensic purposes.

DNA gyrase   DNA gyrase is a type II topoisomerase that introduces negative supercoils into DNA by looping the template so as to form a crossing, then cutting one of the double helices and passing the other through it before releasing the break, changing the linking number by two in each enzymatic step.

DNA ligase   The enzyme that joins the 5′ and 3′ ends of polynucleotide chains by the formation of a phosphodiester bond between them.

DNA methylation   A phenomenon in which a methyl group is added at the 5′ position of cytosine in eukaryotes and 6′ postion of adenosine in prokaryotes, thus repressing expression of regions of the genome. Transcription is prevented when the DNA is methylated and folded into nucleosomes.

DNA polymerases   Enzymes that catalyse the polymerization of deoxy ribonucleotides onto an existing polynucleotide chain using the complementary strand of DNA as a template.

DNA primase   The enzyme that synthesizes short RNA primers complementary to DNA sequences required for initiating synthesis of DNA.

DNA repair   A variety of different mechanisms that remove or modify damaged DNA segments, e.g. pyrimidine dimers, from one strand of double-stranded DNA and its correct resynthesis.

DNase footprinting   It is a technique that detects DNA-protein interaction using the fact that a protein bound to DNA will often protect that DNA from enzymatic cleavage. This makes it possible to locate a protein binding site on a particular DNA molecule.

DNase An enzyme that degrades DNA.

Domain A discrete, independently folded region of a protein. Separate domains indicate different functions in a multifunctional protein.

Donor The source of DNA in a genetic cross. For example, the transducing phage brought into a bacterial cell.

Dot blotting A method to estimate the concentration of a polynucleotide or protein solution by spotting it on a sheet of nitrocellulose or nylon, hybridizing it with a radiolabelled complementary polynucleotide or antibody, visualizing it by autoradiography, then comparing its intensity with that of a similarly treated graded concentration series of the standard polynucleotide or protein.

Double helix (Watson-Crick model) The arrangement in space of two polynucleotide chains in which each chain is wrapped around the other to form two antiparallel spirals. Each strand presents to the other the bases, purine to pyrimidine, with which it can form inter-strand hydrogen bonds.

Double-strand break (DSB) A break in both strands of the DNA backbone, i.e. a break in the 3′-5′ phosphodiester bond, resulting in two exposed double-stranded ends.

Double-strand-break repair model A mechanism proposed for the repair of DNA that has been fragmented at a non-homologous region.

Down-regulation The decrease in the number of hormone receptors on a target cell that occurs after exposure to the hormone. Also, in bacterial nutrition, equivalent to down-shift.

Downstream A sequence located after a particular site relative to the direction of transcription and translation (i.e. located to the 3′ side of a particular site). For example, the lacZ structural gene is located downstream of the lac promoter.

Drosophila melanogaster The fruit fly; the subject of classical genetic studies, because of its rapid generation time and easily observed characteristics.

dsRNA RNA with two complementary strands, similar to the DNA found in all cells. dsRNA forms the genetic material of some viruses.

Duplication A region of DNA that is present in two copies. The DNA is present as an adjacent, direct repeat in tandem duplications. It is also possible for the duplicated DNA to be present at distant sites on a chromosome.

Edman degradation A chemical technique to degrade and cleave amino acid residues sequentially from a protein beginning at the N-terminus, and to identify the residues as they are removed.

Electroblotting An electrophoretic technique for transfer of a protein or nucleic acid from a slab gel, or similar unstable two-dimensional matrix, to a sheet of stable material such as nitrocellulose, for further operations or analysis.

Electron density map In X-ray crystallography, a representation that resembles a geological survey map of planes drawn through a crystal. It shows contours of equal electron density surrounding the individual atoms: steeper around heavier atoms; more shallow around lighter atoms.

Electropherogram A graphical representation of an electrophoretic separation; e.g. for a slab gel, absorbance (ordinate) plotted as a function of distance from the origin (abscissa).

Electrophoresis A technique that separates charged compounds according to their mobilities in an electric field.

Electroporation A method for transferring DNA into cells by exposure to a rapid pulse of high voltage, which causes the transient formation of small pores in the cell membrane.

Elongation factor An accessory protein that is required for transfer of amino acid residues to the polypeptide chain during translation.

Emission spectrum In fluorescence spectroscopy, emission as a function of wavelength, upon excitation at a fixed, shorter wavelength.

Endonuclease An enzyme that makes or breaks a molecule of DNA by hydrolyzing internal phosphodiester bonds.

Enhancer A cis-acting regulatory sequence (generally 50–1500bps in length) that can increase transcription from an adjacent promoter.

Epimer A compound that differs from another by its configuration at only one asymmetrical centre.

Episome   A genetic element that can exist either as an autonomous replicating plasmid or can insert into the bacterial chromosome.

Epistasis   A condition where one gene masks the expression of another gene.

Equilibrium sedimentation ultracentrifugation   A technique for evaluation of the molecular mass of a polymer by determination of the extent of sedimentation at which the force exerted upon a macromolecule in a centrifuge is balanced by its diffusion due to Brownian motion.

Error prone repair   A mechanism for repair of DNA damage that often results in mutations (e.g. SOS repair).

***Escherichia coli* (*E. coli*)**   An enteric bacterium that inhabits the human intestine; much used for experimentation.

Essential amino acid   An amino acid that is not synthesized by an organism at an adequate rate (or at all) from other amino acids or metabolites; therefore one that is a dietary requirement.

Euchromatin   In cytology, the lightly staining regions of chromosomes that contain less-condensed chromatin and are the regions that are transcribed as RNA.

Eukaryote   An organism with a nuclear membrane and membrane bound organelles (e.g. mitochondria), and a mitotic spindle. This group of organisms is more correctly called Eucarya.

Excision repair   A DNA repair system involving removal of a segment of a DNA duplex that contains a thymine or other dimer, formed by ultraviolet-light-induced damage, and resynthesis and ligation of the excised sequence.

Excision   The removal of a section of double-stranded DNA that is faulty due to mutation or incorrect replication.

Exon   A region of a eukaryotic gene that encodes a sequence of amino acids; as opposed to an *intron*, i.e. an intervening region or sequence. It is the sequence in a spliced gene that is retained after removal of the introns to provide the mature mRNA that is translated by the ribosome.

Exonuclease footprinting (DNase footprinting)   A method for identification of a protein-binding region in a double-stranded DNA fragment. This technique is suitable for localization of a binding site to a specific region of a large DNA fragment, whereas *footprinting* gives the exact sequence of the binding site of a smaller fragment.

Exonuclease   An enzyme that digests a molecule of nucleic acid by removing successive nucleotides from the 5' or 3' end.

Exopeptidase   A peptidase that cleaves a protein sequentially, starting at the N-terminal peptide bond (an *aminopeptidase*) or at the C-terminal peptide bond.

Expression cloning   As part of a repetitive process for screening a library, the sampling of cells according to the presence of the product of the gene of interest.

Expression vector   A cloning vector designed so that a foreign gene inserted into the vector will be expressed in the host organism.

Expression   The production of a gene product, i.e. protein or RNA, from a gene; the manifestation of a genotype as a phenotype.

Extragenic   An effect due to a second gene. For example, an extragenic suppressor is a mutation in a second gene that repairs the mutant phenotype.

F⁻ cell   A cell which does not contain the F-factor, and hence able to act as a recipient (female) in a conjugative DNA transfer in matings with F⁺ or Hfr strains.

F⁺ cell   A cell that carries a F-factor as an autonomous plasmid, which enables the cell to act as a donor (male) and transfer the F-factor to a recipient (female) cell.

Flanking sequence   The DNA sequence located on either side of a specific genetic locus.

Fluorescence   The property of a compound or moiety of absorbing ultraviolet or visible light and re-emitting it nearly instantaneously at a longer wavelength.

Fluorophore   A chemical group responsible for the fluorescence of a compound or macromolecule.

Footprinting   A method for identification of a protein-binding site on a DNA molecule.

Freeze-fracture   A technique to prepare cells for scanning electron microscopy; cleavage of a solidly frozen cell, followed by *freeze-etching*, i.e. sublimation of some of the frozen water of the cell and coating the exposed subcellular structures with a film of platinum or carbon.

Functional domain A region of a polypeptide chain or RNA that performs a particular function. Functional domains are often folded into independant structures, separated from the rest of the polymer by a flexible hinge region.

Gel electrophoresis Electrophoresis is the process of separating charged species by subjecting them to a voltage gradient.

Gel filtration chromatography A technique for separation of soluble compounds according to size, based on their relative ability to penetrate the carefully sized pores of the stationary gel phase.

Gel retardation A technique that identifies a DNA fragment that has a bound protein by virtue of its decreased mobility during gel electrophoresis.

Gel shift assay A method to identify by function a band on a polyacrylamide gel. Electrophoresis is performed on a mixture of proteins before and after treatment with a strong ligand that is specific to one component, and note is made of the absence or shift in position of a protein due to prior binding to the ligand.

Gene amplification Multiple copies of a single gene within a cell which remain as tandem repeats or are segregated as satellite DNA. Amplification is generally not inherited.

Gene conversion A genetic event that produces abnormal segregations by non-reciprocal recombination.

Gene duplication The inherited appearance of multiple copies of a gene.

Gene expression The process by which a gene product is made. For genes that encode proteins, the gene must be transcribed into mRNA and then translated into protein. For genes that encode structural RNAs (rRNA, tRNA, etc), the gene must be transcribed into RNA.

Gene fusion A construct that joins the coding region of two open reading frames such that expression of the product results in a chimeric protein or a translational fusion or protein fusion.

Gene library A random collection of DNA fragments that have been inserted into a cloning vector.

Gene mapping Determination of the relative positions of different genes on a DNA molecule.

Gene probe A labelled sequence of single-stranded nucleic acid (DNA or RNA) which can be used to detect the complementary nucleic acid sequence by hybridization techniques.

Gene product The result of gene expression, i.e. an RNA or protein molecule formed by transcription and translation of a gene.

Gene targeting The introduction of a homologous DNA sequence into a specific site in the genome of a cell, either by replacement of the former sequence, i.e. *sequence replacement*, or by insertion into the former sequence, i.e. *sequence insertion*.

Gene therapy The *in vivo* introduction of a functional gene, and its expression, in the germ line cells (*germ line therapy*) or somatic cells (*somatic gene therapy*) of an individual who does not possess the normal gene.

Gene The genetic unit of function. A gene may encode a polypeptide or a molecule of non-translated RNA. In genetics, a unit inferred from the pattern of inheritance.

Genetic code The series of codons of mRNA, each of which specifies a single specific amino acid or a translation stop signal.

Genetic engineering The use of molecular techniques to produce DNA molecules containing new genes or new combinations of genes.

Genetic map The relative position and distance between genes determined from recombination frequencies.

Genetic marker A locus or a DNA sequence that can be recognized and thus used to characterize the larger DNA or chromosome that it is associated with and can be used for genetic selections or screening.

Genetic polymorphism The existence of two or more genetically different classes within a population.

Genetic recombination The process by which a fragment of DNA from one molecule is exchanged with or integrated into another molecule to produce a recombinant molecule(s).

Genome The complete genetic content of a cell or organism, including chromosomes, plasmids, and prophages. The total genetic information of an organism.

Genomic DNA   DNA that has been isolated from a cell and therefore contains introns, as opposed to cDNA.

Genomic library   A collection of clones that represent the total genome of an organism.

Genomics   The comparative analysis of genomic DNA sequences from different organisms. Genomic analysis can provide information about the evolution of genes and can make predictions about the metabolism of an organism.

Genotype   The genetic composition of a cell or organism, usually with reference to a particular set of alleles that may be homozygous or heterozygous for a trait.

Glycoprotein   A protein with one or more carbohydrate moieties attached to it.

Glycosidic bond   The linkage of a sugar hemiacetal or hemiketal through its anomeric carbon with another moiety.

Glycosylation   Post-transcriptional modification of a protein by the addition of a carbohydrate moiety.

GMO   An abbreviation for "genetically modified organism". Commonly used to refer to organisms that have been developed using recombinant DNA approaches, particularly if some foreign DNA remains in the strain.

Green Flourescent Protein (GFP)   An intrinsically fluorescent protein from the jellyfish *Aequorea victoria*. Gene fusions with GFP are commonly used for determining protein localization by fluorescence microscopy.

Growth rate   The rate of increase of cellular density per unit time.

Hairpin loop   A region in a single polynucleotide strand containing adjacent complementary sequences that allows the polynucleotide to fold back upon itself and form a double-stranded section, as in the cloverleaf structure of a tRNA molecule.

Hairpin   A region of single-stranded DNA or RNA that can form base pairs between complementary sequences.

Handshake   An interaction of two identical or nearly identical structures, proteins or nucleic acids, which fit each other in a symmetrical fashion.

Haploid   Only one copy of each chromosome per cell. Prokaryotes are haploid, although more than one copy of a chromosome may be transiently present in the cell, depending on the rate of DNA replication and the growth rate.

Helicases   Proteins that unwind the DNA double helix.

Helix-loop-helix (HLH)   A motif found in some DNA-binding proteins.

Heterochromatin   In cytology, the heavily staining regions of chromosomes that contain condensed chromatin.

Heteroduplex analysis (mismatch analysis)   A method for detection of single base substitutions in DNA fragments.

Heteroduplex DNA   A double-stranded DNA molecule formed by annealing complementary (or partially complementary) single-stranded DNA from two different sources.

Heterogeneous nuclear RNA (hnRNA)   The unprocessed DNA transcripts, which includes pre-mRNAs.

Histone-like proteins   Proteins from bacteria which bind to DNA, and compact the DNA.

Histones   A group of basic proteins from eukaryotes (or Archae) which bind to DNA, forming nucleosomes, and packaging the DNA into chromatin.

HLH protein   One of a group of proteins defined by their characteristic topography. These proteins form DNA-binding heterodimers composed of two proteins, one from a ubiquitous and one from a tissue-specific class of HLH proteins. The heterodimers are involved in regulation of gene expression.

Holliday junction   The cruciform structure formed as an intermediate in homologous genetic recombination.

Holoenzyme   A complex containing all of the subunits required for a functional enzyme. Used to describe enzymes composed of many different protein subunits.

Homeobox   A highly conserved 180-base polynucleotide sequence that controls body part-, organ- or tissue-specific gene expression.

Homologous   (i) Nucleic acid molecules with the same base sequence except for minor differences in alleles; (ii) nucleic acid molecules originating from strains of the same species, thus having at least long stretches of identical DNA base sequences; (iii) gene or protein families having a recognizable common evolutionary origin.

Homology Sequence identity between two nucleotide sequences of protein sequences that denote a common evolutionary origin. Homology is also the similarity of structure or function of proteins that is due to a common evolutionary origin.

Homologous recombination The common cellular mechanism for genetic recombination between DNA molecules with identical or near-identical (homologous) nucleotide sequences.

Homozygous A cell or organism where the same allele is carried by each member of a pair of homologous chromosomes.

Horseradish peroxidase An enzyme that can be complexed to DNA or antibodies for use in some non-radioactive labeling procedures.

Hot-spot Sites at which mutations occur at a much greater frequency than other parts of the genome.

Housekeeping Genes involved in basic functions needed for the sustenance of the cell. Housekeeping genes are constitutively expressed.

Hsp A prefix which denotes a heat-shock protein; e.g. Hsp60 is a 60 kDa heat-shock protein.

Human artificial chromosome (HAC) An artificial assembly of chromosome features, telomere, centromere, etc., that is used to test the minimum features required for replication in transfected cells.

Human leucocyte antigen (HLA) An antigen on the surface of human cells that consists of a generally invariant β_2-microglobulin and a larger polypeptide chain that is characteristic of the individual; responsible for most cases of organ transplant rejection.

Hybridization probe A labelled nucleic acid molecule that can be used to identify complementary or homologous molecules through the formation of stable base-paired hybrids.

Hybridization A technique where a denatured (single-stranded) nucleotide chain (DNA or RNA) is allowed to pair with another single-stranded nucleotide chain.

Hydrogen bond An important cohesive force of biological macromolecules, notably interstrand interactions of bases in the Watson-Crick model of DNA and of amide groups in the α-helix of proteins.

Hydrophilic residue An amino acid residue that has a charged or polar side chain.

Isoelectric Focusing (IEF) A technique to separate proteins based on their isoelectric points.

Immunoaffinity chromatography A variant of affinity chromatography in which an antibody coupled to the stationary phase adsorbs an antigen which is subsequently eluted at a different pH or a higher salt concentration.

Immunofluorescence The visualization of an antigen in a histological section or on a polyacrylamide slab by allowing it to combine with a fluorophore-tagged antibody.

Immunoprecipitation A purification technique that separates antigenic material from a soluble mixture by precipitation with an appropriate antibody; an essential step in radioimmunoassays.

In silico Descriptive of research into bioinformatics databases.

In situ hybridization A technique for gene mapping involving hybridization of a labelled sample of a cloned gene to a large DNA molecule (usually a chromosome), often within a cell.

In situ PCR The performance of PCR on fixed preparations on a microscope slide.

In situ A Latin phrase meaning "in the original place". Commonly used to describe a process that visualizes the position of a biological molecule in a cell.

In vitro mutagenesis A method for mutating DNA outside of a host cell.

In vitro packaging Synthesis of infective phage particles from a preparation of phage capsid proteins and a concatamer of phage DNA molecules.

In vitro Reactions that take place outside of the cell; in a test tube.

In vivo Reactions that take place inside the cell.

Incompatibility The inability of two plasmids to stably coexist in the same cell.

Inducer A chemical or physical agent that turns on gene expression. Usually refers to an agent that alters repressor-operator interactions, often by decreasing DNA-binding.

Induction The switching on of transcription in a repressed system due to the interaction between the inducer and a regulatory protein.

Initiation codon A trinucleotide sequence, AUG, that signals the start of translation of a protein; the codon for methionine in eukaryotes and for *N*-formylmethionine in prokaryotes. It is also called as start codon.

Initiation complex The ribosome charged with mRNA, initiation factors and methionyl-tRNA (in eukaryotes) or *N*-formylmethionyl-tRNA (in prokaryotes, mitochondria and chloroplasts) that is bound to the initiation codon.

Initiation factors Proteins that promote the binding of ribosomes and the initiator tRNA to mRNA to begin the process of translation.

Initiator tRNA The tRNA that recognizes the AUG start codon.

Initiator A site, upstream from a structural gene, for attachment of a protein that stimulates initiation of transcription.

Insertional inactivation A cloning strategy where insertion of a piece of DNA into a vector inactivates a gene carried by the vector.

Insulator A DNA sequence that contributes to limitation of chromatin activation to distinct segments.

Intercalating agent A planer molecule that can insert between two adjacent base pairs in a molecule of double-stranded DNA, distorting the architecture of the double helix.

Intercalating dye A dye which can insert between the bases of nucleic acids. Intercalating dyes may be used to stain DNA or to induce frameshift mutation.

Intercalation The insertion of flat polycyclic molecules between nucleotides in a DNA duplex.

Internal ribosome entry site (IRES) In the 5′-cap-independent initiation of translation, the place on the mRNA that first attaches a ribosome subunit.

Internal signal sequence An internal polypeptide sequence of a protein that is responsible for its targeting to the appropriate locus within a cell after synthesis.

Intron A non-coding polynucleotide sequence that interrupts the coding sequences, the *exons*, of a gene. Introns are common in eukaryotes, but rarer in prokaryotes.

Inversion A DNA rearrangement where a sequence of nucleotides is in the reverse orientation relative to the rest of the molecule.

Inverted repeats A DNA or RNA sequence where the sequence of nucleotides along one strand of DNA is repeated in the opposite physical direction along the other strand.

Ion-exchange chromatography A technique for separation of charged compounds or macromolecules in solution according to their affinities for a positively or negatively charged stationary phase.

Island A large region of DNA that is present on the chromosome of an organism but absent from closely related organisms.

Islet A smaller genetic island that only encodes one or a few gene products.

Isoacceptor One of two or more tRNAs that accept the same amino acid.

Isoelectric point The pH value at which an amphoteric compound is electrically neutral.

Isomerase One of a class of enzymes that rearrange the bonds of their substrates, e.g. an epimerase.

Isoschizomers A pair of restriction endonucleases that recognize the same palindromic sequence, although in some cases, one member of a pair may differ from the other in its response to methylation of the recognition sequence, or may cleave a different phosphodiester bond within the target sequences.

Isotope A variant of an atom, chemically identical but with a different atomic mass; often radioactive.

Junk DNA Regions of DNA that do not encode functional proteins or RNAs and do not function as regulatory sites.

Kinase An enzyme that transfers phosphate from ATP to another molecule.

Kinetics The study of rates of reactions.

Klenow fragment The product of limited proteolysis of *E. coli* DNA polymerase I that retains the polymerase and 3′- to 5′- exonuclease (proof reading) activity but is missing the 5′- to 3′ exonuclease activity.

Labeling The incorporation of a radioactive nucleotide into a nucleic acid molecule.

***Lac* operon** The structural and associated regulatory genes that control the ability of *E. coli* to live on lactose; a subject of classical experimentation.

Ladder The pattern of bands seen on a polynucleotide sequencing gel that differ by integral numbers of nucleotides.

Lagging strand The discontinuous DNA strand that is synthesized at a fork during replication, the direction of synthesis of which is opposite to that of the movement of the fork.

Lambda A temperate phage that infects *E. coli*. Derivatives of phage lambda are widely used as cloning vectors.

Leading strand The strand of newly replicated DNA that is synthesised continuously in the same direction as the replication fork. DNA synthesis proceeds in the 5′ to 3′ direction.

Leaky mutation A nucleotide substitution that changes the amino acid sequence of a protein that results in partial loss of its activity.

Ligand A small molecule or ion that binds to a protein or other structure.

Ligase (DNA ligase) An enzyme that repairs single-stranded discontinuities in double-stranded DNA molecules in the cell. Purified DNA ligase is used in gene cloning to join DNA molecules together.

Ligation The formation of a phosphodiester bond between two adjacent bases separated by a single-strand break. It is catalyzed by DNA ligases.

Linkage group A collection of genes that are inferred to be located together on a single chromosome because of the pattern of their inheritance.

Linkage map A genetic map assembled from recombination data that shows the order of mutant sites and genes along a nucleic acid molecule.

Linkage The tendency of genes located close together on the same DNA molecule to be coinherited.

Linker A synthetic, double-stranded oligonucleotide used to attach sticky ends to a blunt-ended molecule.

Lipid A natural substance that is poorly soluble in water but is soluble in organic solvents; lipids include fatty acids, triacylglycerols, phospholipids, waxes and some hormones and vitamins.

Lipoprotein A complex of lipids and apolipoproteins that is a transport form of lipids in blood. Lipoproteins are characterized by their density, which is determined by the lipid portion, and include high-, low-, and very-low-density lipoproteins.

Locus The position on a chromosome where a particular genetic trait resides. Sometimes used to describe multiple genes that affect the same function.

Long interspersed elements (LINES) Retrotransposons, double stranded DNA sequences, 6-7 kb long, that occur in families of thousands of nearly identical copies which have apparently been randomly inserted into the mammalian genome. Short interspersed elements (SINES) are about 300 bp long.

Long terminal repeat (LTR) A polynucleotide sequence found at each end of an integrated retrovirus genome that contains the signals for expression of the viral genome.

Luminescence Emission of a photon, of the same or lower energy than the energy that excited it, by an excited state of a chemical compound.

Lysate A solution containing progeny phage resulting from the lysis of a population of bacterial cells by infecting phage.

Lysis Disruption of cells with release of the contents.

Lysogen A bacterial cell carrying a phage genome as a repressed prophage.

Lysogenic bacterium A bacterium that hosts a prophage, which may lie dormant even for many generations but eventually may cause lysis.

Lysogenic conversion Expression of particular genes by a prophage that confer a novel phenotype on the host.

Lysogenic cycle The pattern of phage infection that involves integration of the phage DNA into the host chromosome.

Lysogeny The ability of a temperate bacteriophage to maintain itself as a quiescent prophage until induced into the lytic cycle.

Lysozyme An enzyme that hydrolyzes the peptidoglycan within the cell walls of bacteria.

Lytic cycle The development of a bacteriophage, either after infection of a host bacterium or after induction of a prophage, resulting in production and release of free progeny phage particles, and lysis of the host cell.

Lytic phage A phage that can only enter the lytic cycle when it infects a sensitive bacterial cell.

M13 A filamentous, single-stranded DNA phage that infects *E. coli*. The M13 DNA replicates as a circular double-stranded intermediate (called a replicative form) then single-stranded DNA produced by rolling circle replication is packaged into phage particles. Derivatives of M13 are used as cloning vectors.

Macromolecule A compound or complex, usually a polymer such as a protein, nucleic acid or polysaccharide, or a covalent or non-covalent complex of any of these.

Major groove The wider of the two helical spaces on the surface of an A- or B-DNA double helix. The other helical space is the *minor groove*.

Major histocompatibility complex (MHC) The products of genes grouped together on a chromosome that determine whether transplanted tissues will be accepted or rejected. In the human, they are known as *human leucocyte antigen molecules (HLAs)*.

Mapping The creation of an outline of locations of genetic markers within the structures of the chromosomes.

Marker A genetic trait of which one allelic form is selected or screened for following recombination.

Mass spectrometry (MS) It is a powerful technique for characterization of proteins, in which mass of a protein or its fragment is determined. It is also possible to determine the sequence of a part or all of a protein.

Maxam-Gilbert method A technique for sequence analysis of DNA in which four chemical reactions are applied separately, each to cleave the polynucleotide randomly at one of the four bases.

Messenger RNA (mRNA) The transcript of a segment of chromosomal DNA which is a template for protein synthesis.

Methyl transferase Enzymes that catalyze the transfer of methyl groups from one molecule to another.

Methylation analysis Any technique to detect sites at which bases of DNA are methylated, e.g. by analysis of DNA digested by isoschizomeric pairs of restriction endonucleases.

Microarrays Ordered sets of DNA fragments fixed to solid surfaces. The DNA fragments may represent all the open reading frames in a genome, a particular gene family, or any other subset of genes. Microarrays are sometimes called gene chips.

Microinjection A method of introducing new DNA into a cell by injecting it directly into the nucleus.

Microsatellite One of many short, highly polymorphic, non-coding sequences, e.g. poly(TG), found well spaced throughout the genome that can serve as landmarks during physical mapping.

Mismatch repair A mechanism that corrects mismatched base pairs that have escaped correction by the proofreading activities of the DNA polymerases.

Mismatch A defect in the pairing of two complementary DNA sequences where a base in one strand is different from that expected according to complementarity with the other.

Mispairing Improper alignment of two nucleic acid strands.

Missense mutation A mutation that changes a codon for one amino acid to a codon for a different amino acid, resulting in an amino acid substitution in the protein product.

Mitosis The normal process of nuclear division in a eukaryote, whereby nuclear division occurs on a spindle structure without reduction in the chromosome number in the daughter nuclei.

Mobile genetic element A transposon or an insertion sequence; a polynucleotide sequence that can move from a chromosome or plasmid to another chromosome or plasmid.

Module In protein chemistry, a building block from which larger proteins are constructed; a recurring folding pattern in proteins, examples of which may have amino acid sequences that are homologous or non-homologous.

Monocistronic An mRNA that only encodes a single gene product.

Mosaic The substitution of portions of a gene with sequences acquired from another site or another organism via genetic recombination.

Motif   A small portion of a protein (typically less than 20 amino acids) that is homologous to regions in other proteins that perform a similar function.

mRNA   An RNA molecule that includes the coding region(s) and the translation signals for a gene or operon. An intermediate that specifies the amino acid sequence of the encoded polypeptide(s) during translation.

mtDNA   Mitochondrial DNA; a double-stranded polynucleotide composed of a heavy and a light chain, distinguished by the buoyant densities of the separated strands, which are determined by their G+T content.

Mu (μ)   A bacteriophage that reproduces by transposition. Mu infects a variety of enteric bacteria.

Mud   A defective derivative of phage Mu. Mud insertions are commonly used to construct operon and gene fusions in enteric bacteria.

Multifactorial   A phenotype caused by two or more genes.

Multigene family   A number of identical or related genes present in the same organism, usually coding for a family of related polypeptides, e.g. the genes for globins.

Multiplicity of infection (MOI)   The ratio of pathogen to host cells during an infection.

Mutagen   A chemical or physical agent that increases the frequency of mutation, usually by directly damaging the DNA.

Mutagenesis   The formation of mutations.

Mutant   An organism with an altered base sequence in one or several genes. Usually refers to an organism with a mutation that causes a phenotypic difference from the wild-type.

Mutation rate   The number of mutations per cell division.

Mutation   Any heritable alteration in the base sequence of the genetic material.

Mycoplasma   A type of small bacteria that has no cell wall.

Negative regulation   A mechanism of genetic regulation in which a protein or RNA molecule inhibits gene expression.

Nested primers   A second set of PCR primers whose use follows an initial copying of a nucleic acid sequence; the nested primers are specific for sequences internal to the 3'- and 5'-ends of the originally copied sequence.

Nick translation   A method which uses DNA polymerase I to first produce a nick in a DNA duplex, then degrade a stretch of single-stranded DNA using its 5'-exonuclease while synthesizing a new strand in its place.

Nomarski interference optics   A microscope optical system for visualization of differences in refractive indices so as to observe unstained specimens.

Nonsense codon   A codon which does not code for any amino acid, but signals a termination of translation, or punctuation. The three nonsense codons are UAG (amber), UAA (ochre), and UAG (opal). It is also called as *stop codon*.

Nonsense mutation   A mutation which replaces a codon for an amino acid with a codon for chain termination.

Northern blotting   A technique for detection of specific RNA molecules. RNA from a cell is denatured and separated by slab gel electrophoresis, then blotted on to a sheet of nitrocellulose or nylon and hybridized with radiolabelled DNA that is complementary to the desired RNA, whose presence is subsequently indicated by autoradiography.

N-terminal   In a polypeptide sequence, that unique residue which is connected to the linear sequence by its carboxy group, leaving it with a free amino group.

N-terminus   The end of a polypeptide chain that has a free amino acid ($-NH_2$) group.

NTP   Nucleotide triphosphate. Used for the synthesis of RNA or as an energy source.

Nuclear magnetic resonance (NMR)   A technique for studying the microenvironment of certain atoms that can absorb energy in a magnetic field.

Nuclear receptor   A nuclear protein that allows regulation of gene expression by a lipophilic effector molecule. When the effector molecule is bound to the receptor protein, the latter can also bind to a specific DNA sequence.

Nuclear scaffold   The network of protein fibres that underlies the nuclear inner membrane and is exposed upon extraction of soluble proteins from the nucleus.

Nuclease An enzyme which cleaves phosphate-deoxyribose bonds within (endonuclease) or at the end (exonuclease) of a nucleotide sequence.

Nucleic acid hybridization Formation of a double-stranded molecule by base pairing between complementary or homologous polynucleotides.

Nucleic acid DNA or RNA; a macromolecule.

Nucleoid The condensed organization of a prokaryote chromosome inside the cell.

Nucleolus A structure in the nucleus of a cell that is the site of synthesis of rRNA.

Nucleoplasm Between mitoses (interphase) the visibly unstructured matrix of the nucleus, which consists of all but the nucleolus.

Nucleoprotein A protein-DNA or protein-RNA complex.

Nucleoside A purine or pyrimidine base with a sugar attached in a glycosidic linkage; *ribonucleoside* if the sugar is D-ribose and *deoxyribonucleoside* if the sugar is D-deoxyribose.

Nucleosome The repeating structural unit of chromatin that consists of a complex of eight molecules of histones and a DNA double helix wrapped twice around them.

Nucleotide A nucleoside with a phosphate group attached.

Nucleus A membrane-limited structure of eukaryotic cells that contains chromatin and is the site of DNA and RNA synthesis.

Null Completely absent. For example, a null mutation completely disrupts a gene.

Okazaki fragment Fragments of a single-stranded DNA synthesized on the discontinuous site of a DNA replication fork.

Oligomer A small polymer of complexity greater than that of a monomer but less, in common usage, than that of a dodecamer.

Open reading frame (ORF) A stretch of DNA which potentially codes for protein. A length of DNA not interrupted by stop codons. A sequence of in frame codons preceded by a translational initiation codon and terminated by a chain termination triplet.

Operator The DNA sequence where a repressor protein reversibly binds to regulate the activity of one or more closely linked structural genes.

Operon A group of genes, usually metabolically related, that can be controlled co-ordinately from a common regulatory sequence.

Origin of replication (ori) The nucleotide sequence where DNA replication is initiated. Determines the specific position on a DNA molecule where DNA replication begins.

Origin recognition complex (ORC) An aggregate of nuclear proteins involved in initiation of replication and/or repression of transcription (silencing).

Overhang The extension of one polynucleotide strand beyond the terminus of its complementary strand.

Overlapping genes Two genes whose nucleotide sequences partially overlap.

Oxidative phosphorylation The reactions that synthesize the phosphoanhydride bond of ATP by the coupling of that energetically unfavourable reaction with the spontaneous oxidation of metabolites.

PAGE Polyacrylamide gel electrophoresis. Separation of molecules through a polyacrylamide gel matrix in an electric field. Separation may depend upon size and charge of the molecules.

PCR Polymerase chain reaction. A method for amplifying a particular region of DNA by a sequence of denaturation, annealing of specific primers, and synthesis.

Pellet The sedimented portion that accumulates during centrifugation.

Peptidase An enzyme that cleaves the peptide bonds of proteins and peptides.

Peptide bond The amide bond formed by condensation of the α-carboxy group of one amino acid with the α-amino group of another; the bond that joins together the amino acid residues that comprise a protein, peptide or polypeptide.

Peptide A compound formed by incomplete hydrolysis of a protein, or by elimination of the elements of water from between the α-carboxyl and the α-amino groups of α-amino acids to form a linear polymer.

Permease An enzyme system concerned with the transport of specific substances, usually nutrients, through the cytoplasm membrane.

pH A measure of acidity of aqueous solutions. Values below 7 are acidic; values above 7 are alkaline.

Phenotype The appearance or other observable characteristics of an organism. The phenotype expressed by an organism depends upon the particular forms of its genes and the environmental conditions.

Phosphodiester bond The covalent bond joining the 3' hydroxyl of the sugar moiety of one (deoxy) ribonucleotide to the 5' hydroxyl of the adjacent sugar.

Phospholipid A covalent association of phosphoric acid and lipid, e.g. a phosphoacylglycerol, sphingomyelin.

Physical map The linear order of genes and distance between them (usually expressed in base pairs or kilobases). Physical maps are constructed from *in vitro* characterization of the DNA (e.g. by restriction mapping or DNA sequencing).

Plaque A clear area in a lawn of bacterial cells caused by the lysis of infected cells by a phage.

Plasmid A self-replicating extra-chromosomal element, usually a small segment of duplex DNA that occurs in some bacteria; used as a vector for the introduction of new genes into bacteria.

Polycistronic mRNA A single mRNA that contains the information necessary for the production of more than one polypeptide; characteristic of some prokaryotic mRNAs.

Polymorphism Appearance in several forms, e.g. of a protein that is subject to tissue-specific post-translational modification, or of mRNA that is processed through alternative splice sites; also the appearance in a population of more than one structural gene at a single genetic locus.

Polysome A structure that is functional in protein synthesis, formed by several ribosomes attached to a single mRNA molecule.

Positive control The activation of transcription in a bacterium in response to its metabolic state, e.g. synthesis of several enzymes in response to glucose deprivation.

Pre-initiation complex In replication, transcription or translation, a complex that is potentially composed of template, enzymes and initiation factors but that is incomplete for optimal function due to the absence of one or more elements.

Pre-mRNA (heterogeneous nuclear mRNA) The RNA that is the direct transcript from DNA and therefore includes introns before it is processed into mature mRNA.

Pre-mRNA splicing The process by which an intron is excised and exons are religated in the post-transcriptional modification of RNA.

Primary structure The amino acid sequence of a protein or polypeptide, or the nucleotide sequence of a polynucleotide.

Primase An enzyme that synthesizes ribonucleotide primers for lagging strand DNA synthesis of Okazaki fragments. Polymerizes ribonucleotide triphosphates in the 5' to 3' direction.

Primer A short oligonucleotide complementary to a strand of DNA or RNA that is used to initiate synthesis of the complementary DNA strand. The primer provides a 3'OH end which is required by DNA polymerases to initiate synthesis of the complementary DNA.

Primosome A complex of primase and helicase that initiates synthesis of RNA primers on the lagging DNA strand during DNA replication.

Probe A fragment of DNA labeled with radioactivity or non-radioactive ligand and used to hybridize to another DNA molecule to identify complementary base sequences.

Processing Post-transcriptional modification of RNA or post-translational modification of a polypeptide or protein.

Processivity The ability of an enzyme to continue to act on a polynucleotide for a long distance without dissociating.

Prokaryote An organism lacking a nuclear membrane and certain organelles such as mitochondria.

Promoter A sequence on DNA that functions as the RNA polymerase binding site, thus defining the transcription start site.

Proofreading Removal of mismatched base pairs during DNA replication by the 3' to 5' exonuclease activity of DNA polymerase, followed by resynthesis.

Prophage A temperate phage genome whose lytic functions are repressed and which replicates in synchrony with the bacterial chromosome.

Protease An enzyme that degrades proteins to peptides or amino acids.

Protein kinase   An enzyme that uses ATP to phosphorylate a group on a protein, e.g. a serine, threonine or tyrosine hydroxy group.

Protein processing   Limited proteolysis that converts a biologically inactive translation product into an active enzyme, hormone, structural protein, etc.

Protein   A linear polymer of amino acids held together by peptide bonds.

Proteolysis   The degradation of proteins to peptides or amino acids catalyzed by proteases.

Proteome   The repertoire of proteins that exist in a cell or tissue under a particular set of conditions.

Pseudogene   An inactive gene derived from an ancestral active gene. Pseudogenes are often recognized by nonsense or frameshift mutations that disrupt an open reading frame that encodes a functional protein in a related genome. It often occurs without introns.

Puckering   The divergence from planarity of a non-aromatic ring compound, e.g. a sugar or steroid, to accommodate the bond angles of the constituent atoms.

Pulsed field gel electrophoresis (PFGE)   A technique for separation of chromosomes and very large DNA molecules, usually for purposes of genetic mapping.

Purine   One of the two classes of heterocyclic organic bases that are found in nucleic acids and in several other kinds of biological compounds, e.g. coenzymes, nucleotides, sugar derivatives.

Pyrimidine dimers   Covalent bonds formed between two adjacent pyrimidines (e.g. thymidine dimer) on the same strand of DNA induced by Ultraviolet irradiation.

Pyrimidine   One of the two classes of heterocyclic organic bases that are found in nucleic acids and in several other kinds of biological compounds, e.g. nucleotides, sugar and lipid derivatives.

Quadruplex DNA   A form of DNA in which four oligo(G) sequences, either of the same or of different strands, line up either in a parallel, an anti-parallel, or in a fold-back (mixed parallel and antiparallel) pattern.

Quantitative PCR   A method for quantification of a polynucleotide by inclusion with it of a known amount of an easily distinguished template as an internal standard to compensate for variation in efficiency of amplification.

Quaternary structure   The folded structure of a protein including all of the polypeptides required for the intact, fuctional protein.

Quenching   In fluorescence spectroscopy, a decrease in emission due to a variety of causes that dissipate energy of the excited state, e.g. transfer of energy to a non-fluorescent molecule such as molecular oxygen.

Radiation-induced hybrid mapping   A method for location of a human genetic marker within a restricted region of a single chromosome.

Random primers   A set of short oligonucleotides with variable sequences. Within a population of random oligonucleotides, some will anneal to complementary sequences in a DNA or RNA template.

Rate-limiting step   The slowest step of a metabolic pathway or enzymic reaction; the one that determines the rate of appearance of the ultimate product.

Reading frame   The sequence of nucleotides which is read as consecutive triplets during translation of mRNA into protein. A sequence of codons that continues without encountering a stop codon is called an open reading frame (ORF).

RecA protein   The protein encoded by the *recA* gene which is essential for homologous recombination.

Receptor   The binding site for a hormone or neurotransmitter that initiates its action at the cellular level.

Recessive   A genetic trait that is not expressed in a heterozygous or partially heterozygous cell.

Recombinant DNA   A molecule of DNA in which a DNA fragment from a different source has been inserted.

Recombinase   An enzyme that catalyzes genetic recombination.

Recombination   Genetic exchange resulting from a cross-over between two different DNA molecules or different regions of a DNA molecule.

Release factors (RF)   Proteins that facilitate the termination of translation. Translation termination occurs when a ribosome encounters one of the three stop codons (UAA, UAG, or UGA).

Renaturation   The return to native structure from a denatured state; in nucleic acid chemistry, identical to annealing.

Repetitive DNA   Sequences of DNA that are found to be repeated, sometimes thousands of times over.

Replication fork   The site where the two polynucleotides of a parent DNA separate during replication. The daughter strands are attached to the two separated polynucleotides that trail away from the fork as it advances into the parent DNA.

Replication   The process of duplicating a DNA molecule.

Replicon   A DNA molecule that is able to initiate its own replication. A replicon must have an origin of replication and usually also has the necessary regulatory information required for the proper initiation of DNA replication.

Reporter gene   A gene which can be placed downstream of a promoter and expression of the gene followed by a relatively easy assay

Repression   Switching off the expression of a gene or a group of genes in response to a chemical or other stimulus.

Repressor   A gene product that negatively regulates gene expression. Usually refers to a DNA-binding protein that inhibits transcription under certain conditions.

Residue   The part of a single sugar that appears in a polysaccharide; of a single amino acid in a protein; of a single nucleotide in a nucleic acid, etc.; usually the monomer minus the elements of water.

Resolution   In X-ray crystallography, the precision with which atoms are located in space; usually expressed in Å (10^{-10}m).

Restriction endonuclease   One of a group of enzymes that cleave internal phosphodiester bonds of both strands of DNA at specific nucleotide sequences, especially at palindromic sequences. Although restriction endonucleases have specific recognition sites, cleavage may occur at specific or random sites depending on the class of the endonuclease.

Restriction map   A map showing the positions of different restriction sites in a DNA molecule.

Restriction site   A DNA sequence recognized and cleaved by a restriction endonuclease.

Restriction modification   The modification of host DNA to prevent cutting by a restriction endonuclease. The modification is often via methylation of a specific restriction site sequence.

Retrotransposon   A mobile genetic element that transposes to a new location in DNA by first making an RNA copy of itself, then making a DNA copy of this RNA with a reverse transcriptase, and then inserting the DNA copy into the target DNA.

Reverse transcriptase   An enzyme produced by retroviruses that can synthesize a strand of DNA complimentary to an RNA template. Reverse transcriptase can also synthesize DNA from a DNA template.

Reverse transcriptase-PCR (RT-PCR)   A method for amplification of a specific mRNA by prior use of reverse transcriptase to form a cDNA, then use of PCR to amplify.

Reversion   Any mutation that restores the wild-type phenotype of a mutant.

R-factor   A transmissible plasmid that carries genes coding for resistance to several different antibiotics. Also called R-plasmid.

Rho factor   A protein which catalyzes transcription termination at certain sites or when an extended stretch on nontranslated, unstructured RNA is present.

Ribosomal (rRNA)   An RNA molecule that forms part of the structure of a ribosome.

Ribosome binding site   A short nucleotide sequence upstream of a gene which forms the site on the mRNA molecule where the ribosome binds.

Ribosome   An RNA-protein complex responsible for the correct positioning of mRNA and charged tRNAs allowing proper alignment of amino acids during protein synthesis.

Ribozyme   An RNA molecule that owing to peculiarities in its folding, is able to catalyze the interchange of some of its phosphodiester linkages to achieve intramolecular splicing. Ribozymes normally cleave *cis*, i.e. intramolecularly, but can be engineered to cleave a designated RNA target *trans*, i.e. intermolecularly.

Right-handed helix   By analogy with the right hand, a helix, e.g. of a polypeptide, that advances towards the C-terminus in the direction of the extended thumb as the backbone of a chain turns in the direction of the fingers closing on the palm. A *left-handed helix* has the opposite sense.

RNA chaperone   A putative RNA-binding protein that assists RNA in attainment of its biologically active conformation.

RNA polymerase   An enzyme complex that polymerizes RNA from ribonucleotides (NTPs), using one strand of DNA as template (hence called "DNA-dependent RNA polymerase).

RNA   Ribonucleic acid; a macromolecule formed of repeating riboses linked by phosphodiester bonds between the 3-hydroxyl group of one and the 5-hydroxyl group of the next. RNA has several biological functions, most of which depend upon its ability to form sequence-specific interactions with DNA.

RNase   An enzyme that hydrolyzes RNA molecules.

rRNA   Ribosomal RNA; the RNA that forms part of the structure of ribosomes.

Sanger method (dideoxynucleotide sequencing)   A technique for determination of the base sequence of a homogeneous polynucleotide that acts as a template in a replication system.

Satellite DNA   Short repetitive DNA sequences that occur mainly at the ends or in the centre of chromosomes and are therefore suspected of serving structural roles.

Scaffold   Protein engineers use this term to refer to a domain or small protein that is the object of mutation intended to introduce or refine a property, while retaining the folding of the polypeptide backbone.

Scintillation counting   A technique for detection and quantification of radioactivity by capture of the emission from a radiolabel by a primary phosphor which, on emission, excites a secondary phosphor and causes it to emit a pulse of detectable visible light.

Secondary structure   In protein chemistry, the regular folding of a polypeptide in a repeated pattern in nucleic acid chemistry by analogy, the double-helical structure of a polynucleotide and other regular structures seen in RNA foldings.

Selfish DNA   A DNA sequence that does not contribute to the fitness of an organism but is maintained in the genome because it promotes its own replication.

Selfish gene   A gene which confers a selective advantage upon its own propagation, but not on the survival of the individual.

Self-splicing   The activity of the precursor of mature RNA whereby it catalyzes its own splicing.

Sense strand   The strand of DNA that has the same nucleotide sequence as the mRNA.

Sequenator   A device to sequentially degrade a protein to determine its primary structure.

Sequence tagged site (STS)   A landmark in the base sequence of a genome used for placing partial sequences in correct relation to each other; especially as a strategy for co-ordination of the efforts of many laboratories in sequencing the human genome.

Sequence   The ordered linear array of amino acid residues in a protein, of bases in a polynucleotide, or of glycosides in a complex polysaccharide.

Shine-Dalgarno sequence   A polynucleotide sequence of prokaryotic mRNA that is complementary to, and therefore base-pairs with, a sequence at the 3′-end of 16S rRNA, which facilitates positioning of the initiator codon at the peptidyl site of the ribosome.

Shotgun cloning   A cloning strategy that involves the insertion of random fragments of a large DNA molecule into a vector, resulting in a large number of different recombinant DNA molecules.

Shotgun sequencing   An approach to determine the sequence of a large, randomly selected, piece of genomic DNA.

Shuttle vector   A plasmid that has both bacterial and eukaryotic origins of replication and so can propagate in either kind of cell; useful as a form of recombinant DNA for growth in a bacterium and subsequent transfer into eukaryotic cells for expression.

Side chain   The moiety of an amino acid residue in a protein, or of a free amino acid, that is attached to the α-carbon and is unique to each amino acid.

Sigma factor   A protein that functions as a subunit of bacterial RNA polymerases and is responsible for specificity of recognition of promoters. Different sigma factors allow recognition of different promoter sequences.

Signal sequence (leader sequence) The N-terminal portion of a secretory or membrane protein that assists it across the membrane of the rough endoplasmic reticulum, where it is synthesized, but is cleaved from the protein even before the synthesis of the protein is complete.

Signal transduction Conditions that alter the conformation of a protein which regulates expression of other genes.

Signature sequence An insertion or deletion in the coding sequence of a gene which, because it is shared by phylogenetically distant species, is thought to be evolutionarily conserved and therefore can serve to trace evolutionary relationships of species.

Silencing The genetically programmed repression of transcription.

Silent mutation A mutation which changes the nucleotide sequence but does not cause a detectable change in the phenotype.

Similarity Sequence identity between two nucleotide sequences. For example, 85% similarity means that 85 nucleotide positions out of 100 are identical in the two nucleotide sequences.

Simple sequence DNA The highly repetitive noncoding DNA of a satellite that serves a structural role in mitosis; used for genetic characterization by its polymorphism.

Single-stranded DNA binding protein (ssb) The small basic protein that has a high affinity for single-stranded DNA. ssb protein protects single-stranded DNA from nuclease attack and inhibits it from reannealing into double-stranded DNA.

Site-specific recombination Genetic exchange that occurs between particular, short DNA sequences. Unlike RecA mediated recombination, site-specific recombination requires little sequence homology between the two DNA molecules.

Small interspersed repeat element (SINE) A group of recurring species-specific polynucleotide sequences in the human genome, e.g. an *Alu* sequence of 0.3 kb present in approximately 10^6 copies.

Small nuclear RNA (snRNA) A category of polynucleotides, usually from 70 to 500 nucleotides in length.

Small nucleolar RNA (snoRNA) One of several oligonucleotides of the nucleolus that have roles in pre-mRNA processing.

SDS-PAGE A technique to dissociate multimeric proteins and to separate all proteins according to their apparent molecular masses. Proteins are treated with sodium dodecyl sulphate (SDS), an anionic detergent, and separated by electrophoresis on polyacrylamide gels that incorporate SDS.

Southern blot A method for detecting specific DNA fragments separated on an agarose gel.

South-western blotting A technique for identification of DNA-binding proteins and regions of DNA that bind proteins which combines the aspects of both Southern and Western blotting techniques.

Spacer DNA DNA that does not yet have a recognized function.

Species A group of organisms that breed among themselves to produce viable offspring.

Spliceosome A complex of (small nuclear RNA)-protein complexes and other proteins that assemble on a pre-mRNA and catalyze the excision of an intron in a process mechanistically similar to group II self-splicing. It acts presumably by forcing the intron into a loop and bridging the pre-mRNA at its splicing sites.

Splicing The process by which an intron is excised and exons are religated in the post-transcriptional modification of RNA (*cis*- splicing); also the excision of an intein from a precursor protein.

Star activity A secondary nucleotide sequence specificity of an endonuclease. After the canonical specificity is established, a library of potential substrates may be created by substituting at each sequence position, all of the other three nucleotide residues. Each such set is written with an asterisk (star) in the place of the substituted nucleotide residue.

Start (initiation) codon The codon on mRNA where polypeptide synthesis is initiated. The most common start codon is AUG but sometimes GUG or rarely UUG can be used as a start codon.

Strain An organism that is different from other organisms of the same species due to genetic differences.

Stringency Conditions affecting the hybridization of nucleotide sequences. Higher stringency conditions require more base pairing between the two sequences. Higher stringency conditions can be obtained by higher temperatures, higher salt concentrations, or addition of formamide.

Strong promoter An efficient promoter that can direct synthesis of RNA transcripts at a relatively fast rate.

Structural gene The portion of a gene encoding a functional polypeptide or RNA molecule.

Structural protein A protein which fulfill a purely structural role (i.e. not enzymatic). This includes phage capsid proteins, some ribosomal proteins, histone-like proteins, etc.

Subunit One of the identical or non-identical protein molecules that make up a multimeric protein; also one of the ribonucleoprotein complexes that make up the ribosome.

Superfamily A group of genes that are related by evolutionary divergence to an ancestral gene, but that code for products with different functions.

Supernatant fluid The unsedimented portion that remains after centrifugation.

Suppressor gene A mutated gene which produces a product which reverses the effect of a previous specific mutation without actually correcting the original mutation in the DNA can be either intergenic or intragenic.

Tandem array A DNA structure in which a gene and associated sequences are repeated in an immediately adjacent position.

Tandem duplication A DNA sequence that is repeated in direct orientation.

Tandem repeats Multiple adjacent copies of the same sequence.

TATA box A consensus sequence of eukaryotic DNA, named for its sequence on the coding strand, that aligns RNA polymerase II with the start site, about 25 bp downstream from it.

TATA-less promoter A class of transcription promoter sequences that have no TATA box.

Telomere The terminal part of a linear chromosome. Replication of the ends of linear DNA molecules requires specialized enzymes or structures. Often the telomeres have a DNA sequence with a single-stranded end that can fold into a hairpin structure.

Template A single-stranded polynucleotide (or region of a polynucleotide) that can be copied to produce a complementary polynucleotide.

Termination factor A protein that assists in the termination of the action of an RNA polymerase, e.g. the rho factor.

Terminator codon A polynucleotide triplet that signals the limit, at the C-terminus, of protein synthesis.

Terminator A DNA sequence that results in termination of transcription.

Terminus The region of DNA sequences where DNA replication terminates.

Tertiary structure The unique three-dimensional structure of a particular protein or nucleic acid.

Time-of-flight mass spectrometry A variant of MS in which the mass of a molecular ion is calculated from the time of travel from the point of ionization to the detector.

Topoisomerase An enzyme which introduces or removes overwinding or underwinding of the DNA circular duplex by causing a nick, rotating the strands, and then ligating them.

Topoisomers Double-stranded DNA molecules that differ only by their linking numbers.

Topology The nature of supercoiling of a double-stranded DNA molecule; also, in protein chemistry, a formalized array of secondary structures within a molecule.

Trans-acting Descriptive of a controlling effect of a regulatory gene on a structural gene at some distance from it on the same chromosome or on a different chromosome.

Transcript A strand of RNA copied from a DNA template.

Transcription bubble A region where the double-stranded DNA is separated while RNA polymerase is actively transcribing RNA. A short region of RNA-DNA duplex is formed between the newly synthesized RNA and the template DNA in this region.

Transcription factor A protein that binds to a specific DNA sequence and enhances or suppresses transcription of the target gene.

Transcription unit A region of DNA (a gene or an operon) transcribed as a single RNA.

Transcriptional activation   The process of separation of strands of DNA at which replication will commence. Short RNA sequences hold apart the DNA strands to allow a primosome to bind and synthesize primers of DNA synthesis.

Transcriptional silencing   The exercise of genetic control to prevent expression of a structural gene.

Transcription   The synthesis of RNA from a DNA template, catalyzed by RNA polymerase.

Transcriptome   A group of genes that are transcribed under a certain set of conditions.

Transduction   A method of gene transfer between bacteria in which the bacterial donor DNA is carried by a phage. There are two types of transduction: generalized transduction and specialized transduction.

Transfection   The process by which a viral or bacteriophage DNA is introduced into a cell or bacterium.

Transfer RNA (tRNA)   Adaptor molecules which translate the triplet code from the mRNA sequence into the corresponding chain of amino acids. tRNAs are short (about 74–95 bases), single-stranded RNA molecules that contain a high proportion of modified nucleosides.

Transferase   One of a class of enzymes that transfer a chemical group from donor substrates to acceptor substrates, e.g. a kinase, a phosphorylase, a transaminase.

Transformation   The process by which a cell line, that can normally be expected to undergo a limited number of cell divisions before death, becomes immortal; also the process by which an isolated foreign DNA is introduced into a cell or bacterium.

Transgene   DNA that has been experimentally introduced into a transgenic animal.

Trans   Genes located on different DNA molecules present in the same cell (the opposite of *cis*). For example, one copy of a particular gene may reside on the chromosome and the second copy may reside on a plasmid.

Transgenic   Descriptive of an animal that has developed in a surrogate mother from a fertilized ovum that had been injected with a recombinant DNA gene; contrasted with *congenic*, meaning of the same genetic origin.

Transition mutation   A base substitution mutation where a purine is replaced by a different purine, or a pyrimidine is replaced by a different pyrimidine.

Translation   The assembly of amino acids into polypeptides using the genetic information encoded in the molecules of mRNA.

Translesion synthesis   A mechanism that resumes stalled replication due to damage on the template strand. The stalled replicative polymerase is replaced by translesion polymerase(s) that synthesizes a short stretch of DNA across the lesion. Once this occurs, the replicative polymerase resumes DNA synthesis.

Transposable element   A transposon or insertion sequence. An element that can insert in a variety of DNA sequences.

Transposition   The movement of a discrete segment of DNA from one location in the genome to another.

Transposon   One of a class of genes that are capable of moving spontaneously from one chromosome to another, or from one position to another in the same chromosome; also known as a jumping gene or a transposable element.

Triplet   A sequence of three nucleotides. Typically refers to the codons and the corresponding genetic code.

tRNA splicing   The post-transcriptional processing of a tRNA precursor by excision and ligation.

tRNA   Transfer RNA; the RNA that serves in protein synthesis as an interface between mRNA and amino acids.

***Trp* operon**   A region of DNA that, when expressed, results in an RNA transcript that codes for five enzymes that act sequentially on chorismic acid to transform it into tryptophan.

Truncation   To shorten. For example, a truncated protein results if a premature stop codon interrupts the gene.

Two-dimensional electrophoresis   Electrophoretic separation of proteins on a solid support by one technique in one dimension, then by another technique in another. For example, proteins are subjected to isoelectric focusing in a long, thin tubular gel, then the developed gel is placed at the top of a sodium dodecyl sulphate/polyacrylamide slab and subjected to gel electrophoresis in the other direction.

Ultrafiltration A separation procedure in which a solution is forced through a membrane with a pore size that is selected to retain macromolecules of a certain size and to pass smaller ones.

Ultraviolet (UV) spectroscopy A technique which measures absorption of electromagnetic radiation at wavelengths shorter than the visible spectrum, i.e. below 400 nm.

Upstream A sequence located in front of a particular site relative to the direction of transcription and translation. For example, the lac promoter is located upstream of the lacZ structural gene.

UvrABC An enzyme complex that functions as an endonuclease, cutting the DNA on both sides of DNA lesions that distort the double-helix.

Variable number tandem repeat (VNTR) A minisatellite DNA; the repetition of a 35–80 bp sequence, to a size of up to 2 kbp, that is characteristic of an individual. VNTRs are used in forensic science to identify, or exclude, suspects of a crime.

Vector A replicon that is useful for cloning DNA fragments so that they can be amplified or transferred to other cells. Common cloning vectors are derivatives of natural plasmids, phages, or viruses.

Virulence The relative ability of an organism to cause disease.

Virulent phage A bacteriophage which always grows lytically.

Virus A small, infectious, obligate intracellular parasite. The virus genome is composed of either DNA or RNA. Within an appropriate host cell, the viral genome is replicated and uses cellular systems to direct the synthesis of other viral components.

Watson-Crick base pairs The base pairs that are compatible with a DNA double helix; i.e. adenine with thymine, and guanine with cytosine.

Watson-Crick rules The normal base pairing rules for DNA and RNA: A pairs with T or U, and G pairs with C.

Western blotting (immunoblotting) A technique to detect specific proteins. A mixture of proteins is separated by PAGE, blotted on to a plastic sheet and then exposed to a radiolabelled immunoglobulin directed against the desired protein, which is revealed by autoradiography.

Wobble A hypothesis proposed by Francis Crick to explain how one tRNA may recognize two different codons that differ in the third position.

X-gal A common abbreviation for 5-bromo-4-chloro-3-indolyl-β-D-galactoside. It is a sensitive, colour indicator for β-galactosidase.

X-ray crystallography A method for structural analysis of solids that have a repeating unit; dependent upon the ability of electrons to scatter X-rays, and capable of locating in space atoms larger than hydrogen in proteins and nucleic acids as well as in less complex compounds.

Yeast artificial chromosome (YAC) A yeast cloning vector that can accept a relatively large fragment of foriegn DNA, up to 1 Mb, in yeast cells.

Zinc finger A motif of which seven or more may appear in a DNA-binding protein. Each is characterized by two closely spaced cysteine and two histidine residues that serve as ligands for a single Zn^{2+}.

Index

C-value 134
   paradox 136
Carbohydrates 6
   monosaccharides 6
   oligosaccharide 6
   polysaccharides 7
Cell
   chemical constituents 6
   macromolecules 6
   small inorganic molecules 6
   small organic molecules 6
   water 6
Cell-free translation 283
   continuous flow 284
   in ovo system 284
   in vitro system 284
Central dogma of molecular
   biology 2
Centrifugation 58
   centrifugal force 58
   density gradient
     centrifugation 60
   fixed angle rotors 60
   swing out rotors 60
   ultracentifugation 59
Chromatography 81
   affinity 85
   column chromatography 82
   gas chromatography 85
   gel filtration 83
   HPLC 86
   ion exchange 84

   paper chromatography 81
   TLC 81

DNA damages 188
   alkylating agents 188
   UV radiations 189
   X-rays 189
DNA denaturation 136
   hyperchromic shift 136
   melting temperature 136
DNA-protein interactions 105
   ChIP assay 108
   DNase footprinting analysis
     107
   EMSA 105
   filter binding assay 105
DNA recombination 196
   homologous genetic 196
   meiotic 198
DNA repair pathways 189
   base excision repair (BER)
     190
   double strand break repair
     195
   mismatch repair 192
   multi-step repair 190
   nucleotide excision repair
     (NER) 190
   recombination repair 194
   single-step repair 190
DNA replication 172
   in eukaryotes 180

DNA polymerase a 180
DNA polymerase d 180
   lagging strand 173
   leading strand 173
   Okazaki fragments 174
   origin of replication 172
   replicon 172
   semi-conservative mode 173
   terminator for replication 172

Electrophoresis 69
   agarose gel 69
   CHEF 72
   FIGE 72
   formaldehyde-agarose gel 73
   isoelectric focusing 77
   polyacrylamide gel 73
   pulsed-field gel 72
   SDS–PAGE 74
   two-dimensional gel 77
   Western blot analysis 80
Elongation of DNA replication 175
   β ring 175
   DNA polymerase I 175
   DNA polymerase III 175

Genetic code 268
   codons 269
   cracking of 269
   degeneracy 270
   wobble concept 270
Genetic material 132

Genome 132, 140
  DNA 133
  genes 132
  RNA 134
  transforming principle 133
Genome organization 140
  coding sequence 140
  gene clusters 147
  gene families 147
  gene superfamily 147
  non-coding sequences 140
  pseudogenes 147
Genomics 320, 326
  comparative genomics 327
  DNA microarrays 328
  functional genomics 326
  genome annotation 326
  global gene expression 327
  microchips 328
  pharmacogenomics 331
  SAGE 329
  structural genomics 326

Initiation of DNA replication 174
  DnaA 174
  DnaB 174
  DnaC 174
  DnaG 174
  ORC (origin recognition
    complex) 174
  OriC 174

Lipids 8
  cholesterol 11
  fatty acids 8
  monosaturated fatty acids 10
  polyunsaturated fatty acids 10
  saturated fatty acids 10
  triacylglycerol 10

Messenger RNA 258
  3'-poly A tail 260
  3'-UTR 260

5'-cap 260
5'-UTR 260
cap of mRNA 261
internal ribosome entry sites
  (IRES) 262
monocistronic 259
polycistronic mRNA 259
Shine-Dalgarno (S.D.)
  sequence 262
Microscopy 45
  bright field 47
  confocal 52
  cryoelectron 56
  deconvolution 52
  electron 54, 57
  fluorescence 49
  freeze-fracture 57
  green fluorescent protein
    (GFP) 51
  light 45
  nomarski interference 47
  phase contrast and DIC 47
  SEM 58
  TEM 55
Model organisms 3

Nucleic acid modifying enzymes
  chemical nucleases 34
  degrading enzyme 28
  DNA ligases 30
  DNA polymerase I 29
  DNA polymerases 28
  endonucleases 33
  exonucleases 33
  modifying enzymes 28
  phosphatases 32
  polynucleotide kinases 32
  restriction endonucleases 37
  reverse transcritpase 29
  ribonulcease H 35
  RNA polymerases 30
  semi-artificial nucleases 34
  RNA ligase 32

T4 DNA polymerase 29
Taq polymerase 29
terminal deoxynucleotidyl
  transferase 29
Nucleic acids 12
  A and Z forms of DNA 15
  B-DNA 15
  Hoogstein base pairing 16
  nucleoside 14
  nucleotide 14
  phosphodiester bond 14
  purines 12
  pyrimidines 12
  RNA 17
  Watson–Crick model 15

Organelle genome 160
  chloroplast genome 165
  mitochondrial genome 162

Packaging of DNA 153
  chromatin 156
  euchromatin 155
  heterochromatin 155
  MARs 155
  nuclear scaffold 159
  nucleosomal remodeling 159
  nucleosome 156
  SARs 155
  solenoid structure 158
Polymerase chain reaction 96
  oligonucleotide primers 101
  PCR cloning 102
  real time PCR 104
  RT-PCR 102
  Taq polymerase 96
Post-transcriptional events 236
  poly A polymerase 246
  cap 247
  cap0 247
  cap1 247
  cap2 247
  mature rRNA 248

mature tRNA 249
PABP 246
polyadenylation signal 245
processing of mRNA at 3′ end 245
processing of mRNA at 5′ end 247
RNA editing 243
RNaseP 250
RNaseH 249
rrn operon 248
Proteins 17
  amino acids 18
  amino group 17
  carboxyl group 17
  dextro isomers 17
  domains 22
  essential amino acids 20
  functional proteins 17
  isomers dextro 17
  peptide bond 20
  primary structure 20
  quaternary structure 22
  secondary structure 20
  structural proteins 17
  tertiary structure 21
Protein degradation 313
  lysosomal pathway 313
  ubiquitin-proteasome pathway 314
Protein folding 300
  chaperonins 301
  heat shock proteins 301
  molecular chaperones 300
  peptidyl prolyl isomerase 302
  protein disulfide isomerase 302
Protein modifications 305
  attachment of glycolipids 310
  attachment of lipids 307
  glycosylation 305
  GPI anchors 310
  N-myristoylation 307

palmitoylation 309
prenylation 308
protein kinases 311
protein phosphatases 311
protein phosphorylation 311
Protein processing 303
  cleavage of larger precursor proteins 304
  cleavage of initiator methionine 303
  cleavage of signal sequence 304
Protein synthesis 258
  components of 258
  A site 278
  antibiotics 281
  cap-independent initiation 277
  elongation factor 278
  exit site 281
  inhibit 281
  initiation 272
  initiation factors 273
  mechanism of 272
  P site 278
  polysome 283
  release factors 281
  ribosome recycling factor 283
  scanning model 276
  termination 272, 281
  termination codons 281
Protein-protein interactions and interactome 340
  bioinformatics tools and databases 340
  GST-pulldown assay 340
  high throughput protein analysis 340
  library-based interaction mapping 340
Proteomics 332
  antibody arrays 339

antigen arrays 340
ESI 336
general protein arrays 340
LC-MS/MS 338
MALDI 336
mass spectrometry 335
methodologies 333
MS/MS spectrometry 338
objectives 332
Radiolabelling and detection 86
  alpha particle 87
  alpha particle emission 87
  autoradiography 90
  beta particles 88
  gamma rays 87
  Geiger–Muller counters 89
  half-life 88
  isotopes 87
  measurement of radioactivity 89
  radioactive decay 87
  scintillation counters 89
  units of radioactivity 88
Reassociation kinetics 136
  C_0t curve 137
  non-repetitive DNA or unique DNA 139
  repetitive DNA 139
Recombinant DNA techniques 109
  cDNA libraries 110
  colony hybridization 115
  end labelling 118
  gene libraries 109
  hybridization techniques 115
  molecular cloning 109
  molecular probes 117
  nick translation 118
  Northern hybridization 117
  plaque hybridization 116
  random primer labelling 118
  Southern blotting 116

transformation 115
vectors 111
Regulation of protein synthesis 286
global regulation 286
mRNA-binding 288
mRNA-specific regulation 288
multiple ORFs 288
non-coding RNAs 288
poly A tail 288
Regulation of replication 184
Cdc6 185
DAM methylase 184
licensing factor 184
MCM proteins 185
promoter of DnaA 184
SeqA 184
Regulation of transcription 211
by σ factor 218
in Phage λ 219
Replication in eukaryotes 180
DNA polymerase A 180
DNA polymerase D 180
origin recognition complex (ORC) 180
Ribosome 266
catalytic role of rRNAs 268
large and the small subunits 267
peptidyl transferase activity 268
structure of 268

Sequencing techniques 91
DNA sequencing 91
Sanger's dideoxy method 91
sequencing of proteins 93
Spectroscopy 61, 65
Beer-Lambert's law 62
CD spectroscopy 65

dual beam spectrophotometer 63
NMR spectroscopy 64
UV-visible spectroscopy 61
Splicing 237
mechanism of 237
of group I and group II introns 240
tRNA splicing 239
Termination of replication 180
catenates 181
contra-helicase 181
telomerase 182
Ter sequences 180
terminator utilization substances 181
topoisomerase IV 181
Tus 181
Termination of transcription
attenuation 215
establishing lysogeny 221
inducible operon 212
intrinsic terminator 210
negative regulation 212
operon 212
positive regulation 214
repressible operons 215
rho-dependent terminator 210
The Human Genome Project 320
BACs 321
clone-by-clone strategy 321
gene families 324
human genome 323
mapping 320
minisatellite DNA 324
mitochondrial genome 325
nuclear genome 325
repetitive DNA 324
satellite DNA 324
sequencing 320

shotgun sequencing 322
telomeric DNA 324
YACs 321
Transcription 206
activation domain 233
activators 231
elongation 206
initiation 206
RNA polymerase 206
termination 206
transcription factors 206, 227
Transcription in eukaryotes 224
chromatin 236
Class I promoter 226
Class II promoters 225
Class III promoter 226
DNA binding domain 231
insulators 234
mediators 235
nucleosome remodelling agents 236
RNA polymerase I, II, and III 224
Transcription in prokaryotes 207
abortive initiation 208
holoenzyme 207
promoter 207
promoter clearance 208
sigma factor 207
transcription bubble 209
Transfer RNA 263
aminoacyl-tRNA 263
charging of tRNA 265
isoacceptor tRNAs 263
proofreading 266
three dimensional structure of tRNA 264

X-ray crystallography 67
electron-density map 68